Surface Waves in Anisotropic and Laminated Bodies and Defects Detection

NATO Science Series

A Series presenting the results of scientific meetings supported under the NATO Science Programme.

The Series is published by IOS Press, Amsterdam, and Kluwer Academic Publishers in conjunction with the NATO Scientific Affairs Division

Sub-Series

I. Life and Behavioural Sciences	IOS Press
II. Mathematics, Physics and Chemistry	Kluwer Academic Publishers
III. Computer and Systems Science	IOS Press
IV. Earth and Environmental Sciences	Kluwer Academic Publishers
V. Science and Technology Policy	IOS Press

The NATO Science Series continues the series of books published formerly as the NATO ASI Series.

The NATO Science Programme offers support for collaboration in civil science between scientists of countries of the Euro-Atlantic Partnership Council. The types of scientific meeting generally supported are “Advanced Study Institutes” and “Advanced Research Workshops”, although other types of meeting are supported from time to time. The NATO Science Series collects together the results of these meetings. The meetings are co-organized bij scientists from NATO countries and scientists from NATO's Partner countries – countries of the CIS and Central and Eastern Europe.

Advanced Study Institutes are high-level tutorial courses offering in-depth study of latest advances in a field.
Advanced Research Workshops are expert meetings aimed at critical assessment of a field, and identification of directions for future action.

As a consequence of the restructuring of the NATO Science Programme in 1999, the NATO Science Series has been re-organised and there are currently Five Sub-series as noted above. Please consult the following web sites for information on previous volumes published in the Series, as well as details of earlier Sub-series.

http://www.nato.int/science
http://www.wkap.nl
http://www.iospress.nl
http://www.wtv-books.de/nato-pco.htm

Series II: Mathematics, Physics and Chemistry – Vol. 163

Surface Waves in Anisotropic and Laminated Bodies and Defects Detection

edited by

Robert V. Goldstein

Institute for Problems in Mechanics,
Moscow, Russia

and

Gerard A. Maugin

Université Pierre et Marie Curie,
Modélisation en Mécanique, Paris, France

Kluwer Academic Publishers

Dordrecht / Boston / London

Published in cooperation with NATO Scientific Affairs Division

Proceedings of the NATO Advanced Research Workshop on
Surface Waves in Anisotropic and Laminated Bodies and Defects Detection
Moscow, Russia
30 October–2 November 2001

A C.I.P. Catalogue record for this book is available from the Library of Congress.

ISBN 1-4020-2386-3 (PB)
ISBN 1-4020-2385-5 (HB)
ISBN 1-4020-2387-1 (e-book)

Published by Kluwer Academic Publishers,
P.O. Box 17, 3300 AA Dordrecht, The Netherlands.

Sold and distributed in North, Central and South America
by Kluwer Academic Publishers,
101 Philip Drive, Norwell, MA 02061, U.S.A.

In all other countries, sold and distributed
by Kluwer Academic Publishers,
P.O. Box 322, 3300 AH Dordrecht, The Netherlands.

Printed on acid-free paper

Printed in the Netherlands.

Contents

FOREWORD

Among the variety of wave motions one can single out surface wave propagation since these surface waves often adjust the features of the energy transfer in the continuum (system), its deformation and fracture. Predicted by Rayleigh in 1885, surface waves represent waves localized in the vicinity of extended boundaries (surfaces) of fluids or elastic media. In the ideal case of an isotropic elastic half-space while the Rayleigh waves propagate along the surface, the wave amplitude (displacement) in the transverse direction exponentially decays with increasing distance away from the surface. As a result the energy of surface perturbations is localized by the Rayleigh waves within a relatively narrow layer beneath the surface. It is this property of the surface waves that leads to the resonance phenomena that accompany the motion of the perturbation sources (like surface loads) with velocities close to the Rayleigh one; (see e.g., R.V. Goldstein. Rayleigh waves and resonance phenomena in elastic bodies. *Journal of Applied Mathematics and Mechanics (PMM),* 1965, v. 29, N 3, pp. 608-619). It is essential to note that resonance phenomena are also inherent to the elastic medium in the case where initially there are no free (unloaded) surfaces. However, they occur as a result of an external action accompanied by the violation of the continuity of certain physical quantities, e.g., by crack nucleation and dynamic propagation. Note that the aforementioned resonance phenomena are related to the nature of the surface waves as homogeneous solutions (eigenfunctions) of the dynamic elasticity equations for a half-space (i.e. nonzero solutions at vanishing boundary conditions).

Rayleigh waves in a half-space have no dispersion. But Rayleigh and Lamb also discovered surface waves propagating with dispersion. These are eigen waves in an elastic layer symmetric and antisymmetric relatively to its middle plane. The velocities of the symmetric waves belong to the range $c_R < v < c_2$ (c_R, c_2 are the velocities of the Rayleigh and shear waves of an elastic medium), while the antisymmetric waves exist in the velocity range $0 < v < c_R$. Along with the free extended boundaries, interfaces also can serve as wave guides for surface waves. Stoneley (1924) predicted the existence of waves localized along an interface of two half-spaces of joined isotropic materials with different elastic properties, whence so-called Stoneley waves. These waves are related to the case of full contact conditions at the interface (i.e. conditions of continuity of displacement and traction vectors at the interface). Later on, similar types of interface waves were discovered for different types of partial contact conditions at the interface. Among these types of conditions, first of all, the sliding conditions and conditions of possible separation without sliding at the interface need to be mentioned. These interface waves exist not for all combinations of the elastic constants of joined materials and till now the studies on this subject matter are pursued and aimed at searching for more precise conditions for their existence and studying their features.

Evidently, the interface wave picture becomes complicated for a layered medium or a half-space with a coating. These waves were studied starting in the 1950s.

Taking into account the material (media) elastic anisotropy along with the interface induced nonhomogeneity opens up an immense field of study and modeling of the waves of surface and interface types as well as processes of their interaction with bulk waves and waves having combined bulk and surface waves properties (such as so-called leaky waves). Searching for the conditions of their existence and directions which exhibit wave guide properties as well as 'forbidden' directions, permanently attracts the interest of, and demands much effort from, many researchers.

The variety of elastic anisotropic and/or layered natural media, composite materials and structures, as also crystals, makes the study of the wave processes inherent to these objects a rich and active field of research in various branches of engineering and also in some applied sciences such as geophysics. On the other hand, these materials often have complex physical-mechanical properties. This is the case of piezoelectric materials or shape memory ones; more generally, these materials are very sensitive to the action of both mechanical loading and physical fields, (see e.g., G.A. Maugin. *Continuum mechanics of electromagnetic solids.* North-Holland, Amsterdam, 1988). This circumstance leads to an extraordinary vivid activity in solving surface wave problems as well as studying their extended mathematical complexity. Indeed, studying the classical Rayleigh or Stoneley waves is in principle reduced to the search for the transcendental equations or corresponding systems, which we expect can be written in an explicit form. But in the case of anisotropic and nonhomogeneous media sometimes such an explicit form of the appropriate algebraic problem cannot be written explicitly. Nevertheless, qualitative algebraic methods in combination with the ideas of a six-dimensional complex formalism (a remarkably creative step taken by Stroh in 1962) provide successful progress in surface and interface wave analysis.

Along with an evident academic interest, surface and interface waves studies are of utmost importance for applications in Earth Sciences, Materials Sciences and Engineering. In his pioneering paper Rayleigh already pointed to the significance of surface waves "in earthquakes and in the collision of elastic solids". Subsequent field and theoretical studies confirmed this proposition. Moreover, an essential role of the interface waves related to different types of conditions at the interface in the processes of seismic activity was fully demonstrated. The current state of the subject was analyzed in a comprehensive review given by J.R. Rice (New perspectives on crack and fault dynamics. In: *Mechanics for a New Millenium* (Proc. of the 20th Int. Congress of Theoretical and Applied Mechanics, 27 Aug-2 Sept 2000), eds. H. Aref and J.W. Phillips, Kluwer Academic Publishers, 2001, pp. 1-23).

It is now more than forty years that studies and engineering applications of surface waves in active and passive systems aimed at defects and cracks detection have started to blossom. Appropriate devices and systems are now widely used for monitoring the technical state of large scale structures (like pressure vessels, pipelines, etc.) and for materials characterization (see, e.g., R.A. Kline. *Nondestructive characterization of composite materials.* Technomic, Lancaster, PA, 1991). At the same time, the creation of delay lines exploiting the notion of surface waves led to the development of a new generation of devices-transducers in surface acoustic waves (see, e.g., D. Royer and E. Dieulesaint. *Elastic waves in solids.* Springer-Verlag, Berlin, 1999). In turn, these significant engineering applications permanently stimulate further fundamental studies of surface and interface waves. This incentive lies at the basis of the project of the *NATO Advanced Research Workshop* that we organized in Moscow in February 2002.

The Organizing Committee of the Workshop, conscious of its responsibility, gathered the most prominent and active members of the scientific community working in the field of surface wave analysis and applications to discuss current achievements and, what is more important, to find out ways for solving some of the urgent problems, more specifically:

(i) development of an adequate theory capable of analyzing the problem of the "forbidden" directions for genuine surface waves; analyzing non-classical surface waves propagating along forbidden directions (a tutorial lecture given by V.I.Alshits, contributions by V.M. Babich and A.P. Kiselev, S.V. Kuznetsov; V.G. Mozhaev and M. Weihnacht, T.C.T. Ting);

(ii) developing efficient numerical methods and algorithms for the analysis of surface waves (including Love and Lamb waves) propagating in homogeneous and layered media with both arbitrary elastic anisotropy and plasticity, and having a complex internal structure (lectures given by M.V. Ayzenberg- Stepanenko, E.V. Glushkov and N.V. Glushkova, J.D. Kaplunov and M. V. Wilde, A. V. Kaptsov and S. V. Kuznetsov, V.N. Kukudzhanov, R.L. Salganik, I.V. Simonov; S.P. Tokmakova, K.B. Ustinov);

(iii) working out both experimental and theoretical procedures for the identification of material properties, and solitary and dispersed defects by non-destructive testing (lectures given by R.Kline, M.J.S. Lowe, B.N. Pavlakovic, and P. Cawley, A.L. Popov and D.A. Chelyubeev);

(iv) developing efficient analytical and numerical methods for the analysis of surface waves in porous, water saturated media and ice fields (lectures given by P. Adler and V.M. Entov, A. Marchenko);

(v) working out analytical and numerical methods for analyzing interactions of cracks, faults, step-discontinuities and edges with surface waves (lectures given by R.V. Craster and A.V. Shanin, M. Deschamps and O. Poncelet, R.V. Goldstein and A. Marchenko; Y.A. Rossikhin and M.V. Shitikova, D.D. Zakharov);

(vi) improving the theory of crack propagation related to the analysis of velocities of surface waves (lectures given by B. Broberg, G.I. Kanel, S.V. Razorenov, and V.E. Fortov, Y.V. Petrov);

(vii) developing a theory for predicting the behavior of non-linear surface waves (lecture given by D.F. Parker).

Members of the Organizing Committee note with pleasure that the Workshop gave a unique and excellent opportunity to exchange ideas and to elaborate new principles and methods for the analyses of the above stated problems. Moreover, due to the presence of many students at the Workshop, the Workshop helped young people involved in scientific research to get acquainted with the state of the art in surface-wave dynamics. This volume consists of papers presented for publication by most of the lecturers. We express our debt to all contributors for their kind cooperation in the preparation of this book.

In conclusion we like to emphasize that wave dynamics belongs to some of the traditional domains of activity of the Russian Academy of Sciences (former Academy of Sciences of USSR). The Workshop held in Moscow is a tribute to this tradition. We do hope that the materials of this volume will promote further studies of surface waves and related problems.

Last but not least, we acknowledge our debt to NATO that generously supported the organization of this fruitful scientific meeting in the form of an *Advanced Research Workshop* under grant PST.ARW.978081. The NATO Program Director for Physical and Engineering Science and Technology, Dr F. Pedrazzini, has shown his understanding and patience for our delayed organization and our late reports and publication. Our thankful regards go to him. Dr Alexander Kaptsov and Prof. Sergey Kuznetsov in Moscow and Dr Pierre-Yves Lagrée in Paris were most diligent and efficient in preparing these proceedings. Our heartfelt thanks to them are only a minute mark of our appreciation.

The ARW co-chairmen and Editors,

Robert V. Goldstein (Moscow), Gerard A. Maugin (Paris)

Proceedings of the NATO Advanced Research Workshop

Surface Waves in Anisotropic and Laminated Bodies and defects Detection

Held at
Institute for Problems of Mechanics,
Russian Academy of Sciences, Moscow, Russia
7-9 February 2002

Honorary Chairman: Bertram Broberg, Dublin, Ireland
Co-chairmen: Gérard A. Maugin (Paris) and Robert V.Goldstein (Moscow)
Workshop Secretary: Alexander V.Kaptsov (Moscow)

Organizing Committee: Ronald Kline (USA), Gérard A. Maugin (France), Basil M.Babich (Russia), Robert V.Goldstein (Russia)

Scientific Committee: I.D.Abrachams (UK), V.I.Alshits (Russia), V.A.Babeshko (Russia), D.Barnett (USA), L.I.Slepyan (Israel), L.B. Freund (USA), P.Cawley (UK), P.Chadwick (UK), M.Deschamps (France), M Hayes (Ireland), D.M.Klimov (Russia), A.G.Kulikovsky (Russia), I.V.Simonov (Russia), N.F.Morozov (Russia), R.Rannacher (Germany), S.A.Rybak (Russia), T.C.T. Ting (USA).

Local Organizing Committee: A.V.Kaptsov, S.V.Kuznetsov, A.V.Shanin, D.D.Zakharov (Moscow, Russia).

Part I

Theoretical problems concerning propagation of surface wave in elastic anisotropic media

ON THE ROLE OF ANISOTROPY IN CRYSTALLOACOUSTICS

V. I. ALSHITS
Institute of Crystallography, Russian Academy of Sciences, Leninskii pr., 59, 117333 Moscow, Russia

Abstract. The paper presents a short review of some basic theoretical results in the acoustics of anisotropic media. It includes: the general theorems related to phase speeds and polarization vectors of three bulk-wave eigen-modes in arbitrary elastic media; the topological classification of polarization singularities for plane waves; the conditions for the occurrence of the phenomenon of energy concentration; the general criteria for the existence of different classes of acoustic waves in media with surfaces and interfaces; the relations between phase, group and ray velocities of bulk and surface acoustic waves; the theory of resonant reflection/transmission of acoustic waves due to a coupling with a leaky wave branch in various structures; and some other topics. This work was supported by the Russian Foundation for Basic Research (Grant no.01-02-16228).

1. Introduction

The role of anisotropy in crystalloacoustics is not at all reduced just to variations of wave characteristics for different directions of propagation. Anisotropy creates also qualitatively new properties of elastic waves and acoustic phenomena that have not got close analogues in isotropic media. Some of them have already found their practical applications in real devices.

A theoretical description of elastic waves in anisotropic materials is a very non-trivial problem. Impermeability of the wave equations for media of unrestricted anisotropy to explicit analysis has required development of new theoretical methods that allowed obtaining final conclusions without hopeless direct calculations. As a result, during the last few decades, due to contributions of many researchers from different countries, the theory of elastic waves in anisotropic media gradually became an independent branch of modern crystalloacoustics.

R.V. Goldstein and G.A. Maugin (eds.),
Surface Waves in Anisotropic and Laminated Bodies and Defects Detection, 3–68.

2. Elastic Bulk Waves

2.1. BASIC CONCEPTS

2.1.1. *Christoffel's Equation, Acoustical Tensor, its Eigenvectors and Eigenvalues*

Consider an infinite elastic medium of unrestricted anisotropy, free of external forces, in which a plane bulk wave of displacements $\mathbf{u}(\mathbf{x}, t)$ is propagating along the direction specified by a unit vector $\mathbf{m}$,

$$\mathbf{u}(\mathbf{x}, t) = u_0 \mathbf{A} \exp[ik(\mathbf{m} \cdot \mathbf{x} - ct)]. \tag{1}$$

Here u_0 is the scalar amplitude, $\mathbf{A}$ is the unit polarization vector, k is the length of the wave vector, $\mathbf{k} = k\mathbf{m}$, c is the phase speed, $c = \omega/k$, $\mathbf{x}$ is the space coordinate and t is the time. In (1) u_0 and k are supposed to be fixed parameters chosen so that $u_0 k << 1$. The latter condition will allow us to use the equation of linear elastodynamics

$$\rho \frac{\partial^2 u_i}{\partial t^2} = c_{ijkl} \frac{\partial^2 u_l}{\partial x_k \partial x_j}, \tag{2}$$

where ρ is the density of the medium and c_{ijkl} is its elastic moduli tensor. Combining (1) and (2) we obtain the *Christoffel* equation

$$(mm)\mathbf{A} = \rho c^2 \mathbf{A}, \tag{3}$$

where $(mm) \equiv \mathbf{Q}$ is the so-called acoustical tensor with components

$$(mm)_{jk} = m_i c_{ijkl} m_l = Q_{jk}. \tag{4}$$

Christoffel's equation (3) for each direction of the wave normal $\mathbf{m}$ determines the three eigenvectors $\mathbf{A}_\alpha(\mathbf{m})$, α =1, 2, 3, and the three corresponding eigenvalues $\lambda_\alpha(\mathbf{m}) = \rho c_\alpha^2$ as roots of the cubic secular equation

$$\det[\mathbf{Q} - \rho c^2] = 0. \tag{5}$$

By its definition the acoustical tensor $\mathbf{Q}$ is symmetric, real and positive definite. As a result its eigenvalues λ_α are positive (which ensures real speeds c_α) and the eigenvectors $\mathbf{A}_\alpha$ can be chosen real and orthonormal

$$\mathbf{A}_\alpha \cdot \mathbf{A}_\beta = \delta_{\alpha\beta}. \tag{6}$$

It is convenient to express the acoustical tensor $\mathbf{Q}$ in the form of a spectral expansion, as a linear superposition of dyads:

$$\mathbf{Q}(\mathbf{m}) = \rho \sum_{\alpha=1}^{3} c_\alpha^2(\mathbf{m}) \mathbf{A}_\alpha(\mathbf{m}) \otimes \mathbf{A}_\alpha(\mathbf{m}). \tag{7}$$

2.1.2. *Special Directions of Wave Propagation in Crystals*

It is well known that only isotropic media admit propagation along any direction of purely *transverse* waves, polarized arbitrarily in the plane orthogonal to the wave normal,

$$\mathbf{A}_t(\mathbf{m}) \perp \mathbf{m}, \tag{8}$$

and *longitudinal* waves, polarized along the wave normal,

$$\mathbf{A}_l(\mathbf{m}) || \mathbf{m}. \tag{9}$$

In crystals, normally none of the three isonormal waves is purely transverse or longitudinal. Usually, the wave whose polarization makes the most acute angle with $\mathbf{m}$, is called *quasi-longitudinal* while the other two are referred to as quasi-transverse waves. But for some specific orientations $\mathbf{m} = \mathbf{m}_t$ condition (8) may be satisfied for one of the waves. Such a direction $\mathbf{m}_t$ admitting the existence of a purely transverse bulk wave solution is termed a transverse normal. As was proved by Fedorov [1], the equation

$$\mathbf{m}_t \mathbf{Q}^2 \cdot [\mathbf{m}_t \mathbf{Q} \times \mathbf{m}_t] = 0 \tag{10}$$

determines all directions of transverse normals for a given crystal. Eqn.(10) provides only one scalar condition on two unknown variables specifying the orientation of the vector $\mathbf{m}_t$. Therefore one should expect whole lines of solutions of (10) on the unit sphere $\mathbf{m}^2 = 1$. The orientation of the polarization vector $\mathbf{A}_t$ of the corresponding transverse wave and its phase speed c_t are determined by the relations [1]

$$\mathbf{A}_t || \mathbf{m}_t \times \mathbf{Q}\mathbf{m}_t, \tag{11}$$

$$\rho c_t^2 = \frac{|\mathbf{Q}(\mathbf{m}_t \times \mathbf{Q}\mathbf{m}_t)|}{|\mathbf{m}_t \times \mathbf{Q}\mathbf{m}_t|}. \tag{12}$$

Obviously, a longitudinal solution does not generally exist along $\mathbf{m}_t$. But, as was proved by Kolodner [2], for any crystal there always exist at least three solutions $\mathbf{m} = \mathbf{m}_l$ of Eqn.(9), called *longitudinal normals.* By Eqn.(6), along $\mathbf{m}_l$ the existence of two transverse waves is automatically guaranteed, i.e. $\mathbf{m}_l$ is simultaneously a transverse normal as well. On the other hand, it is evident that each of the directions $\mathbf{m}_l$ must be situated at the point of intersection of two lines of solutions for transverse normals. The equation for such points has the form [1]

$$\mathbf{m}_l \times \mathbf{Q}\mathbf{m}_l = \mathbf{0}. \tag{13}$$

In isotropic media for any direction of the wave normal $\mathbf{Qm}$ must be parallel to $\mathbf{m}$, which makes (13) to be an identity. For transverse isotropic media the same situation occurs for any direction in the basal plane and for the wave normals belonging to a cone and making an angle

$$\theta_l = \tan^{-1}\left(\frac{c_{33} - 2c_{44} - c_{13}}{c_{11} - 2c_{44} - c_{13}}\right)^{\frac{1}{2}} \tag{14}$$

with the principal axis, which is also a longitudinal normal. Certainly the cone (14) exists only under the condition

$$\frac{c_{33} - 2c_{44} - c_{13}}{c_{11} - 2c_{44} - c_{13}} > 0. \tag{15}$$

However, in any other crystal including a triclinic one the number of solutions of Eqn.(13) can not exceed 13. In particular, exactly 13 longitudinal normals always exist in cubic crystals. The detailed analysis of numbers of longitudinal normals admitted by various symmetry systems has been accomplished by Khatkevich [3, 4], Brugger [5] and Bestuzheva & Darinskii [6].

The third type of special directions in crystals is related to degeneracy (i.e. coincidence) of the phase speeds of a pair (or all three) of isonormal waves. These directions $\mathbf{m}_d$ are called *acoustic axes.* For the degeneracy direction $\mathbf{m}_d$ where, for instance $c_1(\mathbf{m}_d) = c_2(\mathbf{m}_d)$, Eqn.(7) can be transformed to

$$\mathbf{Q}(\mathbf{m}_d) = \rho\left\{c_1^2(\mathbf{m}_d)\mathbf{I} + \left[c_3^2(\mathbf{m}_d) - c_1^2(\mathbf{m}_d)\right]\mathbf{A}_3(\mathbf{m}_d) \otimes \mathbf{A}_3(\mathbf{m}_d)\right\}. \tag{16}$$

It follows from Eqn.(16) that any vector $\mathbf{A}$ on the plane orthogonal to $\mathbf{A}_3(\mathbf{m}_d)$ is an eigenvector of the acoustical tensor $\mathbf{Q}(\mathbf{m}_d)$, i.e. an allowed orientation for a polarization of the "degenerate" waves propagating along $\mathbf{m}_d$ with the phase speed c_1 (see Fig.1). We shall return below to the theory of acoustic axes, including a classification of polarization singularities in the vicinity of degeneracy directions $\mathbf{m}_d$.

2.2. SOME GENERAL THEOREMS ON PHASE-SPEED BRANCHES $c_\alpha(\mathbf{m})$

The determination of the basic wave parameters, the polarization vectors $\mathbf{A}_\alpha(\mathbf{m})$ and the phase speeds $c_\alpha(\mathbf{m})$, requires a solution of the eigenvalue problem (3), which in turn is reduced to a cubic secular equation (5). Of course, it would be practically impossible to analyse these equations explicitly for arbitrary anisotropy. Nevertheless, as we shall see, one can extract a series of fundamental general properties of the phase speed branches $c_\alpha(\mathbf{m})$ without explicit solution of Eqn.(5). Below we shall number the solutions of Eqns.(3), (5) in accordance with the rule

$$c_1(\mathbf{m}) \leq c_2(\mathbf{m}) \leq c_3(\mathbf{m}), \tag{17}$$

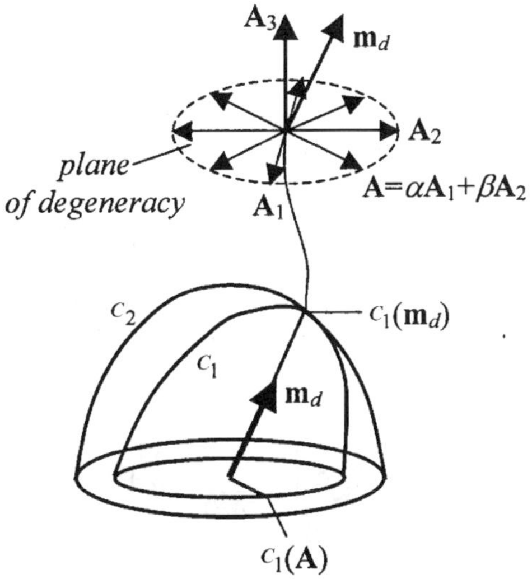

Figure 1. The acoustic axis along $\mathbf{m} = \mathbf{m}_d$ and the corresponding contour in the plane of degeneracy $\mathbf{A} = \alpha\mathbf{A}_1 + \beta\mathbf{A}_2, \alpha^2 + \beta^2 = 1$

and we shall call the $c_1(\mathbf{m})$, $c_2(\mathbf{m})$, $c_3(\mathbf{m})$ functions the *bottom, middle* and *top* branches of the phase speeds, respectively. They correspond to the *inner, intermediate* and *outer* sheets of the phase velocity surface, formed by the ends of vectors $\mathbf{c}_\alpha = \mathbf{m}c_\alpha(\mathbf{m})$, as $\mathbf{m}$ varies over the unit sphere $\mathbf{m}^2 = 1$.

2.2.1. *Invariance Properties of the Combination* $c_1^2 + c_2^2 + c_3^2$ *on the Sphere* $\mathbf{m}^2 = 1$

Let $\mathbf{n}$, $\mathbf{s}$, and $\mathbf{t}$, be three mutually orthogonal unit vectors. Form the unit matrix

$$\mathbf{I} = \mathbf{n} \otimes \mathbf{n} + \mathbf{s} \otimes \mathbf{s} + \mathbf{t} \otimes \mathbf{t}. \tag{18}$$

The convolution of both sides of Eqn.(18) with the tensor c_{ijkl} over the external indices (i and l) may be written, in the notation (4), as

$$\mathbf{V} = (nn) + (ss) + (tt), \tag{19}$$

where $\mathbf{V}$ is a tensor of the second rank,

$$V_{jk} = c_{ljkl}. \tag{20}$$

Making use of the spectral decomposition (7) and taking the trace on both sides of Eqn.(19), one obtains [7-9]:

$$\begin{aligned} \sum_{\alpha=1}^{3} c_\alpha^2(\mathbf{n}) + \sum_{\alpha=1}^{3} c_\alpha^2(\mathbf{s}) + \sum_{\alpha=1}^{3} c_\alpha^2(\mathbf{t}) &= \mathrm{tr}\mathbf{V}/\rho = \\ &= (c_{11} + c_{22} + c_{33} + 2c_{44} + 2c_{55} + 2c_{66})/\rho = \mathrm{const} \end{aligned} \tag{21}$$

irrespective of the orientation of the system {n,s,t}. Thus the following theorem is true.

For any three mutually orthogonal directions of propagation, the sum of the squares of the phase speeds of the bulk elastic waves is constant for a given crystal and equal to the invariant combination $c_{11} + c_{22} + c_{33} + 2c_{44} + 2c_{55} + 2c_{66})/\rho$. (22)

As an evident consequence of this theorem, one can also state that for any two coplanar pairs of the orthogonal vectors $\mathbf{n}\perp\mathbf{s}$ and $\mathbf{p}\perp\mathbf{q}$ $(\mathbf{n}\times\mathbf{s}||\mathbf{p}\times\mathbf{q})$ the following identity [8]

$$\sum_{\alpha=1}^{3}\left[c_\alpha^2(\mathbf{n}) + c_\alpha^2(\mathbf{s})\right] = \sum_{\alpha=1}^{3}\left[c_\alpha^2(\mathbf{p}) + c_\alpha^2(\mathbf{q})\right] \tag{23}$$

must be true.

Let us now approach the problem from another side. It is clear from Eqns.(7) and (4) that the sum $c_1^2 + c_2^2 + c_3^2$ for one arbitrary direction $\mathbf{m}$ is also related to the tensors $\mathbf{Q}$ and $\mathbf{V}$:

$$\sum_{\alpha=1}^{3} c_\alpha^2(\mathbf{m}) = \mathrm{tr}\mathbf{Q}(\mathbf{m})/\rho = \mathbf{m}\cdot[\mathbf{V}/\rho]\mathbf{m}. \tag{24}$$

At this stage of consideration it is convenient to introduce, in the standard way, the characteristic ellipsoid of the tensor $\mathbf{V}/\rho$,

$$\mathbf{r}\cdot(\mathbf{V}/\rho)\mathbf{r} = 1. \tag{25}$$

Its radius-vector $\mathbf{r} = \mathbf{m}r(\mathbf{m})$ has length, which, by (24), may be expressed as

$$r(\mathbf{m}) = \left[\sum_{\alpha=1}^{3} c_\alpha^2(\mathbf{m})\right]^{-\frac{1}{2}}. \tag{26}$$

By its definition (25), the considered ellipsoid (26) gives complete information about the orientational dependence of the combination

$$c_1^2(\mathbf{m}) + c_2^2(\mathbf{m}) + c_3^2(\mathbf{m})$$

on the unit sphere $\mathbf{m}^2 = 1$. Note in particular that in these terms the geometrical meaning of theorem (22) becomes clear: Eqn. (21) is equivalent to a known property of ellipsoid,

$$r^{-2}(\mathbf{n}) + r^{-2}(\mathbf{s}) + r^{-2}(\mathbf{t}) = const \tag{27}$$

for any orientation of the orthogonal system {n,s,t}.

2.2.2. *A General Relation Between* $\max c_1^2(\mathbf{m})$ *and* $\min c_3^2(\mathbf{m})$

Now let us consider any two directions $\mathbf{m}$ and $\mathbf{n}$ (in general they need not be orthogonal to one another). By means of the invariant properties of the tensor c_{ijkl} with respect to transposition of the subscripts, we easily verify that the following convolutions are equal

$$\mathbf{m}(nn)\mathbf{m} = \mathbf{n}(mm)\mathbf{n}. \tag{28}$$

The vector $\mathbf{m}$ can be represented in the form of an expansion in terms of the basis vectors $\mathbf{A}_\alpha(\mathbf{n})$, and the vector $\mathbf{n}$ correspondingly in the system $\mathbf{A}_\alpha(\mathbf{m})$. Then obviously

$$\sum_{\alpha=1}^{3} [\mathbf{m} \cdot \mathbf{A}_\alpha(\mathbf{n})]^2 = 1, \quad \sum_{\alpha=1}^{3} [\mathbf{n} \cdot \mathbf{A}_\alpha(\mathbf{m})]^2 = 1. \tag{29}$$

Substituting the representation of the matrix (mm) from (7) and the analogous representation of (nn) into (28) and remembering that as stipulated, $c_1(\mathbf{m}) \le c_2(\mathbf{m}) \le c_3(\mathbf{m})$, by (29) we have

$$\rho c_1^2(\mathbf{n}) \le \mathbf{m}(nn)\mathbf{m} = \mathbf{n}(mm)\mathbf{n} \le \rho c_3^2(\mathbf{m}). \tag{30}$$

Thus for any propagation directions $\mathbf{m}$ and $\mathbf{n}$,

$$c_1(\mathbf{n}) \le c_3(\mathbf{m}). \tag{31}$$

From (31) we get the following relation between the top and bottom phase speed branches [9]

The largest value of the phase speed in the bottom branch is not greater than the smallest value of the phase speed in the top branch. (32)

In general, this theorem is valid irrespective of whether degeneracy is present or not. It leads to strict limitations on the possible configurations of the sheets of the phase velocity surface. In particular, according to theorem (32) the whole inner sheet must be inside of the sphere of the radius $c_3^{\min} = \min\{c_3(\mathbf{m})\}$ (Fig. 2).

2.2.3. *Relations Between the Phase Speeds of Waves Traveling Along the Directions* $\mathbf{m}$ *and* $\mathbf{A}(\mathbf{m})$

Consider the internal properties of the bottom branch. Suppose that for the propagation direction $\mathbf{m}$ the polarization is $\mathbf{A}_1$, so that by (3)

$$\mathbf{A}_1 \cdot (mm)\mathbf{A}_1 = \rho c_1^2(\mathbf{m}). \tag{33}$$

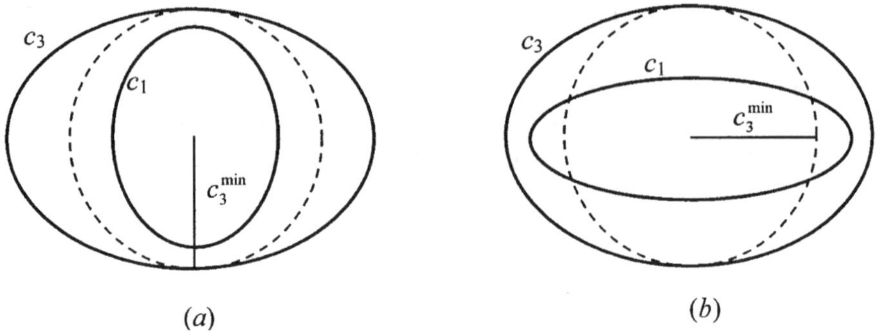

Figure 2. Two schematic phase velocity curves in a plane through the origin intersecting the velocity surface; (a) permissible configuration; (b) forbidden configuration contradicting theorem (32)

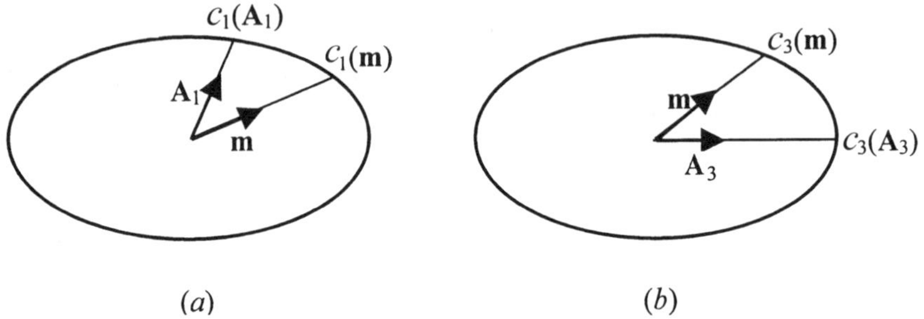

Figure 3. Illustrations to Eqn.(35) - (a) and Eqn.(37) - (b)

On the other hand, by (30),

$$\mathbf{A}_1 \cdot (mm)\mathbf{A}_1 = \mathbf{m} \cdot (A_1 A_1)\mathbf{m} \geq \rho c_1^2(\mathbf{A}_1) \tag{34}$$

where the inequality sign becomes an equality sign if the vector $\mathbf{m}$ happens to be the true polarization for the propagation direction $\mathbf{A}_1$. Combining (33) and (34) we obtain

$$c_1(\mathbf{m}) \geq c_1(\mathbf{A}_1). \tag{35}$$

Relation (35) expresses an important property of the bottom phase-speed branch (Fig. 3a), which may be formulated as follows [9]

> *The polarization vector $\mathbf{A}_1$ of a wave propagating in the direction $\mathbf{m}$ at speed $c_1(\mathbf{m})$ and belonging to the bottom branch points out another propagation direction where the wave speed $c_1(\mathbf{A}_1)$ does not exceed $c_1(\mathbf{m})$.* (36)

Pursuing along this line we obtain an algorithm for the search for a minimum, $c_1^{\min} = \min\{c_1(\mathbf{m})\}$. Of course, if the wave in question is pure longitudinal, this "other" direction simply coincides with the original direction. As is known, purely longitudinal polarization always corresponds to extremal values of the speed $c_\alpha(\mathbf{m})$. But on the bottom branch the minimum does not usually relate to longitudinal polarization.

From (30) together with (34) follows another inequality [9],

$$c_3(\mathbf{m}) \le c_3(\mathbf{A}_3), \tag{37}$$

(Fig. 3b) which enables us to formulate a new statement concerning the top velocity branch and similar to the theorem (36)

The polarization vector $\mathbf{A}_3$ *of a wave propagating in the direction* $\mathbf{m}$ *at speed* $c_3(\mathbf{m})$ *and belonging to the top branch points out another propagation direction where the wave speed* $c_3(\mathbf{A}_3)$ *is not less than* $c_3(\mathbf{m})$. (38)

The relation (37) provides an algorithm for the search for a maximum, $c_3^{\max} = \max\{c_3(\mathbf{m})\}$.

2.2.4. *Relations for Degenerate Phase Speeds*

Now let us consider the degeneracy between the bottom and middle branches in the direction $\mathbf{m}_d$, $c_1(\mathbf{m}_d) = c_2(\mathbf{m}_d)$ (Fig. 1). At the degeneracy point the solution for the polarization vector $\mathbf{A}$ can be represented in the form of any linear combination of the type

$$\mathbf{A} = \alpha\mathbf{A}_1 + \beta\mathbf{A}_2, \quad \alpha^2 + \beta^2 = 1, \tag{39}$$

where $\mathbf{A}_1$ and $\mathbf{A}_2$ are arbitrary orthogonal unit vectors in the plane of "degeneracy" perpendicular to the polarization vector $\mathbf{A}_3(\mathbf{m}_d)$ of the non-degenerate branch. Repeating the arguments, which led us to inequality (35), in this case we again get

$$c_1(\mathbf{m}) \ge c_1(\mathbf{A}). \tag{40}$$

With all possible changes in α and β, the vector $\mathbf{A}$, (39) describes a unit circle in a plane containing $\mathbf{A}_1$ and $\mathbf{A}_2$. From this viewpoint, the content of the inequality (40) may be expressed in the form of the theorem [9]

The phase speed at a degeneracy point between the bottom and middle branches is not less than the greatest speed on the bottom branch on the great circle of directions described by vector $\mathbf{A}$ *(39) on the sphere* $\mathbf{m}^2 = 1$. (41)

See Fig. 1. Quite analogously one can prove the theorem

The phase speed at a degeneracy point between the top and middle branches is not larger than the least speed on the top branch on the great circle of directions described by vector $\mathbf{A}$ *(39) on the sphere* $\mathbf{m}^2 = 1$. (42)

We note that at a point of triple degeneracy vector $\mathbf{A}$ is in general directed in any arbitrary way. In this case the meaning of inequality (40) reduces to the following assertion.

Triple degeneracy can be realized only at a point where the speed on the bottom branch is a maximum and coincides with the minimum velocity on the top branch. (43)

Incidentally, this assertion also follows directly from theorem (32).

2.3. RAY AND GROUP VELOCITIES, NORMAL WAVES, ENERGY CONCENTRATING

In the above considerations we dealt with only one velocity characteristic of bulk wave propagation the phase speed $c = \omega/k$, or in a vector form

$$\mathbf{c} = c\mathbf{m}, \tag{44}$$

which describes the speed and direction of the motion of planes of constant phase. Below we shall introduce two other velocity concepts: the velocity of energy propagation (a ray velocity $\mathbf{v}_r$) and the velocity of the wave packet motion (a group velocity $\mathbf{v}_g$). In contrast to isotropic bodies, in crystals the latter velocities generally do not coincide with the phase speed $\mathbf{c}$. On the other hand, as we shall see, homogeneous bulk waves in arbitrary crystals are characterized by identically equal ray and group velocities, $\mathbf{v}_r = \mathbf{v}_g$. It is less known that the latter identity does not hold for inhomogeneous waves [10-12]. We shall return to this feature of ray and group velocities in the separate section related to the theory of surface waves.

All relations established in this section are equally applicable to all of wave branches. For this reason the subscript α is omitted throughout almost in all expressions.

2.3.1. *Ray and Group Velocities*

The energy flux carried by the elastic wave is described by the elastic Poynting vector,

$$\mathbf{J}(\mathbf{x}, t) = -\dot{\mathbf{u}}(\mathbf{x}, t)\boldsymbol{\sigma}(\mathbf{x}, t), \tag{45}$$

where $\boldsymbol{\sigma}$ is the stress field of the wave,

$$\sigma_{ij} = c_{ijkl}u_{l,k}. \tag{46}$$

The corresponding energy density in the wave field is given by

$$E(\mathbf{x}, t) = \tfrac{1}{2}\left\{\rho[\dot{\mathbf{u}}(\mathbf{x}, t)]^2 + \sigma_{ij}(\mathbf{x}, t)u_{j,i}(\mathbf{x}, t)\right\}. \tag{47}$$

In Eqns.(45), (47), quadratic in $\mathbf{u}$ with (46), one should use the real part of expression (1):

$$\mathbf{u}(\mathbf{x}, t) = u_0 \mathbf{A} \cos\varphi, \quad \varphi = k(\mathbf{m} \cdot \mathbf{x} - ct). \tag{48}$$

In these terms

$$\mathbf{J} = c(ku_0)^2 (AA)\mathbf{m} \sin^2\varphi, \tag{49}$$

$$E = \rho(cku_0)^2 \sin^2\varphi, \tag{50}$$

where we have made use of the Christoffel equation (3). The ratio of the energy flux $\mathbf{J}$ to the energy density E is dimensionally and in the physical meaning a velocity of energy transport,

$$\mathbf{v}_r = \mathbf{J}/E = (AA)\mathbf{m}/\rho c, \tag{51}$$

which is also termed a ray velocity.

The group velocity of the wave is usually defined [1] by the derivative

$$\mathbf{v}_g = \frac{\partial \omega}{\partial \mathbf{k}} = \frac{\partial c}{\partial \mathbf{m}}. \tag{52}$$

As follows from the Christoffel equation (3) the function $c(\mathbf{m})$ is determined by

$$\rho c^2 = \mathbf{m} \cdot (AA)\mathbf{m}. \tag{53}$$

Let us take the gradient with respect to $\mathbf{m}$ on both sides of (53) while making use of the identity

$$\mathbf{m} \cdot \left[\frac{\partial}{\partial m_i}(AA)\right] \mathbf{m} = 2\frac{\partial \mathbf{A}}{\partial m_i}(mm)\mathbf{A} = \rho c^2 \frac{\partial}{\partial m_i}\mathbf{A}^2 = 0, \tag{54}$$

where we have taken into account that $\mathbf{A} \cdot \mathbf{A} = 1$. As a result, we immediately conclude that

$$\mathbf{v}_g = (AA)\mathbf{m}/\rho c, \tag{55}$$

i.e. the group velocity (55) and the ray velocity (51) are identically equal,

$$\mathbf{v}_g = \mathbf{v}_r. \tag{56}$$

Thus, for homogeneous bulk waves we need not distinguish between ray and group velocities and we will often use the common notation $\mathbf{v}$ for these velocities.

Let us derive several useful properties of the group (ray) velocity $\mathbf{v}$. Its universal connection with the phase speed c for the direction $\mathbf{m}$ is easily deduced by comparing Eqns. (51) and (53):

$$\mathbf{v} \cdot \mathbf{m} = c. \tag{57}$$

Hayes [13] established a series of universal connections between the group velocities $\mathbf{v}_\alpha(\mathbf{m})$ $(\alpha = 1, 2, 3)$ of three isonormal waves propagating along $\mathbf{m}$. They follow from Eqn.(51) rewritten in the form

$$\rho c_\alpha(\mathbf{m}) v_{\alpha i}(\mathbf{m}) = c_{ijkl} m_l A_{\alpha j} A_{\alpha k}. \tag{58}$$

Taking into account that the polarizations $\mathbf{A}_\alpha$ $(\alpha = 1, 2, 3)$ form an orthonormal basis, Eqn.(6), satisfying the identity

$$\sum_{\alpha=1}^{3} A_{\alpha j} A_{\alpha k} = \delta_{jk}. \tag{59}$$

we obtain from (58)

$$\rho \sum_{\alpha=1}^{3} c_\alpha(\mathbf{m}) v_{\alpha i}(\mathbf{m}) = c_{ijjl} m_l = V_{il} m_l, \tag{60}$$

or, in an invariant form, with making use of Eqn.(57),

$$\rho \sum_{\alpha=1}^{3} \mathbf{v}_\alpha(\mathbf{m})[\mathbf{v}_\alpha(\mathbf{m}) \cdot \mathbf{m}] = \mathbf{V}\mathbf{m}. \tag{61}$$

The first consequence of the identity (61) is based on the fact that any three coplanar (unit) vectors $\mathbf{m}, \mathbf{p}$ and $\mathbf{q}$ $(\mathbf{m} \cdot [\mathbf{p} \times \mathbf{q}] = 0)$ are linearly dependent, i.e. there exists a linear relation between them, $\mathbf{m} = \beta \mathbf{p} + \gamma \mathbf{q}$. Repeating (61) for $\mathbf{p}$ and $\mathbf{q}$ directions of propagation and combining the results multiplied by β and γ, respectively, with Eqn.(61) one easily obtains

$$\sum_{\alpha=1}^{3} \mathbf{v}_\alpha(\mathbf{m})[\mathbf{v}_\alpha(\mathbf{m}) \cdot \mathbf{m}] = \beta \sum_{\alpha=1}^{3} \mathbf{v}_\alpha(\mathbf{p})[\mathbf{v}_\alpha(\mathbf{p}) \cdot \mathbf{p}] + \gamma \sum_{\alpha=1}^{3} \mathbf{v}_\alpha(\mathbf{q})[\mathbf{v}_\alpha(\mathbf{q}) \cdot \mathbf{q}]. \tag{62}$$

The second consequence of Eqn.(61) directly follows from the symmetry of the matrix $\mathbf{V}$ providing the identity

$$\mathbf{n} \cdot \mathbf{V}\mathbf{m} = \mathbf{m} \cdot \mathbf{V}\mathbf{n}. \tag{63}$$

Taking a scalar product of both sides of Eqn.(61) with the arbitrary unit vector $\mathbf{n}$ and replacing in the result $\mathbf{n} \leftrightarrow \mathbf{m}$ we find, in view of (63), the other universal connection between the six group velocities for two arbitrary directions of propagation:

$$\sum_{\alpha=1}^{3} [\mathbf{v}_\alpha(\mathbf{m}) \cdot \mathbf{n}][\mathbf{v}_\alpha(\mathbf{m}) \cdot \mathbf{m}] = \sum_{\alpha=1}^{3} [\mathbf{v}_\alpha(\mathbf{n}) \cdot \mathbf{m}][\mathbf{v}_\alpha(\mathbf{n}) \cdot \mathbf{n}]. \tag{64}$$

The third connection between the group velocities of isonormal bulk waves, this time for the case of three mutually orthogonal wave normals $\mathbf{n}, \mathbf{s}$ and $\mathbf{t}$, can be deduced by replacing the phase speeds c_α in the identity (21) by corresponding scalar products of the type (57). In these terms Eqn.(21) takes a new appearance [13],

$$\sum_{\alpha=1}^{3} \left\{ [\mathbf{v}_\alpha(\mathbf{n}) \cdot \mathbf{n}]^2 + [\mathbf{v}_\alpha(\mathbf{s}) \cdot \mathbf{s}]^2 + [\mathbf{v}_\alpha(\mathbf{t}) \cdot \mathbf{t}]^2 \right\} = \mathrm{tr}\mathbf{V}/\rho = const. \qquad (65)$$

2.3.2. *The Ray Inclination. Normal Waves.*

If we insert into (51) or (55) the parameters for an isotropic medium the result would be quite predictable: certainly, in this simple case the ray velocities for longitudinal and transverse waves, $\mathbf{v}_{l,t}$, must trivially coincide in both direction and magnitudes with the corresponding phase velocities $\mathbf{c}_{l,t} = c_{l,t}\mathbf{m}$. In crystals, as already mentioned, the situation is more complex. An anisotropy replaces the trivial coincidence between these two velocities by a simple general relation (57).

It follows from Eqn.(57) that the ray velocity vector $\mathbf{v}(\mathbf{m})$ cannot have smaller length than the phase speed $c(\mathbf{m})$. They coincide in magnitude only for the so-called *normal* waves, for which the group velocity is parallel to the wave normal

$$\mathbf{v}(\mathbf{m}) || \mathbf{m}. \qquad (66)$$

For any other directions of wave propagation the length of the group velocity vector $\mathbf{v}$ is larger than the phase speed c. And the ratio v/c is the larger, the larger the angle ψ of inclination of the energy flux $\mathbf{J}$ from the wave normal $\mathbf{m}$:

$$v = c/\cos\psi. \qquad (67)$$

The magnitude v of the ray velocity is, in view of Eqn.(55), given by

$$v = \sqrt{\mathbf{m} \cdot (AA)^2 \mathbf{m}}/\rho c. \qquad (68)$$

Therefore in terms of polarizations the inclination angle ψ is determined by [14]

$$\cos\psi = \frac{c}{v} = \frac{\mathbf{m} \cdot (AA)\mathbf{m}}{\sqrt{\mathbf{m} \cdot (AA)^2 \mathbf{m}}} \qquad (69)$$

where the relation (53) was also used in the derivation.

Let us now derive the equivalent equation for the inclination angle ψ in terms of the phase velocity $c(\mathbf{m})$. Taking into account, that by Eqn.(5) $c(\mathbf{m})$ is a homogeneous function, one may apply to it the Euler theorem, which gives

$$m_i \frac{\partial c}{\partial m_i} \equiv \mathbf{m} \frac{\partial c}{\partial \mathbf{m}} = c. \qquad (70)$$

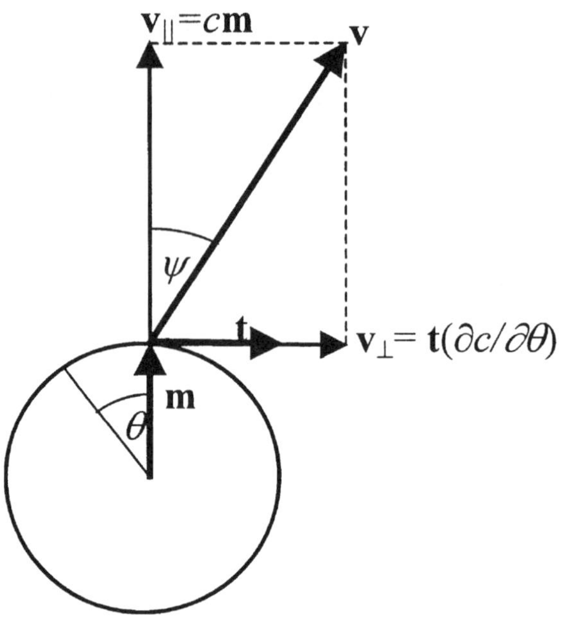

Figure 4. The ray inclination relative to the wave normal

In turn, this allows transformation of Eqn.(52) to the form

$$\mathbf{v} = c\mathbf{m} + (\mathbf{I} - \mathbf{m} \otimes \mathbf{m})\frac{\partial c}{\partial \mathbf{m}}. \tag{71}$$

The last relation represents a decomposition of the vector $\mathbf{v}$ into parts parallel and perpendicular to the wave normal $\mathbf{m}$ (Fig. 4),

$$\mathbf{v} = \mathbf{v}_{||} + \mathbf{v}_{\perp}, \quad \mathbf{v}_{||} = \mathbf{c} = c\mathbf{m}, \quad \mathbf{v}_{\perp} = (\mathbf{I} - \mathbf{m} \otimes \mathbf{m})\frac{\partial c}{\partial \mathbf{m}} = \left.\frac{\partial c}{\partial \mathbf{m}}\right|_{|\mathbf{m}|=1}. \tag{72}$$

In Eqns.(71), (72) the matrix $\mathbf{P}_{\mathbf{m}} = \mathbf{I} - \mathbf{m} \otimes \mathbf{m}$ is obviously a projection operator onto the plane orthogonal to $\mathbf{m} : \mathbf{P}_{\mathbf{m}}\mathbf{m} = \mathbf{0}$. The right-hand side of Eqn.(72) is the gradient in the plane tangent to the sphere $\mathbf{m}^2 = 1$. It may be expressed through the derivative with respect to the orientation of $\mathbf{m}$:

$$\mathbf{v}_{\perp} = \left.\frac{\partial c}{\partial \mathbf{m}}\right|_{|\mathbf{m}|=1} = \mathbf{t}\frac{\partial c}{\partial \theta}, \tag{73}$$

where θ is a polar angle of $\mathbf{m}$ in the plane $\{\mathbf{m}, \mathbf{v}\}$, and $\mathbf{t}$ is the unit vector

$$\mathbf{t} = \frac{\partial \mathbf{m}}{\partial \theta} \perp \mathbf{m} \tag{74}$$

in the same plane (Fig.4). By Eqn.(73) the orientation of this plane is determined by the requirement of the fastest change of the phase speed $c(\mathbf{m})$ for an infinitesimal perturbation $\mathbf{m} \to \mathbf{m} + \delta\mathbf{m}$. The direction $\delta\mathbf{m}_0$ providing the maximum of the difference $c(\mathbf{m} + \delta\mathbf{m}) - c(\mathbf{m})$ marks an orientation of the vectors $\mathbf{t}$ and $\mathbf{v}_\perp$. Combining Eqns.(73) and $(72)_2$ it is easy to derive an alternative formula determining the angle ψ of inclination of the ray relative to the wave normal (Fig.4):

$$\tan\psi = \frac{v_\perp}{v_{||}} = \frac{\partial \ln c}{\partial \theta}. \tag{75}$$

Thus the ray inclination is directly determined by the phase speed sensitivity to the change of orientation of the wave normal. On the other hand, the ray inclination points out the direction $\delta\mathbf{m}_0$ of increased phase speed, the faster the increase the larger is the inclination. In particular, this means that the energy transport for a given velocity branch tends to be concentrated in regions of orientations where the phase velocity surface is furthest away from the origin [14]. Of course, the reverse is also valid, the lesser the sensitivity of the phase speed $c(\mathbf{m})$ to the orientation of $\mathbf{m}$, the closer are the velocities $\mathbf{v}$ and $\mathbf{c}$. For some directions related to the so-called *stationary points* $\mathbf{m}_{st}$ on the unit sphere $\mathbf{m}^2 = 1$ defined by the equation

$$\left.\frac{\partial c}{\partial \mathbf{m}}\right|_{|\mathbf{m}|=1} = 0, \tag{76}$$

the condition (66) is fulfilled and the corresponding waves become normal. It is obvious, that the extremum property (76) represents both a sufficient and necessary condition for the propagation of a normal wave with group velocity $\mathbf{v} = \mathbf{c}$ along $\mathbf{m}_{st}$. In particular, the purely longitudinal waves $(\mathbf{A}||\mathbf{m})$, in view of Eqns.(3) and (55), clearly belong to the class of normal waves $(\mathbf{v}||\mathbf{A}||\mathbf{m}\ )$, which leads to the following consequence:

The waves of purely longitudinal polarization are characterized by extremal values of the phase speeds. (77)

The reverse statement is not valid, i.e. not every extremum of the function $c(\mathbf{m})$ corresponds to a longitudinal normal. Note that transverse waves generally are not normal. However, along symmetry directions, which are always longitudinal normals, transverse waves, as a rule, become normal, with only one exclusion: degenerate transverse waves along a 3 fold symmetry axis.

2.3.3. *Energy Concentration*

Thus, in crystals the energy current of the wave is generally not collinear with the wave vector $\mathbf{k}$. A homogeneous distribution of wave normals $\mathbf{m}$

on the unit sphere will then often result in an orientationally very inhomogeneous distribution of the corresponding rays. The relation between these distributions is very visual on a slowness surface representing a locus of the ends of all slowness vectors $\mathbf{s} = \mathbf{m}/c(\mathbf{m})$ outgoing from the same origin. Since the wave vector $\mathbf{k}$ may be expressed as $\mathbf{k} = \omega\mathbf{s}$, the slowness surface differs from the isofrequency surface

$$\omega(\mathbf{k}) = const, \tag{78}$$

only by scale. Hence the two surfaces possess the same normals. The normal to the isofrequency surface is clearly parallel to the group velocity $\mathbf{v}$, since by (52) and (78)

$$d\omega(\mathbf{k}) = \nabla\omega(\mathbf{k})d\mathbf{k} = \mathbf{v} \cdot d\mathbf{k} = 0 \tag{79}$$

for any $d\mathbf{k}$ belonging to the surface. Thus we can state that the normal direction $\mathbf{n}_S$ at any point $\mathbf{s}_\alpha(\mathbf{m})$ of the slowness surface S is specified by the direction of the group velocity $\mathbf{v}_\alpha = \nabla c_\alpha(\mathbf{m})$ of the wave propagating along $\mathbf{m}$.

More directly group velocity distributions are displayed by a ray surface R, known also as a group velocity surface or a wave surface. This surface is constructed from the radius-vector $\mathbf{v}(\mathbf{m})$ quite analogously to the way the slowness surface S was constructed from the radius vector $\mathbf{s} = \mathbf{m}/c(\mathbf{m})$. One can tell that $\mathbf{v}(\mathbf{m})$ maps the unit sphere $\mathbf{m}^2 = 1$ to the ray surface R. It is remarkable that at any point the normal $\mathbf{n}_R$ to the R surface must be parallel to the corresponding wave normal $\mathbf{m}$.

Both surfaces are shown in Fig.5 for the example of the cubic crystal *Ge*. One can see that a major part of the phase space is made up of *energy concentration* zones of the type COC' (Fig.5a) where the group velocity $\mathbf{v}$ belongs to rather small solid angles of the type BOB' (Fig.5b) around [100] directions. Thus, the energy flux corresponding to the wave packet traveling inside some solid angle $\Delta\Omega_\mathbf{m}$ might be concentrated in a solid angle $\Delta\Omega_\mathbf{v}$ much smaller $\Delta\Omega_\mathbf{m}$. For some directions $\mathbf{m} = \mathbf{m}_c$ the ratio $\Delta\Omega_\mathbf{m}/\Delta\Omega_\mathbf{v}$ may become singular

$$\lim_{\Delta\Omega_\mathbf{m} \to 0} (\Delta\Omega_\mathbf{m}/\Delta\Omega_\mathbf{v}) = \infty. \tag{80}$$

These special directions play a key role in the phonon focusing.

The basic quantitative characteristic of energy concentrating, originally introduced by Maris [15], relates to the left-hand side of Eqn.(80) and is known as the *enhancement factor*

$$A = \left| \frac{d\Omega_\mathbf{s}}{d\Omega_\mathbf{v}} \right|. \tag{81}$$

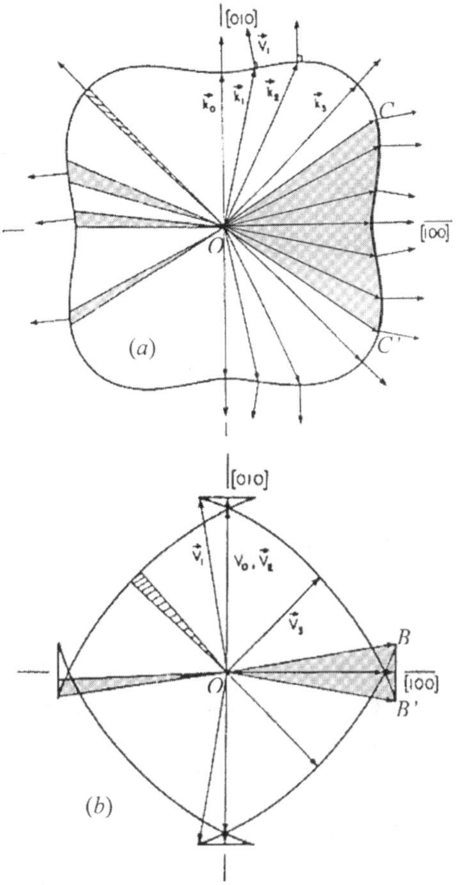

Figure 5. The example of energy concentration for Ge crystal [18] (a) the outer sheet of the slowness surface; (b) the corresponding sheet of the group velocity surface

There are several approaches to the evaluation of the factor A. In particular, Philip & Viswanathan [16] have noted that

$$A = \left| \frac{\sin\theta_s d\theta_s d\varphi_s}{\sin\theta_v d\theta_v d\varphi_v} \right| = \left| \frac{d(\cos\theta_s) d\varphi_s}{d(\sin\theta_v) d\varphi_v} \right| = \frac{1}{|J|}, \tag{82}$$

where J is the Jacobian of the transformation from the coordinates $\cos\theta_s \equiv t_s, \varphi_s$ determining the direction of the slowness vector $\mathbf{s} = \mathbf{m}/c$ to the analogous pair $\cos\theta_v \equiv t_v, \varphi_v$ relating to the orientation of the energy flux or the group velocity $\mathbf{v}$:

$$J = \det \begin{pmatrix} \partial t_v/\partial t_s & \partial t_v/\partial\varphi_s \\ \partial\varphi_v/\partial t_s & \partial\varphi_v/\partial\varphi_s \end{pmatrix}. \tag{83}$$

When both directions, $\mathbf{m}$ and $\mathbf{v}$, are known the further procedure is straightforward. For the details we refer readers to the original papers [16-19]. The alternative method of evaluation of the enhancement factor A is associated with its relation to the Gaussian curvature K of the slowness surface [18-20]:

$$A^{-1} = (v/c^3)|K|. \tag{84}$$

On the other hand, one can show that

$$K = (c/v)\,\det\{(\mathbf{I} - \mathbf{n}_s \otimes \mathbf{n}_s)\mathbf{G}(\mathbf{I} - \mathbf{n}_s \otimes \mathbf{n}_s)\}, \tag{85}$$

where $\mathbf{n}_s$ is the normal to the slowness surface at $\mathbf{s} = \mathbf{m}/c$ and $\mathbf{G}$ is the matrix with the components

$$G_{ij} = \frac{\partial^2 c}{\partial m_i \partial m_j}. \tag{86}$$

As was shown by Fedorov [1], $\mathbf{G}$ has no projection on the wave normal $\mathbf{m}$:

$$\mathbf{Gm} = 0. \tag{87}$$

Therefore $\det \mathbf{G} = 0$ and the inverse matrix $\mathbf{G}^{-1}$ does not exist. However, the adjoint matrix $\bar{\mathbf{G}}$ does exist which allows us to express (85) in a much more compact form

$$K = (c/v)\mathbf{n}_s \cdot \bar{\mathbf{G}}\mathbf{n}_s. \tag{88}$$

Thus, finally

$$A^{-1} = c^{-2}\mathbf{n}_s \cdot \bar{\mathbf{G}}\mathbf{n}_s, \tag{89}$$

or in components

$$\begin{aligned} A^{-1} &= c^{-2}\left\{n_{s1}^2\left(G_{22}G_{33} - G_{23}^2\right) + n_{s2}^2\left(G_{33}G_{11} - G_{31}^2\right) + \right.\\ &+ n_{s3}^2\left(G_{11}G_{22} - G_{12}^2\right) + 2n_{s2}n_{s3}\left(G_{12}G_{13} - G_{11}G_{23}\right) + \\ &\left. + 2n_{s3}n_{s1}\left(G_{23}G_{21} - G_{22}G_{31}\right) + 2n_{s1}n_{s2}\left(G_{31}G_{32} - G_{33}G_{12}\right)\right\}. \end{aligned} \tag{90}$$

The obtained result shows that a vanishing Gaussian curvature K results in an infinite enhancement factor A (81). Solutions of the equation $K = J = 0$ usually form closed lines on the slowness surface, known as *parabolic lines* [18, 21]. Parabolic lines are very common though not a necessary feature of crystals. As was shown by Every [21], for their existence it is sufficient for a crystal to have at least one conical acoustic axis. Parabolic lines on the slowness surface S represent in $\mathbf{k}$ -space the image of the locus points relating to the singulariry (80) of the enhancement factor. The corresponding image in the **r** -space is given by the so-called *caustics* on the ray surface R. As one could expect, caustics display much more complex patterns on R than parabolic lines on S.

Conical points on the slowness surface provide the conditions for a known *conical refraction*, which prevents energy transport within cones of directions about the acoustic axes. This directly corresponds to vanishing A factor at conical points on S where the curvature K diverges. Such singular points on S are displayed on the R surface as empty ellipses or circles of zero wave-intensity and can be regarded as anti- caustics[20-22].

Caustics and anti-caustics are features, which can be directly observed experimentally. Practically all progress with observations of the discussed energy-concentration effects has been made in low temperature experiments with so-called *ballistic phonons* - long wavelength phonons ($10^{11} - 10^{12}$ Hz) having a mean free path comparable with the size of the crystal. Such phonons are usually produced by the *heat-pulse* method, which generally employs a resistive film on the crystal face to create the heat pulse and a fast superconductive bolometer on the opposite face to detect the slight temperature increase when ballistic phonons arrives. The first heat-pulse measurements of phonon focusing in LiF, KCl and Al_2O_3 crystals in the temperature range 1,5-3,5 K were accomplished by Taylor, Maris & Elbaum [23]. A very significant step in the phonon-focusing investigations has been taken by Northrop & Wolfe [17, 18] who developed the alternative phonon imaging method. For the details of this method and for more extensive literature reflecting the further progress in this field, both in theory and in experiments, we refer readers to the review articles [21, 24] and to the book [25].

2.3.4. *Existence, Non-existence and Possible Numbers of Acoustic Axes*
Let us denote the three eigenvalues of the acoustical tensor $\mathbf{Q}$, Eqns.(3)-(5), as $\lambda_\alpha, \lambda_\beta$ and λ_γ, and suppose that along the direction $\mathbf{m} = \mathbf{m}_d$ degeneracy of two of them occurs, say $\lambda_\alpha = \lambda_\beta$. Then, instead of (16), one can write

$$\mathbf{Q}(\mathbf{m}_d) = \lambda_\alpha \mathbf{I} + (\lambda_\gamma - \lambda_\alpha)\mathbf{A}_\gamma \otimes \mathbf{A}_\gamma. \tag{91}$$

In fact, this relation represents one of the forms of the equation for acoustic axes. Taking into account that $Q_{ij} = Q_{ji}$, we are dealing here with a system of six equations with six unknown values: $\lambda_\alpha, \lambda_\gamma$ and the directions of $\mathbf{m}_d$ and $\mathbf{A}_\gamma$. Since each of these six equations is of the degree 2, the system may have at most $2^6 = 64$ solutions. However, it is clear that solutions $(\mathbf{m}_d, \mathbf{A}_\gamma)$, $(\mathbf{m}_d, -\mathbf{A}_\gamma)$, $(-\mathbf{m}_d, \mathbf{A}_\gamma)$ and $(-\mathbf{m}_d, -\mathbf{A}_\gamma)$ are physically equivalent. Therefore among 64 solutions there are only 64:4 $=$ 16 different solutions. Of course, this consideration does not work for isotropic and transversely isotropic media where the above system becomes degenerate and can provide a continuum of solutions (all of the sphere $\mathbf{m}^2 = 1$ for the isotropic case and a circle on the sphere under definite conditions on elastic moduli - for transverse isotropy). Thus the following statement must be valid [26-28]

> *In crystals of unrestricted anisotropy, however not isotropic or transversely isotropic, the total number of acoustic axes does not exceed 16.* (92)

Let us now consider the other forms of equations for acoustic axes. Consider the tensor

$$\mathbf{F} = \mathbf{Q} - \lambda_\alpha \mathbf{I}. \tag{93}$$

In the vicinity of $\mathbf{m}_d$ where $\lambda_\alpha \neq \lambda_\beta$, in view of (7), $\mathbf{F}$ is a planar tensor representing by a sum of two dyads,

$$\mathbf{F}(\mathbf{m}) = (\lambda_\beta - \lambda_\alpha)\mathbf{A}_\beta \otimes \mathbf{A}_\beta + (\lambda_\gamma - \lambda_\alpha)\mathbf{A}_\gamma \otimes \mathbf{A}_\gamma. \tag{94}$$

And at $\mathbf{m} = \mathbf{m}_d$ it becomes linear, i.e. reduces to one dyad,

$$\mathbf{F}(\mathbf{m}_d) = (\lambda_\gamma - \lambda_\alpha)\mathbf{A}_\gamma \otimes \mathbf{A}_\gamma. \tag{95}$$

In accordance with Eqn.(5), in both cases

$$\det \mathbf{F} = 0. \tag{96}$$

And the product $\mathbf{F}$ with the adjoint matrix $\mathbf{F}\bar{\mathbf{F}}$ must vanish:

$$\mathbf{F}\bar{\mathbf{F}} = (\det \mathbf{F})\mathbf{I} = 0. \tag{97}$$

However by [1], in the first case the matrix $\bar{\mathbf{F}}(\mathbf{m}) \neq 0$ being equal to the dyad

$$\bar{\mathbf{F}} = (\lambda_\beta - \lambda_\alpha)(\lambda_\gamma - \lambda_\alpha)\mathbf{A}_\alpha \otimes \mathbf{A}_\alpha. \tag{98}$$

And only at the degeneracy point $\mathbf{m} = \mathbf{m}_d$ where $\lambda_\alpha = \lambda_\beta$ we have

$$\bar{\mathbf{F}}(\mathbf{m}_d) = 0. \tag{99}$$

This is a necessary condition for a degeneracy at $\mathbf{m}_d$. Khatkevich [29] noticed that the corresponding sufficient condition for an acoustic axis is less restrictive and does not require of vanishing all components of the matrix $\bar{\mathbf{F}}$,

$$\bar{F}_{11} = F_{22}F_{33} - F_{23}^2, \quad \bar{F}_{22} = F_{11}F_{33} - F_{13}^2, \quad \bar{F}_{33} = F_{11}F_{22} - F_{12}^2, \tag{100}$$

$$\bar{F}_{12} = F_{13}F_{23} - F_{12}F_{33}, \bar{F}_{13} = F_{12}F_{23} - F_{13}F_{22}, \bar{F}_{23} = F_{12}F_{13} - F_{23}F_{11}. \tag{101}$$

Indeed, for the initial coordinate system chosen so that in the vicinity of $\mathbf{m}_d$

$$Q_{ij} \neq 0, \quad i \neq j, \tag{102}$$

it is sufficient to require

$$\bar{F}_{12} = 0, \quad \bar{F}_{13} = 0, \quad \bar{F}_{23} = 0, \tag{103}$$

in order that the other components (100) of the matrix $\bar{\mathbf{F}}$ automatically vanish. With (93), (101) the system (103) determines both the directions of acoustic axes and the degenerate eigenvalues λ_α. Eliminating λ_α from the system (103), we arrive at the two conditions of Khatkevich [29] solely in terms of the components Q_{ij} :

$$R_1 = (Q_{11} - Q_{22})Q_{13}Q_{23} - Q_{12}(Q_{13}^2 - Q_{23}^2) = 0, \tag{104}$$

$$R_2 = (Q_{11} - Q_{33})Q_{12}Q_{23} - Q_{13}(Q_{12}^2 - Q_{23}^2) = 0. \tag{105}$$

Alshits & Lothe [30] have added to R_1 and R_2 five more components:

$$R_3 = (Q_{22} - Q_{33})Q_{12}Q_{13} - Q_{23}(Q_{12}^2 - Q_{13}^2), \tag{106}$$

$$\begin{aligned} R_4 = (Q_{11} - Q_{22})(Q_{11} - Q_{33})Q_{23} - (Q_{11} - Q_{33})Q_{12}Q_{13}+ \\ +Q_{23}(Q_{12}^2 - Q_{23}^2), \end{aligned} \tag{107}$$

$$\begin{aligned} R_5 = (Q_{22} - Q_{11})(Q_{22} - Q_{33})Q_{13} - (Q_{22} - Q_{33})Q_{12}Q_{23}+ \\ +Q_{13}(Q_{12}^2 - Q_{13}^2), \end{aligned} \tag{108}$$

$$\begin{aligned} R_6 = (Q_{33} - Q_{11})(Q_{33} - Q_{22})Q_{12} - (Q_{33} - Q_{22})Q_{13}Q_{23}+ \\ +Q_{12}(Q_{13}^2 - Q_{12}^2), \end{aligned} \tag{109}$$

$$\begin{aligned} R_7 = (Q_{11} - Q_{22})(Q_{11} - Q_{33})(Q_{22} - Q_{33})+ \\ +(Q_{22} - Q_{33})(Q_{13}^2 - Q_{23}^2) + (Q_{11} - Q_{22})(Q_{13}^2 - Q_{12}^2). \end{aligned} \tag{110}$$

In these terms they formed the seven-component vector

$$\boldsymbol{\xi} = \{R_1, R_2, ..., R_7\} \tag{111}$$

and proved that the equation

$$\boldsymbol{\xi} = 0 \tag{112}$$

represent an invariant criterion of degeneracy valid in an arbitrary coordinate system independently of the condition (102).

Now suppose that we have solved Eqn.(112) and found the direction $\mathbf{m} = \mathbf{m}_d$ of an acoustic axis. The next step should be to find all wave characteristics along $\mathbf{m}_d$. Let us start from the degenerate eigenvalue $\lambda_\alpha = \rho c_\alpha^2$. Here there are two possibilities.

i) If at least one nondiagonal component of tensor $\mathbf{Q}$ does not vanish, say $Q_{23} \neq 0$, then from the condition $\bar{F}_{23} = 0$ with Eqns.(93) and (101) one has the degenerate eigenvalue

$$\lambda_\alpha = Q_{11} - Q_{12}Q_{13}/Q_{23}. \tag{113}$$

ii) If all components Q_{ij} $(i \neq j$) vanish, then along the direction $\mathbf{m}_d$ the condition $R_7 = 0$ must be satisfied, which implies coincidence of some pair of diagonal components, say $Q_{22} = Q_{33}$. The corresponding solution of the system $\bar{F}_{11} = \bar{F}_{22} = \bar{F}_{33} = 0$, as follows from (100), is given by

$$\lambda_\alpha = Q_{22}. \tag{114}$$

Knowing λ_α it is easy to find the nondegenerate eigenvalue λ_γ :

$$\lambda_\gamma = \det \mathbf{Q}(\mathbf{m}_d)/\lambda_\alpha^2 = \mathrm{tr}\mathbf{Q}(\mathbf{m}_d) - 2\lambda_\alpha. \tag{115}$$

The direction of the polarization $\mathbf{A}_\gamma$ of the nondegenerate wave is determined by Eqn.(91),

$$\mathbf{A}_\gamma || \mathbf{FC} = \mathbf{QC} - \lambda_\alpha \mathbf{C}, \tag{116}$$

where $\mathbf{C}$ is an arbitrary real vector. As was already mentioned, the polarizations of the degenerate waves may have any orientations in the plane orthogonal to $\mathbf{A}_\gamma$:

$$\mathbf{A}_\alpha \perp \mathbf{A}_\gamma. \tag{117}$$

Acoustic axes are very common objects in crystals. In fact, until now among practically studied materials there are no examples of crystals free of acoustic axes. And as a rule, a fastest phase-velocity branch remains nondegenerate. The only exclusion found in 1972 by Ohmachi et. al. [31] for TeO_2 crystals, where all three wave branches turn out to be degenerate, has prevented attempts to prove that the latter empirical observation is a general property of bulk eigenwaves in crystals. In the same way, attempts to prove a theorem of obligatory existence of acoustic axes in crystals of unrestricted anisotropy were stopped after Alshits & Lothe [9] in 1979 introduced the example of a thermodynamically stable model crystal without acoustic axes. According to [9], any orthorhombic crystal with the elastic moduli $c_{12} = c_{13} = c_{23} = 0$ and

$$0 < c_{22} < c_{66} < c_{11} < C_1 < C_2 < c_{55} << c_{33} \tag{118}$$

must be completely free of acoustic axes. In (118) $C_{1,2}$ constants are defined by the relations

$$C_1 = \min\left\{c_{44}, \frac{c_{55}(2c_{44} - c_{22})}{c_{55} + 2c_{44} - c_{22}}\right\}, C_2 = \max\left\{c_{44}, \frac{c_{55}(2c_{44} - c_{22})}{c_{55} + 2c_{44} - c_{22}}\right\}. \tag{119}$$

Later an alternative example of orthorhombic medium with nondegenerate phase speed branches was also found numerically [32, 33].

It is worthwhile to mention that acoustic properties of a crystal without acoustic axes must be rather unusual. In particular, in such a medium longitudinal normals are obligatory in all three sheets of a phase velocity surface, including the so-called "quasi-transversal" sheets. And along the latter directions in the "quasi-longitudinal" fastest sheet a purely transversal wave must propagate.

An example of a model medium free of acoustic axes would be impossible for systems of higher symmetry than orthorhombic. Any symmetry axis higher than 2-fold axis must be an acoustic axis. Accordingly, in tetragonal crystals only one acoustic axis along a principal 4-fold axis is obligatory, the other acoustic axes may exist and may not (altogether there could be 1, 5 or 9 acoustic axes in seven different combinations). In trigonal crystals there are possible only two possible variants: 4 (obligatory) or 10 acoustic axes. In cubic crystals 7 obligatory acoustic axes along 4- and 3-fold symmetry axes always exist, and no other degeneracies may occur in this symmetry system. In hexagonal crystals, which are transversely isotropic with respect to elastic properties, apart from 1 obligatory tangent acoustic axis along the principal axis, a cone of acoustic axes, corresponding to the intersection of two sheets of the phase velocity surface, may arise under the condition

$$(c_{66}-c_{44})[(c_{11}-c_{66})(c_{33}-c_{44})-(c_{44}+c_{13})^2]>0. \tag{120}$$

The angle θ_d between the acoustic axis of the cone and the principal axis is given by

$$\theta_d=\tan^{-1}\left[\frac{(c_{11}-c_{66})(c_{33}-c_{44})-(c_{44}+c_{13})^2}{(c_{11}-c_{66})(c_{66}-c_{44})}\right]^{\frac{1}{2}}. \tag{121}$$

The more detailed analysis of acoustic axes in crystals of particular symmetry systems one can find in the original papers [3, 4, 26, 28, 33, 34].

2.3.5. *Geometrical Types of Acoustic Axes and Polarization Singularities Near Degeneracies*

Though orientations of acoustic axes are determined by the same equation (112) and the basic characteristics of the eigenwaves propagating along $\mathbf{m}_d$ are universal, Eqns.(114)-(117), degeneracy directions differ from each other by their neighborhood. They can be classified by geometrical types of contact along $\mathbf{m}_d$ of degenerate velocity sheets or/and by types of singularities of vector polarization fields $\mathbf{A}_{\alpha,\beta}(\mathbf{m})$ around $\mathbf{m}_d$. Geometrical characterization of acoustic axes has a long history. One can find mentioning conical and tangent points of contact and lines of intersection of degeneracy sheets (the latter for transverse isotropy, see Eqn.(121)) already in the papers by Herring [35, 36] and Khatkevich [26, 29], and in the books by Fedorov [1] and Musgrave [37]. Alshits & Lothe [30, 38] (see also [39, 40]) first noticed that geometrical features of degeneracies correlate with definite types of polarization singularities. This observation was elaborated in [41] where a complete classification of acoustic axes was constructed including all possible types of local geometry of the velocity sheets near the degeneracy, and the corresponding polarization singularities (see Fig.6). The developed theory also provides the particular algebraic conditions for any type of degeneracy, without solving the wave equation for arbitrary anisotropy, but using only appropriate convolutions of the elastic moduli tensor c_{ijkl}.

As an example, let us construct the two vectors $\mathbf{p}$ and $\mathbf{q}$,

$$\mathbf{p} = (\mathbf{S}_{11} - \mathbf{S}_{22})\mathbf{m}_d, \quad \mathbf{q} = (\mathbf{S}_{12} + \mathbf{S}_{21})\mathbf{m}_d, \tag{122}$$

where the four matrices $\mathbf{S}_{\alpha\beta}$ are defined similarly to the group velocity (55),

$$\mathbf{S}_{\alpha\beta} = (A_\alpha A_\beta)/2\rho c_d, \tag{123}$$

with c_d and $\mathbf{A}_{1,2}$ being the degenerate phase speed and the arbitrary orthogonal pair of unit vectors in the degeneracy plane, Eqn.(39) and Fig.1. Note in passing that the introduced vector $\mathbf{p}$ has the clear meaning of the semi-difference between the group velocities of two degenerate waves with polarizations $\mathbf{A}_1$ and $\mathbf{A}_2$. It turns out that the vectors $\mathbf{p}$ and $\mathbf{q}$ determine the geometry of the contact of the velocity sheets in the degeneracy point $\mathbf{m}_d$. If the vector product $\mathbf{p} \times \mathbf{q}$ does not vanish, one has a degeneracy of the conical type. It is customary to characterize point singularities in plane distributions of vector fields by a *topological "charge"* - a Poincare index n, which is defined as the angle (in 2π units) of aggregate rotation of vectors around a singular point. In these terms the topological charge of the conical degeneracy at $\mathbf{m}_d$ is given by

$$n = \frac{1}{2}\mathrm{sgn}[\mathbf{m}_d(\mathbf{p} \times \mathbf{q})]. \tag{124}$$

If $\mathbf{p} \times \mathbf{q} = 0$ but $\mathbf{p}$ or $\mathbf{q}$ does not vanish, we arrive at a wedge degeneracy at the point $\mathbf{m}_d$ or along the line passing through $\mathbf{m}_d$. The case $\mathbf{p} = \mathbf{q} = 0$ corresponds to a degeneracy of the tangent type at a point or along a line. In order to distinguish the degeneracy at a point from that along a line and in order to calculate the Poincare index for a wedge-point degeneracy ($n = 0, \pm\frac{1}{2}$) or for a tangent-point degeneracy ($n = 0, \pm 1$), one should use other convolutions of the tensor c_{ijkl} with the vectors $\mathbf{m}_d$ and $\mathbf{A}_\alpha$ ($\alpha = 1$, 2, 3), see [41, 42].

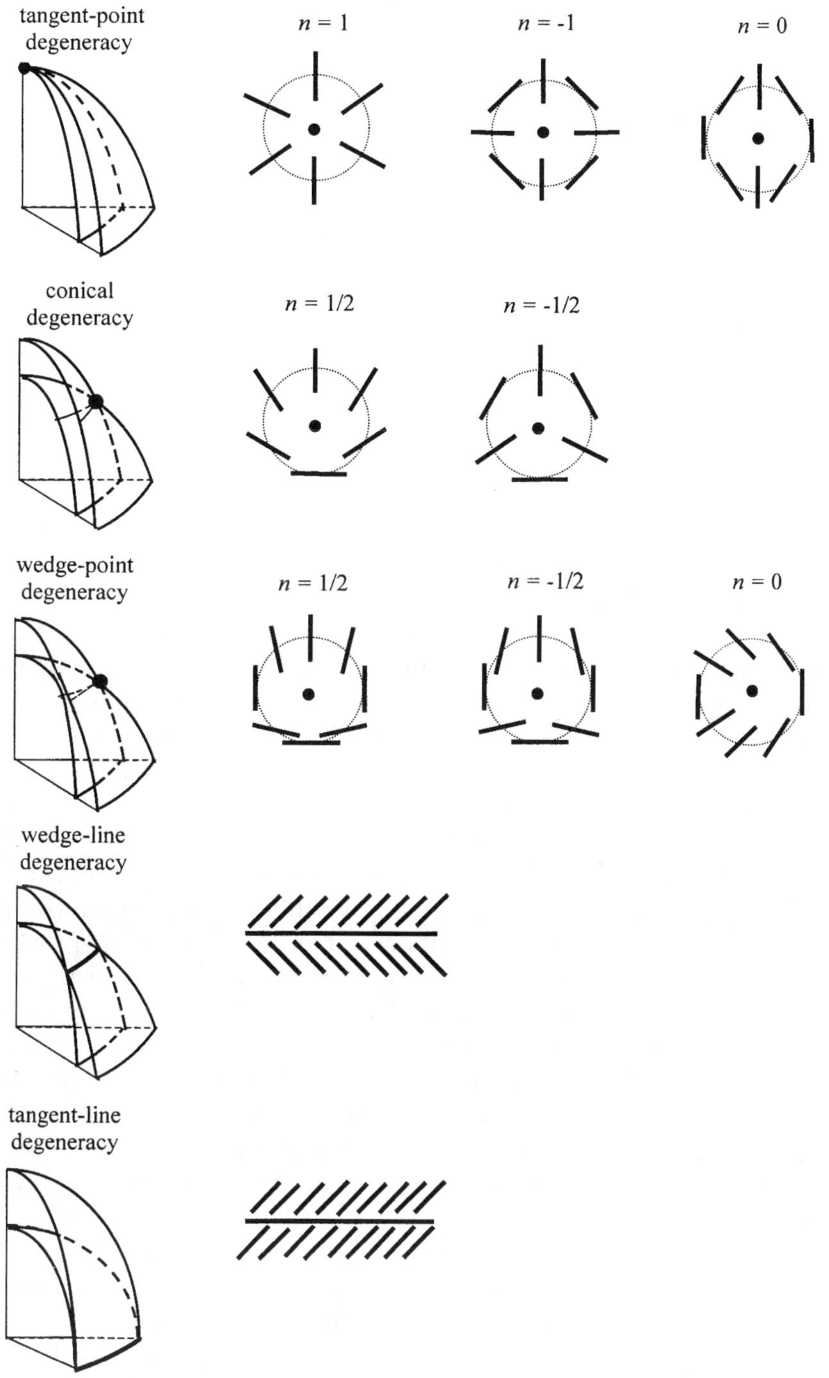

Figure 6. Geometrical and polarization types of degeneracies

It is essential that almost all types of acoustic axes exist in real crystals, only the wedge-point and tangent-line degeneracies are known for model crystals. Conical degeneracies ($n = \pm\frac{1}{2}$) exist practically in all known crystals of orthorhombic, monoclinic and triclinic symmetry systems. It is essential that a 2-fold axis does not obligatory provides a degeneracy, however when it happens to be an acoustic axis the latter must be of tangent type, and its topological charge depending on the material constants may appear of any relevant value $n = 0, \pm 1$. Along a 3-fold axis both in trigonal and in cubic crystals the conical degeneracy with the Poincare index $n = -\frac{1}{2}$ always occurs. A 4-fold axis in tetragonal and cubic crystals is always a tangent acoustic axes with the Poincare index $n = \pm 1$. In particular, for the cubic system the choice of the sign of n is especially simple: $n = \mathrm{sgn}(c_{12} + c_{44})$. An ∞ -fold axis in transverse isotropic crystals is also always a tangent acoustic axis, however its topological charge is definite: $n = 1$. In crystals of this symmetry one can also meet the wedge-line degeneracy, Eqns.(120), (121). In the model crystal, where apart from (120) also the condition $c_{44} = c_{66}$ is satisfied, the two symmetrical wedge lines in accordance with (121) must coalesce into one tangent degeneracy line in the basal plane ($\theta_d = \pi/2$). The appearance of two crossing "meridian" wedge lines with a tangent points at the places of their intersection on the "poles" is predicted [42] for the crystal Hg_2Cl_2 at the temperature of the phase transition of this crystal between the tetragonal and orthorhombic symmetry states. In the same paper [42] the sporadic tangent degeneracy with the Poicare index $n = 0$ is predicted for the monoclinic phase of the crystal $CsDSeO_4$ along the 2-fold axis. The above specific features must arise due to the coincidence $c_{44} = c_{55}$, which retains in the low symmetry phases of both crystals [43,44].

Different types of acoustic axes manifest a different behavior (disappearance, shifting, splitting) under small perturbation of the elastic properties. The analysis of this problem in [30, 41] reveals that only acoustic axes of the conical type are always stable under perturbations, i.e. they can only shift. Unstable points of degeneracy either split in accordance with the rule of conservation of topological charge (the sum of the Poincare indices of the degeneracies after a perturbation must be equal to the index n of the initial degeneracy) or vanish but only if $n = 0$. Such perturbations often arise due to different external influences, like electric fields, mechanical stresses or temperature variations in the vicinity of phase transitions. For instance, at a phase transition from a transverse isotropic to a trigonal crystal the ∞-fold axis ($n = 1$) is replaced by the 3-fold axis ($n = -\frac{1}{2}$). In accordance with the rule of index conservation in addition to the latter degeneracy there must arise also three conical acoustic axes of the index $n = \frac{1}{2}$ (Fig.7).

As was mentioned in [41] and studied in [45], even a conical degeneracy is unstable with respect to the "switching on" a small damping, which

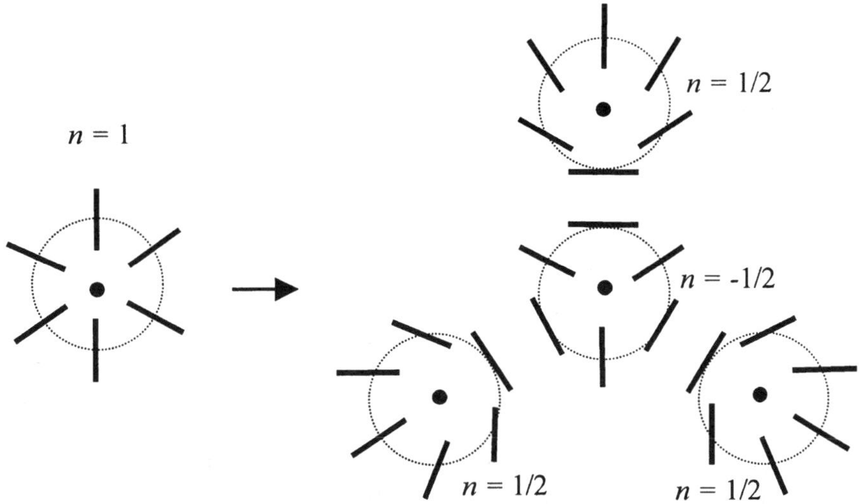

Figure 7. Splitting of the tangent degeneracy along the ∞-fold axis into the four conical degeneracies at the phase transition from the hexagonal to the trigonal symmetry system

is equivalent to a small imaginary perturbation of the phase speed, $v \to v - i\delta v$. As a result of such perturbation, the conical axis split into a pair of singular directions connected on the slowness surface and on the surface of damping by lines of intersection of corresponding sheets. The only two common inversely nonequivalent points of these lines on the unit sphere $\mathbf{m}^2 = 1$ correspond to new positions of acoustic axes. The polarizations of degenerate waves along the two new singular directions are circular. In the vicinity of these points of degeneracy, polarizations are elliptic. The rotation of large semi-axes of these ellipses around each of degeneracy points corresponds to the Poincare index $n = \pm\frac{1}{4}$ in a complete accordance with the rule of index conservation.

The extension of the theory [41] to piezoelectrics was given in [46, 47]. It was found that the classification of acoustic axes does not change apart from renormalization of explicit forms of the algebraic conditions for particular types of degeneracies, due to contributions from piezoelectric moduli. However this contribution may qualitatively change the wave properties near a specific acoustic axis and even change the type of the degeneracy itself, see [47, 48]. Piezoelectric coupling also causes the quasi-static electric field accompanying the elastic waves. Its characteristics depend on the polarization of the elastic wave and therefore also have definite singularities near the directions $\mathbf{m}_d$. These singularities were studied in [46].

Geometrical feature of the degenerate slowness sheets close to a conical point of their contact must create a corresponding cone of group velocities, which in turn may give rise to an internal conical refraction. This phenomenon was first observed by Klerk & Musgrave [49] in Ni crystals along the 3-fold axis parallel to the direction [111]. Quite similarly in the vicinity of the wedge line one can expect the wedge refraction, related to a fan-like distribution of rays, instead of usual cone. Such sort of refraction theoretically predicted by Khatkevich [26, 29] was later experimentally observed by Henneke & Green [50].

When the plane-wave approximation is replaced by packets of elastic waves, one has to answer the following natural question: how do such singular polarization fields in **k** space affect the wave fields in **r** space for beams propagating along directions of acoustic axes? In [51] the phenomenon of internal conical refraction was theoretically investigated from this point of view for acoustic and optical beams. The use of the Lorentzian profile made it possible to obtain the results in an analytical form. The answer on the above question has turned out to be quite non- trivial: the internal conical refraction creates the polarization disclination line in such beam. The wave field vanishes at any point of such disclination and there are two orientational singularities $n = 1$ and $n = -1$ in each cross-section orthogonal to the acoustic axes, $z = const > z_0$.

One general difference between orientational singularities of the wave fields in the **k** and **r** spaces should be pointed out. In **k** space each direction of the wave normal $\mathbf{m} = \mathbf{k}/k$ corresponds to its "own" plane wave with the polarization vector defined apart from sign, and therefore the Poicare index can be half-integer. In **r** space the wave fields must be continuous, and that is why in this case the orientational singularities appear in the vicinity of "zero amplitude" points and can be represented only by integer (or zero) Poincare indices.

3. Surface Waves and Other Inhomogeneous Wave Modes in Anisotropic Media

3.1. SUBSONIC SURFACE WAVES

3.1.1. *Stroh Formalism and Theorems of Existence and Uniqueness for Surface Wave Solutions*

Consider a half-infinite elastic medium of unrestricted anisotropy. Let us choose the coordinate system with the origin at the surface, the y -axis along the internal normal **n** to the surface and the x-axis along the arbitrary direction **m** of wave propagation, Fig.8. The plane specified by the unit vectors **m** and **n** is known as a *sagittal plane*. In the chosen coordinate system the steady-state displacement field of the plane wave can be presented in the

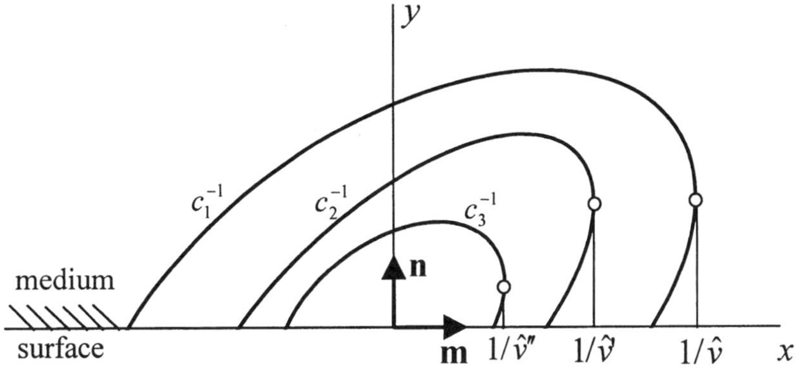

Figure 8. Cross section of slowness surface sheets by a sagittal plane

form of a superposition of partial waves,

$$\mathbf{u}(x - vt, y) = \mathbf{A}(y) \exp[ik(x - vt)], \quad \mathbf{A}(y) = \sum_{\alpha} b_\alpha \mathbf{A}_\alpha \exp(ikp_\alpha y), \tag{125}$$

which have equal x -components of the wave vector, $k_x \equiv k$, and a common tracing speed $v = \omega/k$, but different y-components of the wave vector, $k_y^{(\alpha)} = p_\alpha k$, and polarization vectors $\mathbf{A}_\alpha$. Stroh [52] introduced a powerful sextic formalism for anisotropic steady-state problems. This formalism has played an important role in theoretical crystalloacoustics. In parallel with the displacement field (125), Stroh proposed to consider the so-called traction field represented by a normal projection of the stress tensor,

$$\mathbf{n}\boldsymbol{\sigma}(x-vt, y) = -ik\mathbf{L}(y) \exp[ik(x-vt)], \mathbf{L}(y) = \sum_{\alpha} b_\alpha \mathbf{L}_\alpha \exp(ikp_\alpha y), \tag{126}$$

and formed the unknown 6-component vector function

$$\boldsymbol{\eta}(y) = \begin{pmatrix} \mathbf{A}(y) \\ \mathbf{L}(y) \end{pmatrix} = \sum_{\alpha=1}^{6} b_\alpha \boldsymbol{\xi}_\alpha \exp(ikp_\alpha y), \; \boldsymbol{\xi}_\alpha = \begin{pmatrix} \mathbf{A}_\alpha \\ \mathbf{L}_\alpha \end{pmatrix}. \tag{127}$$

The function $\boldsymbol{\eta}(y)$ yields the equation

$$\frac{\partial \boldsymbol{\eta}}{\partial y} = ik\mathbf{N}\boldsymbol{\eta}, \tag{128}$$

where $\mathbf{N}$ is the 6×6 matrix,

$$\mathbf{N}(v) = -\begin{pmatrix} (nn)^{-1}(nm) & (nn)^{-1} \\ (mn)(nn)^{-1}(nm) - (mm) - \rho v^2 I & (mn)(nn)^{-1} \end{pmatrix} \equiv$$

$$\equiv \begin{pmatrix} N_1 & N_2 \\ N_3 - \rho v^2 I & N_1^t \end{pmatrix}, \tag{129}$$

with the block 3×3 matrices of the type (ab) defined for any vectors $\mathbf{a}$ and $\mathbf{b}$ similarly to Eqn.(4) as $(ab)_{jk} = a_i c_{ijkl} b_l$. It is plain to see from (125)-(127) that the 6-vectors $\boldsymbol{\xi}_\alpha$ and the parameters p_α are just eigenvectors and eigenvalues of the matrix $\mathbf{N}$:

$$\mathbf{N}\boldsymbol{\xi}_\alpha = p_\alpha \boldsymbol{\xi}_\alpha. \tag{130}$$

The Stroh matrix $\mathbf{N}$ is not symmetric, but the matrix $\mathbf{NT}$ is: $\mathbf{NT} = (\mathbf{NT})^t$, where

$$\mathbf{T} = \begin{pmatrix} 0 & \mathbf{I} \\ \mathbf{I} & 0 \end{pmatrix}. \tag{131}$$

Accordingly, the orthogonality and completeness relations for the eigenvectors $\boldsymbol{\xi}_\alpha$ contain $\mathbf{T}$:

$$\boldsymbol{\xi}_\alpha \cdot \mathbf{T}\boldsymbol{\xi}_\beta = \delta_{\alpha\beta}, \quad \sum_{\alpha=1}^{6} \boldsymbol{\xi}_\alpha \otimes \mathbf{T}\boldsymbol{\xi}_\alpha = \mathbf{I}. \tag{132}$$

In terms of decomposition $(127)_2$ of the eigenvectors $\boldsymbol{\xi}_\alpha$ the above relations are transformed to

$$\mathbf{A}_\alpha \cdot \mathbf{L}_\beta + \mathbf{A}_\beta \cdot \mathbf{L}_\alpha = \delta_{\alpha\beta}, \tag{133}$$

$$\sum_{\alpha=1}^{6} \mathbf{A}_\alpha \otimes \mathbf{L}_\alpha = \mathbf{I}, \quad \sum_{\alpha=1}^{6} \mathbf{A}_\alpha \otimes \mathbf{A}_\alpha = 0, \quad \sum_{\alpha=1}^{6} \mathbf{L}_\alpha \otimes \mathbf{L}_\alpha = 0. \tag{134}$$

Eqn.(130) gives rise to a 6-order secular equation with real coefficients,

$$\det(\mathbf{N} - p_\alpha \mathbf{I}) = 0, \tag{135}$$

determining six functions $p_\alpha(v)$, which must be real or form complex conjugate pairs. Here and below, excluding *transonic states*, we suppose all p_α to be non-degenerate. The first transonic state takes place at the so-called *limiting velocity* $\hat{v}$, which is defined as the lowest tracing speed admitting a bulk mode in the superposition (125)-(127), Fig.8. The range $0 < v < \hat{v}$ is called a *subsonic range*. Here all eigenvalues p_α and eigenvectors $\boldsymbol{\xi}_\alpha$ occur in pairs of complex conjugates:

$$p_{\alpha+3} = p_\alpha^*, \quad \boldsymbol{\xi}_{\alpha+3} = \boldsymbol{\xi}_\alpha^*, \quad \alpha = 1, 2, 3. \tag{136}$$

At $v = \hat{v}$ one of the conjugate pairs coalesces into one degenerate eigenvalue $\hat{p}$, which in the *supersonic range* $v > \hat{v}$ splits into a pair of different real parameters. As is clear from Fig.8, in the range $v > \hat{v}$ there must exist at least two other transonic states, $\hat{v}'$ and $\hat{v}''$, where the next pairs of the set (136) are similarly transformed from complex to real state, so that at $v > \hat{v}''$ all partial waves in superposition (125)-(127) are homogeneous bulk waves. On the other hand, in the subsonic range $v < \hat{v}$ three of six inhomogeneous terms of this superposition contain infinitely increasing exponents at $y \to \infty$ and the corresponding three amplitudes b_α must vanish. With the choice of numeration providing Im $p_\alpha > 0$ for $\alpha = 1, 2, 3$, there should be $b_{4,5,6} = 0$. As a result, the solution (125)-(127) describes a 3-partial field localized at the surface, i.e. a surface wave. Its amplitudes b_α are supposed to be found from the boundary conditions. For a free surface the requirement of vanishing traction $\mathbf{n}\boldsymbol{\sigma} = 0$ reduces to the homogeneous equation

$$b_1\mathbf{L}_1 + b_2\mathbf{L}_2 + b_3\mathbf{L}_3 = 0, \tag{137}$$

which may have non-trivial solutions only if the corresponding determinant of the matrix $\{L_{\alpha i}\}$ vanishes,

$$\det\{L_{\alpha i}\} = 0. \tag{138}$$

This equation determines the velocity v_R of the Rayleigh wave.

Let us prove that the complex dispersion equation (138) with one real unknown parameter v is not over-determined being equivalent to one real equation. Indeed, in view of Eqns.$(134)_3$ and (136), the matrix

$$\mathbf{B} = 2i\sum_{\alpha=1}^{3}\mathbf{L}_\alpha \otimes \mathbf{L}_\alpha \tag{139}$$

must be real. Accordingly, Eqn.(138) proves to be equivalent to the real equation [52]

$$\det\mathbf{B} \equiv -8i(\det\{L_{\alpha i}\})^2 = 0. \tag{140}$$

As follows from (140), at any tracing speed v in the subsonic range the real and imaginary parts of $\det\{L_{\alpha i}\}$ must be proportional to each other, namely

$$\det\{L_{\alpha i}\} = e^{\pm i\pi/4}f(v), \tag{141}$$

where $f(v)$ is a purely real function. Taziev [53] has presented a general explicit form for the corresponding real dispersion equation, which is of 27th order over v_R^2.

The above proof of the reality of Eqn.(138) first presented by Stroh has eliminated the natural suspicions of J.L. Synge (1956) that the boundary problem (137) is over-determined and the forbidden directions for surface wave propagation in anisotropic bodies were likely to be the rule rather than the exception (see [54]). However, of course, the reality of an equation does not guarantee an existence of its solutions. In 1973 Barnett et al. [55] proved that the real 3×3 matrix $\mathbf{B}(v)$ (139) is positive definite at $v = 0$ and its eigenvalues monotonic decrease with v growing in the subsonic range, so that two of them simultaneously vanish at $v = v_R$ if a Rayleigh wave exists. On this ground they proved the uniqueness theorem of a solution for a surface wave, when it exists. Later Barnett & Lothe [56-58] and Chadwick & Smith [59] proved also the existence theorem for subsonic surface waves, gradually sharpening its formulation on the basis of establishing new and new mathematical properties of different quantities involved into the theory.

Before presenting the final formulation of this theorem let us return to the concept of a first transonic state $v = \hat{v}$, which plays a key role in the existence conditions. Usually at $v = \hat{v}$ there is only one point of tangency between the vertical line and the outer sheet of the slowness surface, like it is shown in Fig.8. Such configuration is called Type 1 transonic state. The corresponding bulk limiting wave would propagate with the group velocity parallel to the surface, however this might happen only in the exceptional situation when this wave satisfies the condition of a free surface. Such waves, known as exceptional bulk waves [59], occupy only 1- dimensional sub-space (lines) in the 3-D space of all possible surface wave geometries [38], i.e. orientations of the frame $\{\mathbf{m}, \mathbf{n}\}$. The other types of transonic state arise [59] when the vertical line $v^{-1} = \hat{v}^{-1}$ is tangent simultaneously with more than one sheet of the slowness surface at the same point, e.g. at a double or a triple tangent degeneracies (Types 2 and 3 of transonic state), or with only outer sheet but in 2 or 3 points (Types 4 and 6, respectively), or a combination of the Types 2 and 4 (Type 5 transonic state). It is worthwhile to mention that all transonic states but Type 3 may be constructed for real materials [54]. Thus, in these terms the existence theorem may be stated in the form:

1°. The existence of a surface wave at a free surface of an elastic half-space is guaranteed in the subsonic range $v < \hat{v}$ if the first transonic state is not of Type 1, or if it is of Type 1 but the corresponding limiting wave is not exceptional. (142)

2°. If the first transonic state is of Type 1 and the limiting wave at $v = \hat{v}$ is exceptional, then i) at $tr\mathbf{B} < 0$ a surface wave solution does exist in the range $v < \hat{v}$, ii) at $tr\mathbf{B} = 0$ a (2-partial) surface wave solution exists at $v = \hat{v}$, iii) at $tr\mathbf{B} > 0$ in all the range $v \leq \hat{v}$ there is no solutions apart from the exceptional wave ($v = \hat{v}$). (143)

In addition, according to [56, 57], also the non-existence theorem is valid:

A clamped surface cannot support a free localized wave in the elastic half-space. (144)

In other words, it has been proved that the boundary problem $\mathbf{u}_{y=0} = 0$ reducing to the equation

$$b_1\mathbf{A}_1 + b_2\mathbf{A}_2 + b_3\mathbf{A}_3 = 0, \tag{145}$$

in contrast to Eqn.(137), has only a trivial solution $b_{1,2,3} = 0$ because at any v

$$\det\{A_{\alpha i}\} \neq 0. \tag{146}$$

Of course, the conditions of a free or a clamped surface do not represent a complete list of physically possible boundary problems for surfaces waves. The alternative problem of a loaded boundary, $\boldsymbol{\sigma}\mathbf{n} = -\hat{\lambda}\mathbf{u}$, has been considered by Alshits et al [60] for a scalar or a tensor coefficient $\hat{\lambda}$, including the situation when the loading system is characterized by its own eigenfrequency. It was found that in this more general case neither uniqueness nor existence theorems can be formulated: depending on the parameters of the loading system the boundary problem may admit several solutions for surface waves or none. There are indicated the physical situations admitting the existence of three (and even four) surface waves.

However, in practice the mostly frequent situation certainly relates to a free surface condition. For this particular boundary problem the detailed explicit analysis of the surface waves in hexagonal and cubic media was performed in [61-65]. This analysis is in a complete agreement with the above general theorems.

3.1.2. *Extensions to Piezoelectrics and Piezoelectrics - Piezomagnetics - Magnetoelectrics*

The Stroh sextic formalism was later extended [66] so that it could be applied to piezoelectric crystals of unrestricted anisotropy. In piezoelectrics due to the electro-mechanical coupling the wave of elastic displacements is accompanied by the wave of quasi-static electric potential φ^E. The equations of motion are represented by a system

$$\mathrm{div}\boldsymbol{\sigma} = \rho\ddot{\mathbf{u}}, \quad \mathrm{div}\mathbf{D} = 0, \tag{147}$$

where the stress tensor $\boldsymbol{\sigma}$ and the electric displacement vector $\mathbf{D}$ are related to the elastic displacement $\mathbf{u}$ and the electric potential φ^E by the constitutive equations

$$\sigma_{ij} = c_{ijkl}u_{k,l} + e_{lij}\varphi^E_{,l}, D_j = e_{jlk}u_{k,l} - \varepsilon_{jl}\varphi^E_{,l}. \tag{148}$$

Here e_{lij} and ε_{jl} are the piezoelectric moduli and electric permittivity tensors, respectively.

Let us add to (125), (126) the similar relations

$$\varphi^E(x-vt, y) = \Phi^E(y)\exp[ik(x-vt)], \Phi^E(y) = \sum_\alpha b_\alpha \Phi^E_\alpha \exp(ikp_\alpha y), \tag{149}$$

$$\mathbf{n}\cdot\mathbf{D}(x-vt, y) = -ikD^n(y)\exp[ik(x-vt)], D^n(y) = \sum_\alpha b_\alpha D^n_\alpha \exp(ikp_\alpha y), \tag{150}$$

and introduce the 4-vectors

$$\mathbf{U}(y)=\begin{pmatrix}\mathbf{A}(y)\\ \Phi^E(y)\end{pmatrix},\mathbf{V}(y)=\begin{pmatrix}\mathbf{L}(y)\\ D^n(y)\end{pmatrix},\mathbf{U}_\alpha=\begin{pmatrix}\mathbf{A}_\alpha\\ \Phi^E_\alpha\end{pmatrix},\mathbf{V}_\alpha=\begin{pmatrix}\mathbf{L}_\alpha\\ D^n_\alpha\end{pmatrix}, \tag{151}$$

and the 8-vectors

$$\boldsymbol{\eta}(y)=\begin{pmatrix}\mathbf{U}(y)\\ \mathbf{V}(y)\end{pmatrix},\ \boldsymbol{\xi}_\alpha=\begin{pmatrix}\mathbf{U}_\alpha\\ \mathbf{V}_\alpha\end{pmatrix}. \tag{152}$$

The 8-component unknown function $\boldsymbol{\eta}(y)$ is organized so that it again satisfies Eqn.(128), where this time the $\mathbf{N}$ matrix (129) is a 8×8 matrix, because the block matrices of the type (ab) are in this case 4×4 matrices,

$$(ab)_{JK}=a_i c_{iJKl} b_l,\quad i,l=1,2,3;\quad J,K=1,2,3,4. \tag{153}$$

Here c_{iJKl} is the tensor defined by

$$c_{iJKl}=\begin{cases} c_{ijkl}\ for\ J,K=j,k=1,2,3,\\ e_{lij}\ for\ J=j=1,2,3,\ K=4,\\ e_{ilk}\ for\ J=4,\ K=k=1,2,3,\\ -\varepsilon_{il}\ for\ J=4,\ K=4.\end{cases} \tag{154}$$

The 8-vectors $\boldsymbol{\xi}_\alpha$ and parameters p_α ($\alpha = 1,\ldots,8$), as before, are found as eigenvectors and eigenvalues of the matrix $\mathbf{N}$, Eqn.(130). The set of eigenvectors $\boldsymbol{\xi}_\alpha$ is again orthonormal and complete (see Eqn.(132) where one should increase the upper limit of summation from 6 to 8 and consider $\mathbf{T}$ (131) as 8×8 matrix). The set p_α is given by roots of Eqn.(135), which form at $v<\hat{v}$ the four complex conjugate pairs. Accordingly, a subsonic surface wave in a piezoelectric medium consists generally of four partial waves. The further consideration, including the proof of reality of secular equations related to basic boundary problems, is very similar to the above speculations. On this basis Lothe & Barnett [67, 68] have extended their existence analysis for the case of piezoelectric half-spaces. They limited themselves to a consideration of only the most restrictive Type 1 transonic state. The following boundary problems have been solved: the surface was supposed to be mechanically either free of tractions, $\mathbf{L}(0)=0$, or clamped, $\mathbf{A}(0)=0$; and electrically either closed, $\Phi^E(0)=0$, or open, $D^n(0)=0$, or adjoined to an isotropic dielectric medium with the permittivity ε, $D^n(0)+i\varepsilon\Phi^E(0)=0$.

It was shown that for all considered electrical boundary conditions at the clamped surface the surface wave solutions are forbidden, i.e. the non-existence theorem (144) remains valid. The other three boundary problems,

$$\begin{array}{lll} i) & \mathbf{L}(0)=0, & D^n(0)+i\varepsilon\Phi^E(0)=0\\ ii) & \mathbf{L}(0)=0, & D^n(0)=0;\\ iii) & \mathbf{L}(0)=0, & \Phi^E(0)=0,\end{array}$$

are less restrictive. Namely:

The problem i) has no more than two surface wave solutions. The problem ii) has no more than one surface wave solution. The problem iii) has at least one surface wave solution (provided that the limiting wave is not exceptional) but not more than two solutions; if the limiting wave is exceptional then there could be two, one or none surface wave solutions. (155)

Examples of coexistence in piezoelectrics of two surface waves, of the Rayleigh-type and of the Bleustein-Gulyaev type [69, 70], are very well known.

Surface waves in media possessing the piezomagnetic effect, instead of piezoelectric coupling, may be described quite similarly to the case of piezoelectrics [71-75]. The same theorems (144) and (155) remain valid after simple rephrasing of the electrical quantities D^n, Φ^E and ε in i), ii) and iii) by the corresponding magnetic equivalents B^n, Φ^H and μ where B^n and Φ^H are the amplitudes of the normal projection of the magnetic induction $\mathbf{B}$ and of the magnetic potential and μ is the magnetic permeability of the adjoined medium. Here we must stress that, in spite of the mentioned formal similarity in the description of surface waves in piezoelectrics and piezomagnetics, physically their properties in those two cases are not completely symmetric. For example, the surface of a piezoelectric half-space is electrically closed when it is metallized, i.e. is electrically screened, which is described by the condition $\Phi^E(0) = 0$. The physically equivalent situation of magnetically closed surface is reached by a superconducting (or superdyamagnetic) coating of the surface, which provides screening of the induction field and is described by the condition $B^n(0) = 0$. In particular, in [74, 75] a transformation of surface waves in a piezomagnetic medium is studied when the coating at its surface changes the state from superconducting to normal, i.e. when the boundary condition $B^n(0) = 0$ is replaced by $B^n(0) + i\mu\Phi^H(0) = 0$.

In addition to purely piezoelectric and piezomagnetic media there are also crystal classes where simultaneously piezoelectric, piezomagnetic and magnetoelectric couplings coexist. An analysis of the surface wave existence in such media cannot be accomplished in the framework of the above 8-dimensional formalism and calls for a further extension of the theory. Such extension developed by Alshits et al [76] has led to a 10-dimensional version of the Stroh-like formalism. This time one arrives at the 10×10 $\mathbf{N}$ matrix (129), where the block matrices of the type (ab) are defined similar to (153) however with $J, K = 1, \ldots, 5$ in accordance with a more complex structure of the tensor c_{iJKl}. The latter now must combine in addition to (154) the piezomagnetic and magnetoelectric moduli and the permeability tensor. In this case at $v < \hat{v}$ the 10-D eigenvectors $\boldsymbol{\xi}_\alpha$ and the eigenvalues p_α ($\alpha = 1, \ldots, 10$) of the $\mathbf{N}$ matrix form pairs of complex conjugates. Therefore the corresponding surface wave solutions in the range $v < \hat{v}$ are generally expected to be 5-partial. We shall not go into further details and just present the final results.

Let us divide all possible combinations of mechanical, electrical and magnetic boundary conditions into four groups omitting everywhere the argument $y = 0$.

$$
\begin{aligned}
&(a)\mathbf{A} = D^n = B^n = 0 \text{ or } \mathbf{A} = \Phi^E = B^n = 0 \text{ or } \mathbf{A} = D^n = \Phi^H = 0;\\
&(b)\mathbf{A} = \Phi^E + D^n/i\varepsilon = \Phi^H + B^n/i\mu = 0\\
&\qquad\qquad\qquad \text{or } \mathbf{A} = \Phi^E = \Phi^H = 0 \text{ or } \mathbf{L} = D^n = B^n = 0;\\
&(c)\mathbf{L} = \Phi^E + D^n/i\varepsilon = \Phi^H + B^n/i\mu = 0\\
&\qquad\qquad\qquad \text{or } \mathbf{L} = \Phi^E = B^n = 0 \text{ or } \mathbf{L} = D^n = \Phi^H = 0;\\
&(d)\mathbf{L} = \Phi^E = \Phi^H = 0.
\end{aligned}
$$

The corresponding four theorems can be stated.

The problem (a) has no surface wave solutions.
The problem (b) has no more than one surface wave solution.
The problem (c) has no more than two surface wave solutions.
The problem (d) has at least one surface wave solution (156)
(provided that the limiting wave is not exceptional) but not more than two solutions; if the limiting wave is exceptional then there could be two, one or none surface wave solutions.

In principle, varying ε and μ in $(b)_1$ and $(c)_1$ from 0 to ∞ we can obtain all possible physical situations including all the other problems in the presented list (a), (b), (c), (d). As follows from the consideration in [76], the decrease of ε and/or μ is not favorable for an existence of surface wave solutions. In particular, tending in $(b)_1$ and $(c)_1$ ε or μ to zero we obtain the problems, which should be added to the previous groups (a) and (b), respectively:

$$(a)_{4,5}\ \mathbf{A} = D^n = \Phi^H + B^n/i\mu = 0 \text{ or } \mathbf{A} = \Phi^E + D^n/i\varepsilon = B^n = 0;$$

$$(b)_{4,5}\ \mathbf{L} = D^n = \Phi^H + B^n/i\mu = 0 \text{ or } \mathbf{L} = \Phi^E + D^n/i\varepsilon = B^n = 0.$$

On the other hand, after the limit ε or $\mu \to \infty$ we obtain the problems belonging to the same groups (b) and (c), respectively. And at ε and $\mu \to \infty$ we have the transitions $(b)_1 \to (b)_2$, $(c)_1 \to (d)$.

It is noteworthy that here we meet for the first time the cases $(b)_{1,2}$ when a clamped surface may admit the existence of a surface wave solution. In [77] such solutions were explicitly found for a model thermodynamically stable crystal. The other nontrivial situation was indicated in [78], where a new branch of surface magnetoelectroelastic waves, existing completely due to magnetoelectric coupling, was theoretically predicted in a piezomagnetic material adjoining a superconducting or a superdiamagnetic material without mechanical contact. This corresponds to the above boundary problems $(c)_2$ and $(b)_5$. When the magnetoelectric interaction is "switched off", these surface waves either disappear or transform into bulk waves.

3.1.3. *Exceptional Bulk Waves and Quasi-bulk Surface Waves*

As we have seen, the situations, when a limiting bulk wave related to a limiting speed $\hat{v}$ proves to be exceptional, play a crucial role in the theory of surface waves in anisotropic media, because these particular situations turn out to be exclusive for the existence of surface wave solutions in the subsonic range $v < \hat{v}$. From this point of view, it appears to be important to find out the conditions for the appearance of exceptional wave solutions and especially to establish the dimension of the sub-space occupied by such solutions in the 3D space of all possible orientations of the frame $\{\mathbf{m}, \mathbf{n}\}$ specifying a surface wave geometry. This will allow us to answer the question how characteristic is it for the Rayleigh wave to exist in an arbitrary direction of an anisotropic medium. The other motivation for a wide interest in exceptional waves has purely applicational origin: the devices based on bulk waves are often more preferable compared with their surface wave analogues. The theory of exceptional waves has been developed by Alshits & Lothe [38].

Formally, a bulk wave is defined as exceptional with respect to the family of parallel planes with the normal $\mathbf{n}$ if the wave produces no tractions on these planes, $\sigma_{ij}n_j = 0$. With this definition, the energy flux of exceptional wave $\mathbf{J} = -\dot{\mathbf{u}}\boldsymbol{\sigma}$ together with its ray velocity $\mathbf{v} = \mathbf{J}/E$, Eqns.(45), (51), must be parallel to the surface, because $\mathbf{J} \cdot \mathbf{n} = -\dot{u}_i\sigma_{ij}n_j = 0$. For the bulk wave (1) the stress field $\sigma_{ij} = c_{ijkl}\partial u_k/\partial x_l$ can be presented in the form

$$\sigma_{ij}(\mathbf{x}, t) = iku_0\mu_{ij}\exp[ik(\mathbf{m}\cdot\mathbf{x} - ct)], \tag{157}$$

where

$$\mu_{ij} = c_{ijkl}m_lA_k. \tag{158}$$

In view of the symmetry of the elastic moduli tensor, $c_{ijkl} = c_{jikl}$, the tensor μ_{ij} is also symmetric. In terms of (158) the Christoffel equation (3), (4) can be rewritten as

$$m_i\mu_{ij} = \rho c^2 A_j. \tag{159}$$

Multiplying scalarly both sides of this equation by the vector $\mathbf{n}$ and taking into account that

$$\mu_{ij}n_j = 0, \tag{160}$$

we obtain an important relation

$$\mathbf{A}\cdot\mathbf{n} = 0. \tag{161}$$

Thus, the following important property of exceptional waves is deduced [38].

An exceptional bulk wave is polarized in the plane with respect to which the wave is exceptional. (162)

For a given surface wave geometry there are several possible exceptional waves. For instance, in Fig.8 three such waves relating to the tracing speeds $\hat{v}$, $\hat{v}'$ and $\hat{v}''$ might be exceptional. However, in the particular case when all these transonic waves have the same wave normals, i.e. equal $p_\alpha : \hat{p} = \hat{p}' = \hat{p}''$, they cannot be simultaneously exceptional, because polarization vectors of three isonormal waves are mutually orthogonal and cannot be simultaneously parallel to the same plane, as it is required by theorem (162).

The wave normal $\mathbf{m}$ of exceptional wave can be found from the condition for the occurrence of non-trivial solutions of Eqn.(160) for the components n_j of the normal to the exceptional plane:

$$\det \boldsymbol{\mu} = 0. \tag{163}$$

This equation determines the admitted directions $\mathbf{m}$ for propagation of exceptional waves. Of course, before solving Eqn.(163) one should find the polarization distribution $\mathbf{A}(\mathbf{m})$ for a considered wave branch from the Christoffel equation. The orientation of the unit vector $\mathbf{m}$ is completely specified by the two spherical angles $\varphi_\mathbf{m}$ and $\theta_\mathbf{m}$. So the scalar equation (163) fixes a relation between $\varphi_\mathbf{m}$ and $\theta_\mathbf{m}$,

$$F(\varphi_\mathbf{m}, \theta_\mathbf{m}) = 0, \tag{164}$$

which is a line on the unit sphere of directions $\mathbf{m}^2 = 1$. For any $\mathbf{m}$ at this line the direction of the normal $\mathbf{n}$ to the corresponding free surface is specified by the only parameter the angle $\varphi_\mathbf{n}$ in the plane orthogonal to $\mathbf{m}$. The components n_j and consequently the angle $\varphi_\mathbf{n}$,

$$\varphi_\mathbf{n} = f(\varphi_\mathbf{m}, \theta_\mathbf{m}), \tag{165}$$

may be found from system (160). Thus, in the 3D space $\{\varphi_\mathbf{m}, \theta_\mathbf{m}, \varphi_\mathbf{n}\}$ the exceptional wave orientations must generally occupy a 1D sub-space (164), (165), i.e. lines on the sphere $\mathbf{m}^2 = 1$.

The theorem of existence of such lines of exceptional wave solutions (EWS) for media of unrestricted anisotropy was proved in [38]. The intimate relationships between exceptional waves and acoustic axes were established. According to [38], the direction $\mathbf{m}_d$ of any acoustic axis determines the wave normal of at least one exceptional wave and the properties of EWS lines on the unit sphere of directions depend on the type of degeneracy along

$\mathbf{m}_d$. For instance, the EWS lines diverge from and end on the points of conical degeneracy, passing in these points from one degenerate branch to another. On the other hand, the discussed lines necessarily pass through points of genuine tangent degeneracy ($n = \pm 1$) in one or in both degenerate branches. Only in the case of a sporadic tangent degeneracy ($n = 0$), it may happen that in the vicinity of $\mathbf{m}_d$ there are no other solutions for exceptional waves. However, even in this case a line of EWS necessarily exists in one of degenerate branches. And, finally, in media free of acoustic axes the lines of solutions for exceptional waves must present in all velocity sheets.

Thus, lines of EWS must be an important characteristic feature of the acoustic identity of most crystals (except those of transverse isotropic media where due to the symmetry the whole sphere $\mathbf{m}^2 = 1$ is exceptional). Indeed, such EWS lines were found numerically for a series of crystals [79, 80]. Some extension of the results [38] was accomplished in [80, 81], where the equations were derived for the orientation of the polarization of the exceptional waves propagating along the acoustic axes of different types, and also for the direction of EWS lines at $\mathbf{m}_d$. Analysis of these equations allowed finding the number of exceptional waves that can propagate along different acoustic axes and correspondingly the number of the lines of solutions for exceptional waves going from degeneracy points or crossing them.

Lyubimov & Sannikov [82, 83] have noticed that at a perturbation of the propagation geometry $\{\mathbf{m}, \mathbf{n}\}$ of an exceptional bulk wave the latter transforms either to a quasi-bulk (deep penetrating) surface wave or to a non-physical solution with increasing amplitude into the depth. Lothe & Alshits [61] explained such a behavior of perturbed solutions basing on the theorem (143) (see Fig.9) and introduced the following criterion of existence of quasi-bulk solutions under the perturbation of exceptional wave configuration.

If for a given propagation geometry the limiting wave is exceptional and at $v < \hat{v}$ there is no surface wave solutions then, by theorem (142), under any perturbation destroying exceptionality of the limiting wave there must arise a quasi-bulk surface wave solution. (166)

And the other statement is valid.

If in an exceptional situation there is a surface wave at $v < \hat{v}$ then, by the uniqueness theorem, at any perturbation destroying exceptionality of the limiting wave the initial bulk wave solution must disappear transforming into a non-physical solution. (167)

Later on Darinskii [84] found that independently of the symmetry of the crystal the penetration depth for a quasi-bulk wave is inversely propor-

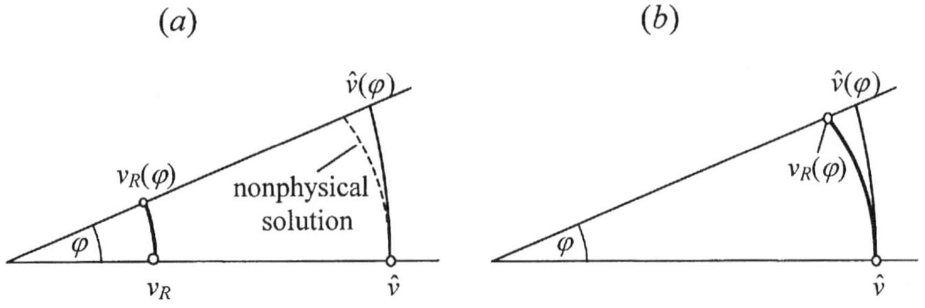

Figure 9. Existence (a) and nonexistence (b) of the Rayleigh wave solution at $\nu < \hat{\nu}$, when the limiting wave is exceptional, and the corresponding nonexistence (a) and existence (b) of the quasi-bulk surface wave branch at $\nu < \hat{\nu}$

tional to the square of angles specifying the deflection from the geometry of propagation corresponding to the existense of an exceptional wave. And the value of the "gap" between the tracing speed of a quasi-bulk wave and the associated perturbed limiting velocity $\hat{v}$ is proportional to the fourth power of the values of these angles. It should be added that in general an exceptional wave may transform into a quasi-bulk wave under perturbations of a different nature. Besides the well-known example piezoelectric coupling, which brings up the Bleustein-Gulyaev wave solutions [69, 70, 61, 85, 86], it may be surface cohesion forces, corrugation of the free surface, non-linear elastic properties *etc.*: see e.g., reviews [87-89]. Apart from this, a Rayleigh wave can become a quasi-bulk due to a strong elastic anisotropy, for instance, at phase transition [90, 91].

3.1.4. *Group, Ray and Tracing Velocities of Surface Waves*

As was shown in sec.2.3.1, homogeneous bulk elastic waves in unbounded anisotropic media have identically equal ray and group velocities, $\mathbf{v}_r = \mathbf{v}_g \equiv \mathbf{v}$, that is related to the phase speed c by a simple relation $\mathbf{v} \cdot \mathbf{m} = c$. For surface acoustic waves in crystals the situation is not so simple [10-12]. In this case we have a ray velocity, which in general must depend on the distance y from the surface, $\mathbf{v}_r = \mathbf{v}_r(y)$, a group velocity $\mathbf{v}_g$ independent of y and therefore not coinciding with $\mathbf{v}_r$, and a scalar tracing speed $v = \omega/k$ ($k = k_x = \mathbf{k} \cdot \mathbf{m}$), which plays a role of a phase speed for surface waves. Below we shall establish some relationships between these velocities for the case of surface waves in a piezoelectric half-space.

The ray velocity $\mathbf{v}_r$ is defined similar to $(51)_1$,

$$\mathbf{v}_r = \mathbf{J}(y)/E(y), \tag{168}$$

where the energy density $E(y)$ is a sum of the average kinetic (E_{kin}), strain (E_{str}) and electric (E_{el}) energies [12]:

$$E = E_{kin} + E_{str} + E_{el}, \tag{169}$$

$$E_{kin} = \frac{1}{T}\int_0^T \frac{\rho}{2}\left(\frac{\dot{\mathbf{u}}+\dot{\mathbf{u}}^*}{2}\right)^2 dt = \frac{\rho}{4}\dot{\mathbf{u}}\cdot\dot{\mathbf{u}}^* = \frac{\rho\omega^2}{4}\mathbf{A}(y)\cdot\mathbf{A}^*(y), \tag{170}$$

$$E_{str} = \frac{1}{8}\left(\sigma_{ij}u^*_{i,j} + \sigma^*_{ij}u_{i,j}\right) = E_{kin} + \frac{1}{8}ik\frac{\partial}{\partial y}\left(\mathbf{A}(y)\cdot\mathbf{L}^*(y) + \mathbf{A}^*(y)\cdot\mathbf{L}(y)\right), \tag{171}$$

$$E_{el} = -\frac{1}{8}\left(\varphi^E_{,l}D^*_l + \varphi^{E*}_{,l}D_l\right) = -\frac{1}{8}ik\frac{\partial}{\partial y}\left(\Phi^E(y)D^{n*}(y) - \Phi^{E*}(y)D^n(y)\right). \tag{172}$$

In the above relations we have used Eqns.(125), (126), (149) and (150). The energy flux $\mathbf{J}(y)$ in (168) is a sum of the strain and electrical components:

$$\mathbf{J} = \mathbf{J}_{str} + \mathbf{J}_{el}, \tag{173}$$

$$\mathbf{J}_{str} = -\frac{1}{4}\left(\boldsymbol{\sigma}\dot{u}^* + \boldsymbol{\sigma}^*\dot{\mathbf{u}}\right), \qquad \mathbf{J}_{el} = -\tfrac{1}{4}\left(\varphi^E\dot{\mathbf{D}}^* + \varphi^{E*}\dot{\mathbf{D}}\right). \tag{174}$$

Let us first prove that the ray velocity $\mathbf{v}_r(y)$ of the surface wave is parallel to the surface, i.e. $\mathbf{v}_r\cdot\mathbf{n} = 0$, or $\mathbf{J}\cdot\mathbf{n} = 0$. Basing again on Eqns.(125), (126), (149)-(151) it is easy to find

$$\mathbf{J}\cdot\mathbf{n} = \frac{\omega k}{4}\sum_{\alpha,\beta=1}^{4} b_\alpha b^*_\beta\left(\mathbf{U}_\alpha\cdot\mathbf{V}^*_\beta + \mathbf{U}^*_\beta\cdot\mathbf{V}_\alpha\right)\exp\left[ik\left(p_\alpha - p^*_\beta\right)y\right]. \tag{175}$$

Accepting a customary choice of numeration $\mathbf{V}^*_\beta = \mathbf{V}_{\beta+3}$, $\mathbf{U}^*_\beta = \mathbf{U}_{\beta+3}$, and taking into account that in analogy with (133) the orthogonality relation

$$\mathbf{U}_\alpha\cdot\mathbf{V}_\beta + \mathbf{U}_\beta\cdot\mathbf{V}_\alpha = \delta_{\alpha\beta} \tag{176}$$

is valid, we conclude that $\mathbf{J}\cdot\mathbf{n} = 0$ and

$$\mathbf{n}\cdot\mathbf{v}_r(y) = 0. \tag{177}$$

Now let us define the group velocity $\mathbf{v}_g$ for the surface wave (125), (149)

$$\begin{pmatrix}\mathbf{u}(\mathbf{r},t)\\ \varphi^E(\mathbf{r},t)\end{pmatrix} \equiv \mathbf{U}(\mathbf{r},t) = \mathbf{U}(y)\exp[i(\mathbf{k}\cdot\mathbf{x} - \omega t)], \tag{178}$$

where the vectors $\mathbf{k}$ and $\mathbf{x}$ belong to the plane xz parallel to the surface, $\mathbf{x} = x\mathbf{m} + z\mathbf{t}$, $\mathbf{k} = k\mathbf{m}$. As usual, the group velocity is determined by the gradient

$$\mathbf{v}_g = \frac{\partial\omega}{\partial\mathbf{k}} = \frac{\partial\omega}{\partial k}\mathbf{m} + \frac{\partial\omega}{\partial k_z}\mathbf{t} = \mathbf{v}^{||}_g + \mathbf{v}^{\perp}_g. \tag{179}$$

Thus, by its definition the group velocity of a surface wave is also parallel to the surface,

$$\mathbf{v}_g \cdot \mathbf{n} = 0. \tag{180}$$

The electro-elastic field (178) is supposed to obey the basic equations of motion,

$$div\, \boldsymbol{\sigma} = \rho \ddot{\mathbf{u}}, \quad div \mathbf{D} = 0, \tag{181}$$

and therefore the energy equation

$$\frac{1}{2}\frac{\partial}{\partial t}\left[\rho \dot{u}_k^2 + \sigma_{kl} u_{k,l} - D_l \varphi^E_{,l}\right] = \frac{\partial}{\partial x_l}\left(\dot{u}_k \sigma_{kl} - \varphi^E \dot{D}_l\right). \tag{182}$$

Form now a wave packet adding to the wave $\mathbf{U}(\mathbf{r}, t)$ (178) another solution $\tilde{\mathbf{U}}^*(\mathbf{r}, t)$ of Eqns.(181), (182), which differs from $\mathbf{U}(\mathbf{r}, t)$ by complex conjugation and slight changes of wave vector ($\tilde{\mathbf{k}} = \mathbf{k} + \delta\mathbf{k}$) and frequency ($\tilde{\omega} = \omega + \delta\omega$, $\delta\omega = \mathbf{v}_g \cdot \delta\mathbf{k}$). The sum $\mathbf{U}(\mathbf{r}, t) + \tilde{\mathbf{U}}^*(\mathbf{r}, t)$ certainly also satisfies Eqns.(181) and (182). On inserting into (182) the sum $\mathbf{U}(\mathbf{r}, t) + \tilde{\mathbf{U}}^*(\mathbf{r}, t)$, instead of $\mathbf{U}(\mathbf{r}, t)$, we obtain

$$\delta\omega E(y) - \delta\mathbf{k} \cdot \mathbf{J}(y) = -ik\omega \frac{1}{4}\frac{\partial}{\partial y}\left[\mathbf{U}(y) \cdot \tilde{\mathbf{V}}^*(y) + \tilde{\mathbf{U}}^*(y) \cdot \mathbf{V}(y)\right]. \tag{183}$$

After integrating Eqn.(183) over y from 0 to ∞, we obtain

$$\delta\omega \bar{E} - \delta\mathbf{k} \cdot \bar{\mathbf{J}} = ik\omega \frac{1}{4}\left[\mathbf{U}(0) \cdot \tilde{\mathbf{V}}^*(0) + \tilde{\mathbf{U}}^*(0) \cdot \mathbf{V}(0)\right]. \tag{184}$$

where $\bar{E}$ and $\bar{\mathbf{J}}$ are the total energy and the total energy flux in the wave field per unit area of the surface,

$$\bar{E} = \int_0^\infty E(y)dy \quad \bar{\mathbf{J}} = \int_0^\infty J(y)dy. \tag{185}$$

For the surface mechanically free ($\boldsymbol{\sigma}\mathbf{n} = \tilde{\sigma}^*\mathbf{n} = 0$, i.e. $\mathbf{L}(0) = \tilde{\mathbf{L}}^*(0) = 0$) and electrically closed ($\varphi^E = \tilde{\varphi}^{E*} = 0$, i.e. $\Phi^E(0) = \tilde{\Phi}^{E*}(0) = 0$) or open ($\mathbf{D} \cdot \mathbf{n} = \tilde{\mathbf{D}}^* \cdot \mathbf{n} = 0$, i.e. $D^n(0) = \tilde{D}^{n*}(0) = 0$) the right-hand side in (184) vanishes and we obtain [10, 12]

$$\mathbf{v}_g = \bar{\mathbf{v}}_r \equiv \frac{\bar{\mathbf{J}}}{\bar{E}}, \tag{186}$$

which replaces the identity (56). In view of (171), (172), for the chosen boundary conditions,

$$\bar{E}_{str} = \bar{E}_{kin}, \quad \bar{E}_{el} = 0, \quad \bar{E} = 2\bar{E}_{kin} \ . \tag{187}$$

With this observation $\bar{\mathbf{v}}_r$ in (186) is equal

$$\bar{\mathbf{v}}_r = \frac{\bar{\mathbf{J}}}{2\bar{E}_{kin}}. \tag{188}$$

If the piezoelectric half-space with a free surface is adjoined (without a mechanical contact) to a dielectric space (e.g. vacuum), then the considered electro-elastic surface wave (178) is accompanied in the adjoined medium by a quasi-static wave of potential $\varphi' = \Phi'(y)\exp[ik(x-vt)]$, where $\Phi'(y) \propto \exp(|k|y)$. In this case the electric boundary conditions reduce to the continuities

$$\Phi^E(0) = \Phi'(0), \quad D^n(0) = D^{n\prime}(0). \tag{189}$$

As a result, the above Eqns.(186)-(188) retain their form if to put there, for the total energy $\bar{E}$ and the total energy flux $\bar{\mathbf{J}}$ of the wave field per unit area of the surface, the integrals

$$\bar{E} = \int_{-\infty}^{\infty} E(y)dy, \quad \bar{\mathbf{J}} = \int_{-\infty}^{\infty} J(y)dy. \tag{190}$$

The vanishing $\bar{E}_{el}$ is automatically provided by the conditions (189).

Let us now return to Eqn.(184) and rearrange its right-hand side by introducing the surface impedance matrix $\mathbf{Z}$:

$$\mathbf{V}(0) = i\mathbf{Z}\mathbf{U}(0). \tag{191}$$

Taking into account that the matrix $\mathbf{Z}$ is Hermitian [67] (i.e. $Z^*_{ij} = Z_{ji}$), we obtain

$$\delta\omega\bar{E} - \delta\mathbf{k}\cdot\bar{\mathbf{J}} = \omega\delta L^{\Phi}_{U_0}, \tag{192}$$

where L^{Φ} is the Φ -Lagrange function [67]

$$L^{\Phi} = -\frac{1}{4}k\mathbf{U}_0^*\cdot\mathbf{Z}\mathbf{U}_0, \quad \delta L^{\Phi}_{\mathbf{U}_0} = -\frac{1}{4}k\mathbf{U}_0^*\cdot\delta\mathbf{Z}\mathbf{U}_0, \tag{193}$$

$\mathbf{U}_0 \equiv \mathbf{U}(0)$ and the subscript $\mathbf{U}_0$ at δL^{Φ} indicates that the variation of L^{Φ} is effected at $\mathbf{U}_0 = const$. For a given crystal with a fixed orientation of the surface, the impedance matrix $\mathbf{Z}$ depends only on the tracing speed $v = \omega/k$ and the direction of the wave propagation $\mathbf{m} = \mathbf{k}/k$ [10, 67]. Therefore, by (193) the function $L^{\Phi}_{\mathbf{U}_0}(\omega, \mathbf{k})$ must have the structure

$$L^{\Phi}_{\mathbf{U}_0}(\omega, \mathbf{k}) = kF\left(\frac{\omega}{k}, \frac{\mathbf{k}}{k}\right). \tag{194}$$

Accordingly,

$$\delta L^{\Phi}_{\mathbf{U}_0} = \left(\frac{\partial L^{\Phi}_{\mathbf{U}_0}}{\partial \omega}\right)_{\mathbf{k}} \delta\omega + \left(\frac{\partial L^{\Phi}_{\mathbf{U}_0}}{\partial \mathbf{k}}\right)_{\omega} \cdot \delta\mathbf{k} =$$
$$= \left(\frac{\partial F}{\partial v}\right)_{\mathbf{m}} \delta\omega + \left\{\left[F - v\left(\frac{\partial F}{\partial v}\right)_{\mathbf{m}}\right]\mathbf{m} + \left(\frac{\partial F}{\partial m_z}\right)_{v} \mathbf{t}\right\} \cdot \delta\mathbf{k}. \tag{195}$$

Combining (195) with (194) one can present the right-hand side of Eqn.(192) in the form

$$\omega\delta L^{\Phi}_{\mathbf{U}_0} = \omega \left(\frac{\partial L^{\Phi}_{\mathbf{U}_0}}{\partial \omega}\right)_{\mathbf{k}} \delta\omega + v\left\{\left[L^{\Phi}_{\mathbf{U}_0} - \omega\left(\frac{\partial L^{\Phi}_{\mathbf{U}_0}}{\partial \omega}\right)_{\mathbf{k}}\right]\mathbf{m} + \left(\frac{\partial L^{\Phi}_{\mathbf{U}_0}}{\partial \theta}\right)_{\omega} \mathbf{t}\right\} \delta\mathbf{k}. \tag{196}$$

where θ is the angle specifying the orientation of the propagation direction $\mathbf{m}$ in the surface plane xz. Taking into account the identities [67]

$$L^{\Phi} = \bar{E}_{kin} - \bar{E}_{str} + \bar{E}_{el}, \quad \omega\left(\frac{\partial L^{\Phi}_{\mathbf{U}_0}}{\partial \omega}\right)_{\mathbf{k}} = 2\bar{E}_{kin} \tag{197}$$

and Eqn.(187) we obtain the important relation [10, 12]

$$\bar{\mathbf{v}}_r \cdot \mathbf{m} = v. \tag{198}$$

Basing on this relation together with Eqns.(179), (186) one can easily extend the basic properties (72), (73) of the group velocity of bulk waves for the case of surface waves:

$$\mathbf{v}_g = \frac{\partial \omega}{\partial k}\mathbf{m} + \frac{\partial \omega}{\partial k_z}\mathbf{t} = \mathbf{v}_g^{\|} + \mathbf{v}_g^{\perp}, \quad \mathbf{v}_g^{\|} = v\mathbf{m}, \quad \mathbf{v}_g^{\perp} = \frac{\partial v}{\partial \theta}\mathbf{t}. \tag{199}$$

The inclination angle ψ of the vector $\mathbf{v}_g$ from the sagittal plane is given by the expression [12]

$$\tan\psi = \frac{\partial \ln v}{\partial \theta}, \tag{200}$$

which is naturally also very similar to the corresponding relation for bulk waves (see Eqn.(75) and Fig.4). It is plain to see that Eqns.(199), (200) retain their form for the alternative electric boundary problem (189) related to the case of an adjoined dielectric half-space (e.g. vacuum).

Summarizing, we can state that the group velocity $\mathbf{v}_g$ of a surface wave behaves very similar to its counterpart for bulk waves (Fig.10 (a)). However this cannot abolish the fact that the physical energy current characterized by the ray velocity $\mathbf{v}_r$ of a surface wave in an anisotropic medium have generally different directions and magnitudes at different distances y from

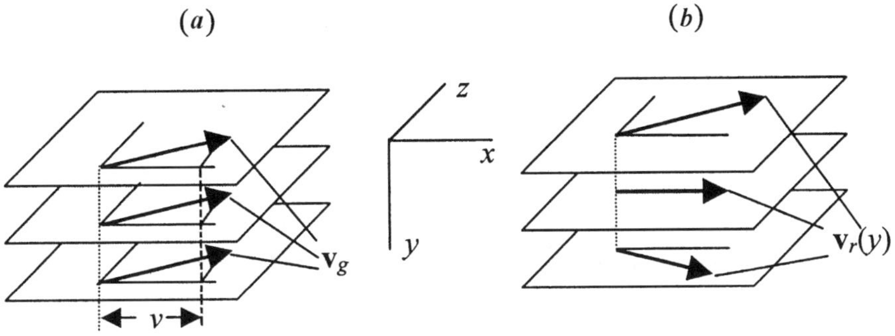

Figure 10. The group (a) and the ray (b) velocity distributions for a surface wave

the surface, Eqn.(168), Fig.10 (b). And the group velocity $\mathbf{v}_g$ coincides not with the ray velocity $\mathbf{v}_r$ but with the integral characteristic $\bar{\mathbf{v}}_r$ defined by the ratio (186) of the total energy flux $\bar{\mathbf{J}}$ per unit area of the surface, Eqns.$(185)_2$ or $(190)_2$, to the total energy density $\bar{E}$, Eqns.$(185)_1$ or $(190)_1$.

3.2. SUPERSONIC SURFACE WAVES, LEAKY WAVES, SIMPLE REFLECTION

In this section we shall limit ourselves to a consideration of presumably purely elastic effects and only in the first transonic range $\hat{v} < v < \hat{v}'$ (see Fig.8), where there are two pairs of complex roots, p_1, p_1^*, p_2, p_2^*, continued from the subsonic range, and two real roots, p_i and p_r, which have developed from $\hat{p}$. The two bulk waves corresponding to the two real roots will be denoted by i and r, the subscript i being used for the incident wave, and r for the reflected wave. With four physically acceptable partial waves, $1, 2, i$ and r, we can always form the solution of the reflection problem. Only two partial waves are available for the construction of pure surface waves, and usually they can exist only for some specific orientations of the surface and the direction of propagation. However, with a little admixture of the reflected wave r, solutions are possible, and these are the leaky surface waves.

3.2.1. *The Reflection Problem in the Transonic Range*

In the first transonic range $\hat{v} < v < \hat{v}'$ the fixed tracing speed v for the reflection problem specifies the incident and reflected waves. The inverse speed v^{-1} is then the projection on the surface of the inverse phase speeds of the incident and reflected waves (see Fig.11). So that

$$\mathbf{u}^{(i)}(\mathbf{r}, t) = \mathbf{a}^{(i)} \exp[ik(x + p_i y - vt)], \;\; \mathbf{u}^{(r)}(\mathbf{r}, t) = \mathbf{a}^{(r)} \exp[ik(x + p_r y - vt)], \tag{201}$$

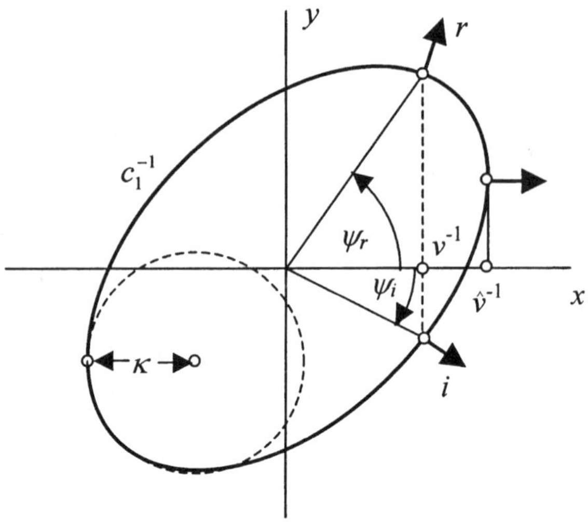

Figure 11. Incident wave, reflected wave and limiting wave, related to the slowness surface

where $\mathbf{a}^{(i,r)} = b_{i,r}\mathbf{A}_{i,r}$ and

$$p_i = \tan\psi_i, \quad p_r = \tan\psi_r. \tag{202}$$

The vector amplitudes $\mathbf{a}^{(i,r)}$ and the corresponding tractions $\mathbf{l}^{(i,r)} = b_{i,r}\mathbf{L}_{i,r}$ of the bulk waves (201) can be chosen real. Taking into account that the projections of the energy flux on the normal $\mathbf{n}$,

$$\mathbf{J}^{(i,r)} \cdot \mathbf{n} = -\tfrac{1}{2}\mathbf{n}\cdot\boldsymbol{\sigma}^{(i,r)}\dot{\mathbf{u}}^{(i,r)*} = -\tfrac{1}{2}k\omega\mathbf{a}^{(i,r)}\cdot\mathbf{l}^{(i,r)}, \tag{203}$$

for the incident and reflected waves must be negative and positive, respectively, it is plain to see that

$$\mathbf{a}^{(i)}\cdot\mathbf{l}^{(i)} > 0, \quad \mathbf{a}^{(r)}\cdot\mathbf{l}^{(r)} < 0. \tag{204}$$

And accordingly, due to the normalization (133) $2\mathbf{A}_i\cdot\mathbf{L}_i = 2\mathbf{A}_r\cdot\mathbf{L}_r = 1$, the vectors $\mathbf{A}_i$ and $\mathbf{L}_i$ must be purely real, and the vectors $\mathbf{A}_r$ and $\mathbf{L}_r$ must be purely imaginary.

The boundary problem for the reflection at a free surface in terms of the Stroh formalism reduces to a requirement of linear dependence, similar to (137),

$$\mathbf{L}(0) = b_1\mathbf{L}_1 + b_2\mathbf{L}_2 + b_i\mathbf{L}_i + b_r\mathbf{L}_r = 0. \tag{205}$$

Certainly, four vectors are always linearly dependent. Eqn.(205) describes the situation that a wave i is incident and is reflected as a wave r and that the local surface modes $\alpha = 1, 2$ enter by the condition of free surface. Thus, for any speed v, there must be a solution for the unknown amplitudes b_β, and the Stroh formalism determines the solution readily as follows [92]:

$$b_\beta = C[\mathbf{L}_\beta \mathbf{L}_1^* \mathbf{L}_2^*]. \tag{206}$$

Here the constant C is given by the known amplitude b_i of the incident wave and $[\dots]$ denotes the mixed vector product, or the determinant of the matrix constructed from the components of the vectors enclosed in brackets. The derivation of (206) is based on the identity $(134)_3$. Indeed, the substitution of (206) into Eqn.(205), in view of $(134)_3$ gives

$$\mathbf{L}(0) = -C\{\mathbf{L}_1^*[\mathbf{L}_1^*\mathbf{L}_1^*\mathbf{L}_2^*] + \mathbf{L}_2^*[\mathbf{L}_2^*\mathbf{L}_1^*\mathbf{L}_2^*]\} = 0. \tag{207}$$

Quite similarly, basing on Eqn.$(134)_1$, one can find

$$\mathbf{A}(0) = b_1\mathbf{A}_1 + b_2\mathbf{A}_2 + b_i\mathbf{A}_i + b_r\mathbf{A}_r = C\left(\mathbf{L}_1^* \times \mathbf{L}_2^*\right). \tag{208}$$

Multiplying (205) with (208) and making use of the orthogonality (133) we obtain the following useful relation among b_α [92],

$$b_1^2 + b_2^2 + b_i^2 + b_r^2 = 0. \tag{209}$$

The solution (206) and the identity (209) may be generalized to cases of more complicated boundary problems [93-96]. For instance, the reflection-transmission problem on the interface between two elastic media $j = 1, 2$ (the incident mode is supposed to belong to the medium 1),

$$\sum_{\alpha=1}^{4} b_\alpha^{(1)} \boldsymbol{\xi}_\alpha^{(1)} = \sum_{\beta=1}^{3} b_\beta^{(2)} \boldsymbol{\xi}_\beta^{(2)}, \tag{210}$$

yields the solution [96, 97]

$$b_{\alpha,\beta}^{(j)} = C[\boldsymbol{\xi}_{\alpha,\beta}^{(j)} \boldsymbol{\xi}_5^{(1)} \boldsymbol{\xi}_6^{(1)} \boldsymbol{\xi}_4^{(2)} \boldsymbol{\xi}_5^{(2)} \boldsymbol{\xi}_6^{(2)}]. \tag{211}$$

And the property (209) is transformed to

$$\sum_{\alpha=1}^{4} (b_\alpha^{(1)})^2 = \sum_{\beta=1}^{3} (b_\beta^{(2)})^2. \tag{212}$$

Another example of an extension of the above theory to the case of a bicrystal is the reciprocity relation, which is a generalization of the analogous property on a fluid-fluid [98] and fluid-isotropic solid [99] interfaces. Designate by $T_{\beta\alpha}(v) = b_\beta^{(j)}/b_\alpha^{(1)}$ $(j = 1, 2)$ the coefficient of transformation of the incident α -mode in the medium 1, into the mode of any β -branch, which is either bulk or inhomogeneous for the given speed v and propagates either in the medium 1 or in the medium 2. The inverse transformation of the incident β -mode into the α -mode is then defined by the coefficient $T_{\alpha\beta}(v) = \tilde{b}_{\alpha}^{(1)}/\tilde{b}_{\beta}^{(j)}$. The reciprocity relation reads [96, 97]

$$|T_{\alpha\beta}(v)| = |T_{\beta\alpha}(v)|. \tag{213}$$

For a more extensive study of the reciprocity relations see [100].

Let us now return to the reflection problem in the anisotropic half-space. A particular and interesting situation arises in the limiting case $\hat{v} \leftarrow v$ where ψ_i and ψ_r coalesce to $\hat{\psi}$, and accordingly, $p_i, p_r \rightarrow \hat{p}$, $\mathbf{a}_i, \mathbf{a}_r \rightarrow \hat{\mathbf{a}}$, and $\mathbf{l}_i, \mathbf{l}_r \rightarrow \hat{\mathbf{l}}$ (the corresponding normalized $\mathbf{L}_i$ and $\mathbf{L}_r$ diverge). This is the case of grazing incidence since the ray direction of the limiting wave is parallel with the surface (Fig.11). The necessary fourth partial solution required for satisfying the condition of free surface can be obtained as follows. Since $\mathbf{u}^{(i)}$ and $\mathbf{u}^{(r)}$ (201) are partial solutions, also

$$\mathbf{u}_4 = \frac{\mathbf{u}^{(r)} - \mathbf{u}^{(i)}}{\Delta\psi}, \Delta\psi = \psi_r - \psi_i, \tag{214}$$

is a partial solution. With (202) in (201) going to the limit $\Delta\psi \rightarrow 0$ as $\hat{v} \leftarrow v$, one obtains from (214) the 4-th partial solution [92]

$$\mathbf{u}_4(\mathbf{r},t) = \left[\left(\frac{\partial \mathbf{a}}{\partial \psi}\right)_{\hat{\psi}} + \frac{iky\mathbf{a}(\hat{\psi})}{\cos^2\hat{\psi}}\right] \exp\left[ik\left(x + y\tan\hat{\psi} - \hat{v}t\right)\right], \tag{215}$$

which is an inhomogeneous wave. In the surface $y = 0$, this wave exerts a force with amplitude

$$-\boldsymbol{\sigma}_4\mathbf{n} = ik\left(\frac{\partial \mathbf{l}(\psi)}{\partial \psi}\right)_{\hat{\psi}}. \tag{216}$$

The condition of free surface can now be satisfied by a superposition

$$b_1\mathbf{L}_1 + b_2\mathbf{L}_2 + \hat{b}\hat{\mathbf{l}} + b\left(\frac{\partial \mathbf{l}}{\partial \psi}\right)_{\hat{\psi}} = 0. \tag{217}$$

At grazing incidence, this requirement replaces (205). The solution of (217) could be obtained by limiting procedures starting with (206). We will not undertake such a discussion here.

The construction leading to (215) is similar to the Achenbach-Gautesen [101] construction of inhomogeneous surface wave solution. However one should notice that the combined solution containing the partial wave (215) in its literal form is divergent at infinity $y \rightarrow \infty$ together with its derivatives. Nevertheless, this inhomogeneous solution has the definite physical sense as a limiting asymptotic form describing the wave field at grazing reflection ($\Delta\psi << 1$) not very far from the surface ($ky << 1/\Delta\psi$), when indeed

$\sin(\Delta\psi ky)/\Delta\psi \approx ky$. Such particular solutions arise for the Lamb waves in elastic plates at the short-wavelength range when an infinite number of dispersion curves $v_n(k)$ for wave-guided modes tend to the level $\hat{v}$ [102], independently of the "status" of the limiting wave, exceptional or non-exceptional:

$$v_n(k) \approx \hat{v} + \frac{1}{2\kappa}\left(\frac{\pi n}{kd}\right)^2, \quad n = 1, 2, ... \tag{218}$$

Here $\kappa \sim \hat{v}^{-1}$ is the radius of curvature of the external slowness sheet at $v^{-1} = \hat{v}^{-1}$ (Fig.11), d is the thickness of the plate, and for each number n the wavelength is supposed to be short enough in order the dispersion curve $v_n(k)$ would be close to the asymptotic level $\hat{v} : v_n - \hat{v} << \hat{v}$, i.e.

$$kd >> \pi n. \tag{219}$$

On the other hand, from the elementary geometrical consideration in the vicinity of $\hat{v}$ one has

$$v_n - \hat{v} \approx \frac{1}{2\kappa}\left(\frac{\Delta\psi}{2\cos^2\hat{\psi}}\right)^2, \tag{220}$$

which together with (218) provides the estimate $\Delta\psi \sim \pi n/kd$. Thus, the wave-guided solutions in the asymptotic range (219) contain the inhomogeneous component (215) in the pre-surface zones of the thickness

$$\Delta y << d/\pi n. \tag{221}$$

Looking at Eqn.(217) one would think that rather than the exceptional case $\hat{\mathbf{l}} = 0$, the less restrictive situation

$$b_1\mathbf{L}_1 + b_2\mathbf{L}_2 + \hat{\mathbf{l}} = 0, \tag{222}$$

could also represent the limiting case of a surface wave of infinite penetration at $\hat{v}$. This would extend the definition of an exceptional limiting wave for a larger class of solutions than we studied above. However, as was rigorously proved in [92], Eqn.(222) cannot be satisfied at $\hat{\mathbf{l}} \neq 0$, and at $\hat{\mathbf{l}} = 0$ there must be either $b_1 = b_2 = 0$ or $b_1\mathbf{L}_1 + b_2\mathbf{L}_2 = 0$. In the latter case a two-partial surface wave would coexist with the exceptional bulk wave at $v = \hat{v}$, which is not ruled out by the theorem (143) (see the case ii).

More generally, let us ask whether for some reason one should expect linear dependence between the other three vectors in (217) when $\hat{\mathbf{l}} = 0$,

$$b_1\mathbf{L}_1 + b_2\mathbf{L}_2 + b\left(\frac{\partial\mathbf{l}}{\partial\psi}\right)_{\hat{\psi}} = 0. \tag{223}$$

The answer to this question is also negative. As was proved in [92], the third term in (223) may not belong to the same plane as $\mathbf{L}_1$ and $\mathbf{L}_2$. Thus, an extension of the concept of an exceptional wave in elastic half-space is not possible. On the other hand, extended exceptional waves representing a superposition of a single bulk mode with localized modes may arise in elastic plates [103] and in piezoelectric half-spaces [104, 86].

In this section we have limited ourselves to an analysis of reflection only in the first transonic range $\hat{v} < v < \hat{v}'$. That is why we missed in our consideration some features of reflection specific for a more "fast" region $v > \hat{v}'$, for instance, the phenomenon of mode conversion, similar to a Brewster reflection in optics, when the incident and outgoing waves belong to different sheets of the slowness surface (see [105]).

3.2.2. *Supersonic Surface Waves and Simple Reflection*

In contrast to isotropic media, where the Rayleigh surface waves are always subsonic ($v_R < \hat{v} = c_t$), in crystals there are known supersonic surface waves with the speed v_R belonging to the first transonic range $\hat{v} < v < \hat{v}'$. For instance, the two-partial surface wave propagating and polarized in a symmetry plane may be faster than the independent bulk SH-wave representing an exceptional limiting wave. Such wave is termed a symmetric supersonic surface wave [106, 107]. However, symmetry is not a necessary condition for the existence of supersonic surface waves. In general (beyond symmetry planes), such two-partial solutions are called secluded supersonic surface waves [108]. The term implies that in the general case the dispersion equation for a speed of such wave is not real as for the subsonic surface waves. So, it appears natural that an existence of such non-symmetric solutions was considered over some time to be doubtful. However, now secluded supersonic surface waves are not only theoretically studied [108-112] but also observed experimentally, e.g. in [113] by means of Mandelstamm-Brillouin light scattering.

For the supersonic surface wave the boundary condition of a free surface reduces to

$$b_1\mathbf{L}_1 + b_2\mathbf{L}_2 = 0, \tag{224}$$

instead of Eqn.(137). In the given case, quite similarly to (209) we have

$$b_1^2 + b_2^2 = 0, \tag{225}$$

which transforms (224) into

$$\mathbf{L}_1 \pm i\mathbf{L}_2 = 0. \tag{226}$$

But then,

$$\mathbf{L}_1 \otimes \mathbf{L}_1 + \mathbf{L}_2 \otimes \mathbf{L}_2 = 0, \tag{227}$$

and also, from (227)

$$\mathbf{L}_1^* \otimes \mathbf{L}_1^* + \mathbf{L}_2^* \otimes \mathbf{L}_2^* = 0. \tag{228}$$

Combining (227) and (228) with the completeness relation $(134)_3$, we conclude that also

$$\mathbf{L}_i \otimes \mathbf{L}_i + \mathbf{L}_r \otimes \mathbf{L}_r = 0. \tag{229}$$

But this is the condition for linear dependence between $\mathbf{L}_i$ and $\mathbf{L}_r$,

$$\mathbf{L}_i \pm i\mathbf{L}_r = 0, \tag{230}$$

or the condition for simple reflection, i.e. reflection without inhomogeneous partial modes. This observation was first made in [92]. Thus, the following statement is true.

When a two-component supersonic surface wave exists, the conditions for simple reflection are also satisfied. (231)

However, the converse is not true: simple reflection does not secure the existence of a supersonic surface wave. Simple reflection does however, via $(134)_3$, secure that the reduced matrix

$$\mathbf{B}_r = 2i(\mathbf{L}_1 \otimes \mathbf{L}_1 + \mathbf{L}_2 \otimes \mathbf{L}_2) \tag{232}$$

is purely real (compare with (139)).

Let us answer the question, what is the subspace of tracing speed v and orientations of the frame $\{\mathbf{m}, \mathbf{n}\}$ specified by three angles φ_j where the conditions for simple reflection are fulfilled. Note that, since $\mathbf{L}_i$ and $i\mathbf{L}_r$ are both real vectors, Eqn.(230) imposes only two conditions upon v and φ_j 's. This means that simple reflection occupies a 2D subspace in the above 4D space. We shall call it a simple reflection surface S_{sr}. The two signs in Eqn.(230) specify two different kinds of simple reflection [109-111]. The first one forms a surface $S_{sr}^{(1)}$, which can continuously go from the transonic range to the subsonic region crossing at a transonic state whose limiting wave belongs to the line of exceptional waves, i.e. it satisfies the condition of free surface. The second kind of simple reflection corresponds to a surface $S_{sr}^{(2)}$, where the incident and outgoing waves are antiphase waves and eliminate each other in the surface at grazing incidence. As was proved in [111], two-partial Rayleigh waves may belong only to the first type subspace $S_{sr}^{(1)}$. We shall suppose that at normalization the signs of $\mathbf{L}_i$, $\mathbf{L}_r$ were chosen so that the surface $S_{sr}^{(1)}$ is described by

$$\mathbf{L}_i + i\mathbf{L}_r = 0. \tag{233}$$

Let us now evaluate the dimensionality of the sub-space of solutions for supersonic surface waves. In contrast to the pair $\mathbf{L}_i$, $i\mathbf{L}_r$ of real vectors in Eqn.(233), the both vectors in (226) are generally complex: $\mathbf{L}_\alpha =$

$\mathbf{L}'_\alpha + i\mathbf{L}''_\alpha, \alpha = 1, 2$. In these terms Eqn.(226) is equivalent to the two real equations,

$$\mathbf{L}'_1 = \pm\mathbf{L}''_2, \ \mathbf{L}'_2 = \mp\mathbf{L}''_1. \tag{234}$$

Taking into account that this system implies four scalar equations with respect to the four real variables, v and φ_j 's, it appears that only isolated surface wave solutions are possible. However, it can be demonstrated that the real and imaginary components of the vectors $\mathbf{L}_{1,2}$ are linear dependent. Indeed, according to [108], throughout the space $S_{sr}^{(1)}$ they belong to the same plane with the normal

$$\mathbf{A} = \mathbf{A}_i + i\mathbf{A}_r. \tag{235}$$

On the other hand, one can prove that inside of $S_{sr}^{(1)}$ the above conditions (234) reduce to the following scalar equation:

$$\mathbf{A} \cdot (\mathbf{L}'_1 \times \mathbf{L}''_2) = 0. \tag{236}$$

Let us prove that the requirement

$$\mathbf{L}'_1 || \mathbf{L}''_2 \tag{237}$$

equivalent to Eqn.(236) provides the vanishing matrix $\mathbf{B}_r$ (232). As was shown above, $\mathbf{B}_r$ is real in S_{sr}, which transforms (232) into the two equations: $\mathrm{Im}\mathbf{B}_r = 0$, $\mathbf{B}_r = \mathrm{Re}\mathbf{B}_r$, i.e.

$$\mathbf{L}'_1 \otimes \mathbf{L}'_1 + \mathbf{L}'_2 \otimes \mathbf{L}'_2 - \mathbf{L}''_1 \otimes \mathbf{L}''_1 - \mathbf{L}''_2 \otimes \mathbf{L}''_2 = 0, \tag{238}$$

$$\mathbf{B}_r = -2(\mathbf{L}'_1 \otimes \mathbf{L}''_1 + \mathbf{L}''_1 \otimes \mathbf{L}'_1 + \mathbf{L}'_2 \otimes \mathbf{L}''_2 + \mathbf{L}''_2 \otimes \mathbf{L}'_1). \tag{239}$$

It is clear seen from (238) that by (237) we also have $\mathbf{L}'_2 || \mathbf{L}''_1$. Inserting into (238) and (239)

$$\mathbf{L}'_1 = \alpha\mathbf{L}''_2, \quad \mathbf{L}'_2 = \beta\mathbf{L}''_1, \tag{240}$$

one obtains

$$\alpha^2 = 1, \quad \beta^2 = 1, \tag{241}$$

and

$$\mathbf{B}_r = -2(\alpha + \beta)(\mathbf{L}''_1 \otimes \mathbf{L}''_2 + \mathbf{L}''_2 \otimes \mathbf{L}''_1). \tag{242}$$

In accordance with [111] in the sub-space $S_{sr}^{(1)}$ there must be

$$\alpha = \pm 1, \quad \beta = \mp 1, \quad \alpha + \beta = 0, \tag{243}$$

which provides both (234) and vanishing $\mathbf{B}_r$. Thus, in conclusion it is sufficient that v and φ_j, j = 1,2,3, satisfy in all three independent conditions in order for $\mathbf{L}_1$ and $\mathbf{L}_2$ to become collinear. These conditions are formed by Eqn.(233), that is equivalent to two scalar real relations as already mentioned, and by one additional condition (236). Summarizing, we can make the following statement.

In crystals of unrestricted anisotropy there is a one-dimensional space of two-partial supersonic surface wave solutions determined by Eqns.(233), (236) and represented by a line in the simple reflection surface $S_{sr}^{(1)}$. (244)

For a more rigorous and detailed analysis of this delicate problem we address readers to the paper by Darinskii et al [112].

Supersonic surface waves may exist also on a free surface of a piezoelectric medium and on an interface between two half-infinite media. But it should be noted that in piezoelectrics the supersonic surface wave and the solution of the reflection problem existing at the same speed v, may contain the common surface partial modes (but not a single one) [93, 94]. So in this case there is no such direct relation between supersonic surface waves and simple reflection as in the purely elastic case.

3.2.3. *Leaky Waves and Resonance Reflection*

Thus, in the transonic range, $\hat{v} < v < \hat{v}'$, pure surface wave solutions are not possible except for certain orientations of the surface and the sagittal plane. However, branches of leaky surface waves may exist [10, 92, 114] (below we shall follow [92]): with a small admixture of the reflected wave r, the condition of a free surface

$$b_1\mathbf{L}_1 + b_2\mathbf{L}_2 + b_r\mathbf{L}_r = 0 \tag{245}$$

may be almost satisfied for a speed v_l, and regarding the partial waves as analytic functions of a complex speed, the boundary condition (245) may be exactly satisfied for some complex speed

$$v_l - iv_l'. \tag{246}$$

Since the wave r carries energy away from the surface, the solution must be decaying with time, i.e. $v_l' > 0$. Of course, such a pseudo-surface mode can have a physical meaning only when this decay is small enough during one period, i.e. until $v_l' << v_l$.

Define $q(v)$ as

$$q(v) = [\mathbf{L}_r\mathbf{L}_1\mathbf{L}_2]. \tag{247}$$

When $\mathbf{L}_r, \mathbf{L}_1$ and $\mathbf{L}_2$ are linearly dependent so that the boundary condition (245) of free surface can be satisfied, $q = 0$, i.e. (246) is the zero of $q(v)$. Note that for complex root (246), in which case $\mathbf{L}_r$ is not purely imaginary, it does not follow from this that (206) generalized to complex v gives zero b_r ; (206) correctly gives the r-admixture. However, it can be shown, that $b_i = 0$ as it should be. For real v, the function $q(v)$ can be decomposed into a real and imaginary part

$$q(v) = X(v) + iY(v). \tag{248}$$

Let more generally $X(v)$ and $Y(v)$ be analytical continuation of the $X(v)$ and $Y(v)$ on the real axis. Then (248) is a valid expression also for complex v. Certainly, for a complex argument, $X(v)$ and $Y(v)$ are also complex; only on the real axis does (248) decompose $q(v)$ into its real and imaginary parts.

With $v_l - iv_l'$ a zero of $q(v)$, we now obtain in an expansion to first order in v_l',

$$q(v_l - iv_l') = X(v_l) + iY(v_l) - iv_l'\frac{\partial X}{\partial v_l} + v_l'\frac{\partial Y}{\partial v_l} = 0. \tag{249}$$

For this equation to agree in both its real and imaginary parts, v_l must be such that both expressions

$$v_l' = -\frac{X(v_l)}{\partial Y/\partial v_l} = \frac{Y(v_l)}{\partial X/\partial v_l} \tag{250}$$

give the same v_l', or

$$X\frac{\partial X}{\partial v_l} + Y\frac{\partial Y}{\partial v_l} = 0. \tag{251}$$

Let us define $Q(v)$ as

$$Q(v) = [X(v)]^2 + [Y(v)]^2. \tag{252}$$

For real v, $Q(v)$ is then simply

$$Q(v) = |q(v)|^2. \tag{253}$$

From (251) we see that on the real axis $Q(v)$ has a minimum at v_l,

$$\frac{\partial Q}{\partial v_l} = 0. \tag{254}$$

This equation determines v_l. Either of expressions (250) can be used to determine v_l'.

Equation (252) defines $Q(v)$ as an analytical function of v. Thus, also, in view of (254),

$$Q(v_l - iv_l') = Q(v_l) - \tfrac{1}{2}(v_l')^2\frac{\partial^2 Q}{\partial v_l^2} = 0, \tag{255}$$

which gives an alternative expression for v_l',

$$v_l' = -\left(\frac{2Q(v_l)}{\partial^2 Q/\partial v_l^2}\right)^{1/2}. \tag{256}$$

Eqn.(256) is not identical with (250), but is equivalent to what one obtains including second-order correction terms $\sim (v_l')^2$ in (249) and is thus actually more accurate than the first order result (250). Also, (256) allows for a very simple expression for $Q(v)$ on the real axis in the neighborhood of v_l. By (254)

$$Q(v) = Q(v_l) + \tfrac{1}{2}\frac{\partial^2 Q}{\partial v_l^2}(v - v_l)^2. \tag{257}$$

But by (256), this can be rewritten as

$$Q(v) = \tfrac{1}{2}\frac{\partial^2 Q}{\partial v_l^2}[(v - v_l)^2 + v_l'^2], \tag{258}$$

which clearly displays the behavior of a minimum at v_l of width $\sim v_l'$ and of height $\sim (v_l')^2$.

Typically, leaky wave branches $v_l(\phi)$ arise in the vicinity of a particular direction ϕ_0 where $v_l' = 0$ and $Q(v_l) = 0$, i.e. a pure undamped two-component surface wave exists. Here ϕ defines any of angles φ_j ($j = 1,2,3$) specifying the surface orientation and the propagation direction. Physically, as discussed after (246), one expects that always $v_l' > 0$. Changing the direction of ϕ slightly, one therefore expects

$$v_l' = \gamma(\phi - \phi_0)^2 + \text{ higher order terms}, \tag{259}$$

i.e. the first order term in $(\phi - \phi_0)$ must be absent. The rigorous proof of this property one can find in [92].

An interesting situation occurs in the particular case when at $\phi = \phi_0$ the pure solution proves to be an exceptional bulk wave. As is shown by Darinskii [84] (see also [115]) in a general manner, a leaky wave branch may arise from the exceptional wave only close to the second transonic state when $v_l(\phi_0) = \hat{v}'$ (Fig. 8). The exceptional limiting wave ($v = \hat{v}$) at any perturbation of geometry can transform either into a quasi-bulk surface wave branch, or into a non-physical solution with the amplitude exponentially increasing at $y \to \infty$ (see theorem (167)).

As was shown in [92, 116], a leaky wave branch evokes resonant behavior of the reflection for incidence angles relating to tracing speeds v close to $v_l(\phi)$. This is rather expectable, because for small $\Delta\phi = \phi - \phi_0$, $\Delta v = v - v_l(\phi)$, i.e. close to the pure point where $\mathbf{L}_1(\phi_0)||\mathbf{L}_2(\phi_0)$, all coefficients b_β (206) are proportional to the small parameter $\mathbf{L}_1^*(\phi) \times \mathbf{L}_2^*(\phi)$. So, their ratios might manifest some peculiarities depending on the relations between Δv and $\Delta\phi$. Now we are in order to exhibit the mentioned resonances.

As we know, the complex speed $v_l - iv_l'$ (246) is a zero point of the analytically extended function $b_i(v)$. So, for $\Delta v << v_l$ (in view of (259) for small $\Delta\phi$ always $v_l' << v_l$) one has to main order

$$b_i(v) \approx \left(\frac{\partial b_i}{\partial v}\right)_{v_l} [v - v_l + iv_l']. \tag{260}$$

The same complex speed $v_l - iv_l'$ is also a zero of the function $q(v)$ (247). Taking into account that for real v the vector $\mathbf{L}_r$ is purely imaginary, it is easily seen from (206), (247) and (248) that

$$b_r(v) = -Cq^*(v) = -C[X(v) - iY(v)]. \tag{261}$$

Accordingly, a zero point of this function is given not by $v_l - iv_l'$, but by its complex conjugation $v_l + iv_l'$. Therefore in analogy with (260)

$$b_r(v) \approx \left(\frac{\partial b_r}{\partial v}\right)_{v_l} [v - v_l - iv_l']. \tag{262}$$

As a result, the reflection coefficient $R = b_r/b_i$ near a pure point of supersonic surface wave solution is described by the expression

$$R = \frac{(\partial b_r/\partial v)_{v_l}}{(\partial b_i/\partial v)_{v_l}} \frac{v - v_l - iv_l'}{v - v_l + iv_l'} = \exp\left\{i[\chi_0 + \chi(v, \Delta\phi)]\right\}, \tag{263}$$

where

$$\chi = 2\arctan\left[\frac{v_l'}{v_l - v}\right]. \tag{264}$$

Taking into account that here v_l' is a small parameter (259), the phase χ proves to be very sensitive to the speed v (i.e. to the incidence angle) in the vicinity of v_l. On the other hand, such a sensitivity of the phase of the reflection coefficient is an indication on a possible nonspecular reflection of the corresponding acoustic beam [98]. And indeed, the pronounced anomalies due to the coupling with the leaky wave branch at reflection of two-dimensional and three-dimensional Gausian beams at the free surface of an elastic halfspace have been theoretically found and investigated in [116, 117].

As we have seen, the expansions (260), (262) of the coefficients $b_{i,r}$ in Δv and $\Delta\phi$ do not contain linear terms in $\Delta\phi$, because in view of (259) $v_l' \propto (\Delta\phi)^2$. On the other hand, as was shown in [92], there is no reason for the coefficients b_α, $\alpha = 1, 2$, not to contain such a linear term:

$$b_\alpha(v) \approx \left(\frac{\partial b_\alpha}{\partial v}\right)_{v_l} (v - v_l) + \left(\frac{db_\alpha}{d\phi}\right)_{\phi_0} \Delta\phi, \tag{265}$$

where

$$\frac{d}{d\phi} = \frac{\partial}{\partial\phi} + \frac{\partial v_l}{\partial\phi}\frac{\partial}{\partial v}. \tag{266}$$

Below we shall suppose that the maximum variations of Δv are limited from both sides:

$$\gamma(\Delta\phi)^2 << \max|v - v_l| << \bar{\gamma}\Delta\phi, \tag{267}$$

with

$$\bar{\gamma} = \frac{(db_\alpha/d\phi)_{\phi_0}}{(\partial b_\alpha/\partial v)_{v_l}}. \tag{268}$$

Under the above condition the first term in (265) may be omitted and we obtain the excitation coefficient $E_\alpha = b_\alpha/b_i$ of the localized mode α in the form

$$E_\alpha = \frac{(db_\alpha/d\phi)_{\phi_0}}{(\partial b_i/\partial v)_{v_l}} \frac{\sqrt{v_l'/\gamma}}{v - v_l + iv_l'}. \tag{269}$$

This dependence reveals the typical resonance behavior with the center at the leaky wave branch $v = v_l(\phi)$, where the amplitude $|E_\alpha| \propto 1/|\Delta\phi|$, and with the width $|v - v_l| \sim v_l' \propto (\Delta\phi)^2$. Physically this resonance of reflection at $v = v_l$ with the leaky eigenmode is rather understandable: the incident "pumping" wave just compensates the leakage of energy from the surface in the eigenwave which also eliminates an imaginary addition to the speed v.

Since [92] the resonance reflection of the considered type in the vicinity of a leaky wave branch were studied in multiple publications [95, 96, 118-133] for various elastic, piezoelectric and piezomagnetic structures. These different schemes might be important for applications, however the basic features of considered resonances are very similar to the presented picture. So, we shall not go into details, referring readers to the above references.

4. Conclusions

Thus, as we have seen, anisotropy in crystalloacoustics is not at all reduced just to more complex theoretical descriptions of properties and phenomena, which are very well known and compactly described for isotropic media. Indeed, the properties of elastic waves in isotropic bodies are very simple, but this simplicity has its drawback. For instance, the group velocities of bulk waves in such bodies are directed along the wave normals and therefore are uniformly distributed on the sphere of propagation directions. The corresponding group and phase velocity surfaces coincide, both having a trivial spherical form. But this simplicity eliminates such important physical effects as energy concentrating and phonon focusing. In isotropic media there are no individual acoustic axes: the whole sphere is degenerate. But then in these media there are also no such phenomena, as conical and wedge refraction (both internal and external). Isotropy excludes piezoelectric and piezomagnetic couplings and at the same time the Bleusten-Gulyaev surface waves and their piezomagnetic analogs together with a series of beautiful physical effects based on those couplings. In isotropic media there are no supersonic surface waves and therefore also no leaky waves and the above described resonant reflection. More examples could be given to support this argument.

On the other hand, as we tried to demonstrate, anisotropy in crystalloacoustics is a beautiful world of elegant mathematics, non-trivial theorems, unexpected peculiarities, singularities and topological catastrophes. In the above considerations we have limited ourselves to the analysis of acoustic wave properties only in infinite and half-infinite anisotropic media. In fact, the same methods have proved to be very efficient in many other acoustic problems. For those who are interested in the theory of interfacial waves in anisotropic bicrystalline structures we can recommend the original publications [134-137]. For the theory of fundamental and wave-guided eigenmodes in anisotropic plates we address readers to the papers [138-142, 103, 143-147, 102]. The anisotropic layer-substrate structure is also rather popular in recent theoretical studies of fundamental properties of elastic waves (see, e.g., [129, 130, 148-150]).

5. Acknowledgements

The author is very grateful to the collaborators with whom many of the above presented or mentioned results were obtained, especially to J. Lothe, V.N. Lyubimov, A.L. Shuvalov and A.N. Darinskii for the pleasure of a joint work during many years. Separate thanks to G.A. Maugin, who initiated this paper, for his patience, understanding and helpful constructive editor's remarks. This work was supported by the Russian Foundation for Basic Research (Grant no.01-02-16228).

References

1. Fedorov, F.I. (1968) *Theory of Elastic Waves in Crystals*, Plenum Press, New York.
2. Kolodner, I.I. (1966) Existence of longitudinal waves in anisotropic media, *J. Acoust. Soc. Amer.* **40**, No. 3, 730-731.
3. Khatkevich, A.G. (1964) Special directions for elastic waves in crystals, *Sov. Phys. Crystallogr.* **9**, No. 5, 579-582.
4. Khatkevich, A.G. (1977) Classification of crystals by acoustic properties, *Sov. Phys. Crystallogr.* **22**, No. 6, 701-705.
5. Brugger, K. (1965) Pure modes for elastic waves in crystals, *J. Appl. Phys.* **36**, 759-768.

6. Bestuzheva, N.P. and Darinskii, B.M. (1993) Longitudinal normals of elastic waves in crystals, *Crystallogr. Rep.* **38**, No. 5, 592-597.
7. Fitzgerald, E.R. and Wright, T.W. (1967) Invariance of sound velocity sums in crystals, *Phys. Stat. Sol.* **24**, No. 1, 37-44.
8. Hayes, M. (1972) A universal connection for waves in anisotropic media, *Arch. Ration. Mech. Anal.* **46**, 105-113.
9. Alshits, V.I. and Lothe, J. (1979) Elastic waves in triclinic crystals. II. Topology of polarization fields and some general theorems, *Sov. Phys. Crystallogr.* **24**, No. 4, 393-398.
10. Ingebrigtsen, K.A. and Tonning, A. (1969) Elastic surface waves in crystals, *Phys. Rev.* **184**, No. 3, 942-951.
11. Hayes, M.A. and Musgrave, M.J.P. (1979) On the energy flux and group velocity, *Wave Motion* **1**, No. 1, 75-82.
12. Alshits, V.I., Kessenikh, G.G. and Lothe, J. (1982) Energy flux, group and phase velocities of acoustic waves in piezoelectrics, *Ferroelectrics* **42**, 103-108.
13. Hayes, M. (1974) Simple universal connections between energy velocities in anisotropic elastic media, *J. Acoust. Soc. Am.* **56**, No. 1, 1-3.
14. Sirotin, Yu.I. and Shaskolskaya, M.P. (1982) *Fundamentals of Crystal Physics*, "Mir" Publishers, Moscow.
15. Maris, H.J. (1971) Enhancement of heat pulses in crystals due to elastic anisotropy, *J. Acoust. Soc. Am.* **50**, No. 3, 812-818.
16. Philip, J. and Viswanathan, K.S. (1978) Phonon magnification in cubic crystals, *Phys. Rev. B* **17**, No. 12, 4969-4978.
17. Northrop, G.A. and Wolfe, J.P. (1979) Ballistic phonon imaging in solids a new look at phonon focusing, *Phys. Rev. Lett.* **43**, No. 19, 1424-1427.
18. Northrop, G.A. and Wolfe, J.P. (1980) Ballistic phonon imaging in germanium, *Phys. Rev. B* **22**, No. 22, 6196-6212.
19. Lax, M. and Narayanamurti, V. (1980) Phonon magnification and the gaussian curvature of the slowness surface in anisotropic media: detector shape effects with application to GaAs, *Phys. Rev. B* **22**, No. 10, 4876-4897.
20. Every, A.G. (1981) Ballistic phonons and the shape of the ray surface in cubic crystals, *Phys. Rev. B* **24**, No. 6, 3456-3467.
21. Every, A.G. (1994) Thermal phonon imaging, in T. Pashkiewicz and K. Rapcewicz (eds.), *Die Kunst of Phonons*, Plenum Press, New York, pp. 55-72.
22. Every, A.G. (1986) Formation of phonon-focusing caustics in crystals, *Phys. Rev. B* **34**, No. 4, 2852-2862.
23. Taylor, B., Maris, H.J. and Elbaum, C. (1969) Phonon focusing in solids, *Phys. Rev. Lett.*, **23**, No. 8, 416-419.
24. Every, A.G., Kim, K.Y. and Sachse, W. (2001) Point-source/point-receiver methods, In M. Levy (ed.), *Handbook of Elastic Properties of Solids, Liquids, and Gases*, Vol.I: Dynamic Methods for Measuring the Elastic Properties of Solids, Academic Press, New York, pp. 87-108.
25. Wolfe, J.P. (1993) *Heat Pulses and Phonon Imaging*, Springer, Berlin.
26. Khatkevich, A.G. (1962) The acoustic axes in crystals, *Sov. Phys. Crystallogr.* **7**, No. 5, 601-604.
27. Holm, P. (1992) Generic elastic media, *Physica Scr.* **T44**, 122-127.
28. Darinskii, B.M. (1994) Acoustic axes in crystals, *Crystallogr. Rep.* **39**, No. 5, 697-703.
29. Khatkevich, A.G. (1962) Internal conical refraction of elastic waves, *Sov. Phys. Crystallogr.* **7**, No. 5, 742-745.

30. Alshits, V.I. and Lothe, J. (1979) Elastic waves in triclinic crystals. I. General theory and the degeneracy problem, *Sov. Phys. Crystallogr.* **24**, No. 4, 387-392.
31. Ohmachi, Y., Uchida, N. and Niizeki, N. (1972) Acoustic wave propagation in TeO2 single crystal, *J. Acoust. Soc. Am.* **51**, 164-168.
32. Ledbetter, H.M. and Kriz, R.D. (1982) Elastic-wave surfaces in solids, *Phys. Stat. Sol. (b)* **114**, 475-480.
33. Boulanger, Ph. and Hayes, M. (1998) Acoustic axes for elastic waves in crystals: theory and applications, *Proc. Roy. Soc. Lond. A*, **454**, 2323-2346.
34. Musgrave, (1981) On the elastodynamic classification of orthorhombic media, *Proc. R. Soc. Lond. A* **374**, 401-429.
35. Herring, C. (1937) Accidental degeneracy in the energy bands of crystals, *Phys. Rev.* **52**, 365-373.
36. Herring, C. (1954) Role of low-energy phonons in thermal conduction, *Phys. Rev.* **95**, 954- 965.
37. Musgrave, (1970) *Crystal Acoustics*, Holden-Day, Cambridge.
38. Alshits, V.I. and Lothe, J. (1979) Elastic waves in triclinic crystals. III. Existence problem and some general properties of exceptional surface waves, *Sov. Phys. Crystallogr.* **24**, No. 6, 644-648.
39. Alshits, V.I. and Shuvalov, A.L. (1984) Polarization fields of elastic waves near the acoustic axes, *Sov. Phys. Crystallogr.* **29**, 373-378.
40. Alshits, V.I. and Shuvalov, A.L. (1984) On the effect of external influencies on the bulk elastic wave properties near acoustic axes, *Ferroelectric Letters* **1**, 151-154.
41. Alshits, V.I., Sarychev, A.V. and Shuvalov, A.L. (1985) Classification of degeneracies and analysis of their stability in the theory of elastic waves in crystals, *Sov. Phys. JETP* **62**, 531-539.
42. Shuvalov, A.L. (1998) Topological features of the polarization fields of plane acoustic waves in anisotropic media, *Proc. R. Soc. Lond. A* **454**, 2911-2947.
43. Cao Xuan An, Hauret, G. and Chapelle, J.P. (1977) Brillouin scattering in Hg2Cl2. *Solid State Commun.* **24**, 443-445.
44. Lushnikov, S.G., Prohorova, S.D., Sinii, I.G. and Smolenskii, G.A. (1987) Elastic properties of the cesium deuterium selenite crystal in the monoclinic phase, *Sov. Phys. Solid State* **29**, 280-284.
45. Alshits, V.I. and Lyubimov, V.N. (1998) Elastic waves in absorptive anisotropic media: perculiarities of wave surfaces and singularities in the polarization fields, in A. Radowicz (ed.) *Proceedings of II Workships on Dissipation in Physical Systems*, Politechnika Swietokrziska, Kielce, Poland, pp.15-44.
46. Alshits, V.I., Lyubimov, V.N., Sarychev, A.V. and Shuvalov, A.L. (1987) Topological characteristics of singular points of the electric field accompanying sound propagation in piezoelectrics, *Sov. Phys. JETP* **66**, No. 2, 408-413.
47. Alshits, V.I. and Shuvalov, A.L. (1988) On acoustic axes in piezoelectric crystals, *Sov. Phys. Crystallogr.* **33**, No. 1, 1-4.
48. Every, A.G. and McCurdy, A.K. (1987) Phonon focusing in piezoelectric crystals, *Phys. Rev. B* **36**, 1432-1447.
49. Klerk, J. and Musgrave, M. (1955) Internal conical refraction of transverse elastic waves in a cubic crystal, *Proc. Phys. Soc.* **68**, No. 2, 81-88.
50. Henneke, E.G. and Green, R.E. (1969) Light-wave / elastic wave analogies in crystals, *J. Acoust. Soc. Amer.* **45**, No. 6, 1367-1373.
51. Alshits, V.I., Darinskii, A.N. and Shuvalov, A.L. (1989) Disclinations in a vector field of polarizations in acoustic and optical beams under conical refraction conditions, *Sov.Phys. JETP* **69**, No. 6, 1105-1108.

52. Stroh, A.N. (1962) Steady state problems in anisotropic elasticity, *J. Math. Phys.*, **41**, 77- 103.
53. Taziev, R.M. (1989) Dispersion relation for acoustic waves in an anisotropic elastic half- space, *Sov. Phys. Acoustics* **35**, 535-538.
54. Barnett, D.M. (2000) Bulk, surface, and interfacial waves in anisotropic linear elastic solids, *Int. J. Solids Struct.* **37**, 45-54.
55. Barnett, D.M., Lothe, J., Nishioka, K. and Asaro, R.J. (1973) Elastic surface waves in anisotropic crystals: a simplified method for calculating Rayleigh velocities using dislocation theory, *J. Phys.F: Metal Physics* **3**, 1083-1096.
56. Barnett, D.M. and Lothe, J. (1974) Consideration of the existence of surface wave (Rayleigh wave) solutions in anisotropic elastic crystals, *J. Phys.F: Metal Physics* **4**, 671- 676.
57. Lothe, J. and Barnett, D.M. (1976) On the existence of surface wave solutions for anisotropic half-spaces with free surface, *J. Appl. Phys.* **47**, 428-433.
58. Barnett, D.M. and Lothe, J. (1985) Free surface (Rayleigh) waves in anisotropic elastic half-spaces, *Proc. R. Soc. Lond. A* **402**, 135-152.
59. Chadwick, P. and Smith, G.D. (1977) Foundations of the theory of surface waves in anisotropic elastic materials, in C.-S. Yih (ed.) *Advances in Applied Mechanics*, Academic Press, New York, pp. 303-376.
60. Alshits, V.I., Lyubimov, V.N. and Shuvalov A.L. (1994) Elastic surface waves at the loaded boundary of a crystal, *JETP* **79**, No. 3, 455-465.
61. Lothe, J. and Alshits, V.I. (1977) Existence criterion for quasi-bulk surface waves, *Sov. Phys. Crystallogr.* **22**, No. 5, 519-525.
62. Alshits, V.I. and Lothe, J. (1978) Surface waves in hexagonal crystals, *Sov. Phys. Crystallogr.* **23**, No. 5, 509-515.
63. Chadwick, P. and Smith, G.D. (1982) Surface waves in cubic elastic media, in H.G. Hopkins and M.J. Sewel (eds.), *Mechanics of Solids*, Pergamon Press, Oxford, pp.47-100.
64. Chadwick, P. (1989) Wave propagation in transversely isotropic elastic media. I.Homogeneous plane waves, II. Surface waves, *Proc. R. Soc. Lond. A* **422**, 23-66, 67-101.
65. Alshits, V.I., Gorkunova, A.S. and Shuvalov, A.L. (1994) Parameters of surface and bulk transonic acoustic waves in hexagonal crystals for the γ-geometry, *Crystallogr. Reports* **39**, No. 6, 879-886.
66. Barnett, D.M. and Lothe, J. (1975) Dislocations and line charges in anisotropic piezoelectric insulators, *Phys. Stat. Sol. (b)* **67**, 105-111.
67. Lothe, J. and Barnett, D.M. (1976) Integral formalism for surface waves in piezoelectric crystals. Existence considerations, *J. Appl. Phys.* **47**, 1799-1807.
68. Lothe, J. and Barnett, D.M. (1977) Further development of the theory for surface waves in piezoelectric crystals, *Physica Norvegica* **8**, 239-254.
69. Bleustein, L. (1968) A new surface wave in piezoelectric materials, *Appl. Phys. Lett.* **13**, No. 12, 412-413.
70. Gulyaev, Yu.V. (1969) Surface electroacoustic waves in solids, *JETP Letters*, **9**, No. 1, 63-65.
71. Gulyaev, Yu.V., Kouzavko, Yu.A., Oleinik, I.N. and Shavrov, V.G. (1984) New surface magnetoacoustic waves caused by piezomagnetism, *Sov. Phys. JETP* **60**, 674-676.
72. Kosevich, Yu.A. (1985) Existence of surface magnetoacoustic waves in crystals, *Sov. Phys. Solid State* **27**, 193-195.
73. Kaganov, M.I. and Kosevich, Yu.A. (1986) Shear surface magnetosonic waves due

to piezomagnetic interaction in crystals, *Poverkhn., Fiz. Khim. Mekh. (USSR)* **5**, 148-154.

74. Alshits, V.I. and Lyubimov, V.N. (1989) Crystals with a superconducting coating: Surface and bulk acoustic waves, *Sov. Phys. Solid State* **31**, No. 3, 99-101.
75. Alshits, V.I. and Lyubimov, V.N. (1989) Transformation of acoustic waves on the interface between a piezoelectric or piezomagnetic crystal and a superconductor, *Sov. Phys. Solid State* **31**, No. 12, 63-64.
76. Alshits, V.I., Darinskii, A.N. and Lothe, J. (1992) On the existence of surface waves in half-infinite anisotropic elastic media with piezoelectric and piezomagnetic properties, *Wave Motion* **16**, 265-283.
77. Alshits, V.I., Darinskii, A.N., Lothe, J. and Lyubimov, V.N. (1994) Surface acoustic waves in piezocrystals: an example of surface wave existence with clamped boundary, *Wave Motion* **19**, 113-123.
78. Alshits, V.I. and Lyubimov, V.N. (1994) Surface acoustic waves caused by a magnetoelectric interaction in piezomagnetic crystals, *JETP* **79**, No. 2, 364-368.
79. Naumenko, N.F., Bondarenko, V.S. and Perelomova, N.V. (1983) Features of volume and surface acoustic wave propagation in paratellurite crystals, *Sov. Phys. Solid State* **25**, No. 9, 1512-1513.
80. Alshits, V.I., Lyubimov, V.N., Naumenko, N.F., Perelomova, N.V. and Shuvalov, A.L. (1985) Exceptional elastic body waves in crystals of various symmetris, *Sov. Phys. Crystallogr.* **30**, No. 2, 123-126.
81. Alshits, V.I., Lyubimov and Shuvalov, A.L. (1987) Exceptional body waves in the neighborhoods of acoustic axes of various types, *Sov. Phys. Crystallogr.* **32**, No. 4, 487-491.
82. Lyubimov, V.N. & Sannikov, D.G. (1973) Surface quasi-bulk and Rayleigh elastic waves in crystals, *Sov. Phys. Solid State* **15**, No. 6, 830-831.
83. Lyubimov, V.N. & Sannikov, D.G. (1975) Surface quasi-bulk waves in the neighborhood of exceptional directions and surfaces in crystals, *Sov. Phys. Solid State* **17**, No. 2, 300-302.
84. Darinskii, A.N. (1998) Quasi-bulk Rayleigh waves in semi-infinite media of arbitrary anisotropy, *Wave Motion* **27**, No. 1, 79-93.
85. Lyubimov, V.N., Alshits, V.I. and Lothe, J. (1980) Body waves and quasi-body surface waves in a semi-infinite piezoelectric medium, *Sov. Phys. Crystallogr.* **25**, No. 1, 16-21.
86. Alshits, V.I. and Lyubimov, V.N. (1985) Two types of exceptional body waves in piezoelecrics and the sectors of existence of Bleustein-Gulyaev waves, *Sov. Phys. Crystallogr.* **30**, No. 3, 252-256.
87. Gulyaev, Yu.V. (1986) Shear surface acoustic waves in solids, in *Proc. Intern. Symp. "Surface Waves in Solids and Layered Structures"*, VINITI, Novosibirsk, V.1, pp.357-358.
88. Gulyaev, Yu.V. (1998) Review of shear surface acoustic waves in solids, *IEEE Trans. Ultrason. Ferroel. Freq. Control* **45**, No. 4, 935-938.
89. Maugin, G.A. (1988) Shear horizontal surface acoustic waves in solids, in: *Recent Developments in Surface Acoustic Waves*, eds. D.F.Parker and G.A.Maugin, pp. 158-172, Springer series on Wave Phenomena, Vol. 7, Springer-Verlag, Berlin.
90. Royer, D. and Dieulesaint, E. (1984) Rayleigh wave velocity and displacements on orthorhombic, tegragonal, hexagonal, and cubic crystals, *J. Acoust. Soc. Am.* **76**, 1438-1444
91. Kosevich, Yu.A. and Syrkin, E.S. (1985) Existence criterion and properties of

deeply penetrating Rayleigh waves in crystals, *Sov. Phys. JETP* **62**, No. 6, 1282-1286.
92. Alshits, V.I. and Lothe, J. (1981) Comments on the relation between surface wave theory and the theory of reflection, *Wave Motion* **3**, 297-310.
93. Alshits, V.I., Darinskii, A.N. and Shuvalov, A.L. (1989) Theory of reflection of acoustoelectric waves in a semiinfinite piezoelectric medium. I. Metallized surface, *Sov. Phys. Crystallogr.* **34**, No. 6, 808-812.
94. Alshits, V.I., Darinskii, A.N. and Shuvalov, A.L. (1990) Theory of reflection of acoustoelectric waves in a semiinfinite piezoelectric medium. II. Non-metallized surface, *Sov. Phys. Crystallogr.* **35**, No. 1, 1-6.
95. Alshits, V.I., Darinskii, A.N. and Shuvalov, A.L. (1991) Theory of reflection of acoustoelectric waves in a semiinfinite piezoelectric medium. III. Resonance reflection in the vicinity of the leaky wave branch, *Sov. Phys. Crystallogr.* **36**, No. 2, 145-152.
96. Alshits, V.I., Darinskii, A.N. and Shuvalov, A.L. (1993) Acoustoelectric waves in bicrystal media in conditions of a rigid contact or a vacuum gap at an interface, *Crystallogr.Reports* **38**, No. 2, 147-158.
97. Alshits, V.I., Darinskii, A.N. and Shuvalov, A.L. (1992) Elastic waves in infinite and semi- infinite anisotropic media, *Physica Scripta* T**44**, 85-93.
98. Brekhovskikh, L.M. (1980) *Waves in Layered Media*, 2nd ed. Academic Press, New York.
99. Qu, J., Achenbach, J.D. and Roberts, R.A. (1989) Reciprocal relations for transmission coefficients: theory and applications, *IEEE Trans. Ultrason. Ferroel. Freq. Control* **36**, No. 2, 280-286.
100. Shuvalov, A.L. and Lothe, J. (1997) The Stroh formalism and the reciprocity properties of reflection-transmission problems in crystal piezo-acoustics, *Wave Motion* **25**, 331-345.
101. Achenbach, J.D. and Gautesen, A.K. (1977) Elastic surface waves guided by the edge of a slit, *J. Sound and Vibr.* **53**, No. 3, 407-416.
102. Alshits, V.I., Deschamps, M. and Maugin, G.A. (2003) Elastic waves in anisotropic plates: short-wavelength asymptotics of the dispersion branches vn(k), *Wave Motion*, **37**, No. 3, 273-292.
103. Alshits, V.I., Lothe, J. and Lyubimov, V.N. (1983) Surface and exceptional body elastic waves in a plate of a hexagonal crystal, *Sov. Phys. Crystallogr.* **28**, No. 4, 374-378.
104. Filippov, V.V. (1984) On exceptional bulk waves in piezoelectric crystals, *Doklady Akad. Nauk BSSR* **28**, No. 5, 412-414.
105. Lyubimov, V.N. and Alshits, V.I. (1982) Brewster reflections of elastic waves in hexagonal crystals, *Sov. Phys. Crystallogr.* **27**, No. 5, 512-516.
106. Chadwick, P. (1990) The behavior of elastic surface waves polarized in a plane of material symmetry. I. General analysis, *Proc. R. Soc. Lond.* A **430**, 231-240.
107. Barnett, D.M., Chadwick, P. and Lothe, J. (1991) The behavior of elastic surface waves polarized in a plane of material symmetry. I. Addendum, *Proc. R. Soc. Lond.* A **433**, 699-710.
108. Gundersen, S.A., Wang, L. and Lothe, J. (1991) Secluded supersonic elastic surface waves, *Wave Motion* **14**, 129-143.
109. Lothe, J. (1991) Body waves in anisotropic elastic media, in J.J. Wu, T.C.T. Ting and D.M. Barnett (eds.), *Modern Theory of Anisotropic Elasticity and Applications*, SIAM, Philadelphia, pp. 173-185.
110. Wang, L. and Lothe, J. (1992) Simple reflection in anisotropic elastic media and

its relation to exceptional waves and supersonic surface waves, *Wave Motion* **16**, 89-112.
111. Lothe, J. and Wang, L. (1995) Self-orthogonal sextic formalism for anisotropic elastic media: spaces of simple reflection and two component surface waves, *Wave Motion* **21**, 163-181.
112. Darinskii, A.N., Alshits, V.I., Lothe, J., Lyubimov, V.N. and Shuvalov, A.L. (1998) A criterion for the existence of two-component surface waves in elastically anisotropic media, *Wave Motion* **28**, 241-257.
113. Aleksandrov, V.V., Velichkina, T.S., Mozhaev, V.G., Potapova, Ju.B., Khmelev, A.K. and Yakovlev, I.A. (1992) New data concerning surface Mandelstamm-Brilouin light scattering from the basal plane of germanium crystal, *Phys. Lett.* A **162**, No. 5, 418-422.
114. Darinskii, A.N. (1997) On the theory of leaky waves in crystals, *Wave Motion* **25**, No. 1, 35-49.
115. Alshits, V.I. and Lyubimov, V.N. (1983) Leaky elastic waves in hexagonal crystals, *Sov. Phys. Crystallogr.* **28**, No. 2, 130-135.
116. Alshits, V.I., Darinskii, A.N., Kotowski, R.K. and Shuvalov, A.L. (1988) Analog of Schoch effect in reflection of acoustic beams from free boundaries of a crystal, *Sov. Phys. Crystallogr.* **33**, No. 3, 318-325.
117. Alshits, V.I., Darinskii, A.N., and Shuvalov, A.L. (1990) Resonant reflection of "three- dimensional" acoustic beam from the free boundary of a crystal, *Sov. Phys. Crystallogr.* **35**, No. 4, 476-478.
118. Alshits, V.I., Darinskii, A.N., and Shuvalov, A.L. (1992) Resonant reflection and refraction of sound at a liquid-crystal interface, *Sov. Phys. Solid State*, **34**, No. 8, 1337-1346.
119. Alshits, V.I., Darinskii, A.N., and Shuvalov, A.L. (1992) Effect of resonant reflection of acoustic waves on the 180-domain wall, *Ferroelectrics*, **126**, 323-328.
120. Alshits, V.I., Darinskii, A.N., and Radowicz, A. (1995) Resonance reflection of elastic waves at the interface between two crystals with sliding contact. I. Plane waves in structures with arbitrary anisotropy; II. Plane waves and acoustic beams in structures with hexagonal symmetry, *Cristallogr. Reports* **40**, No. 3, 364-389.
121. Alshits, V.I. and Lyubimov, V.N. (1995) Magnetoelectroelastic resonances in layered piezocrystalline structutes, *Phys. Solid State* **37**, No. 6, 1014-1020.
122. Alshits, V.I., Lyubimov, V.N. and Radowicz, A. (1995) Resonant excitation of Rayleigh waves in a bicristalline piezoelectric structure, *Cristallogr. Reports* **40**, No. 4, 538-542.
123. Darinskii, A.N. (1995) Resonance phenomena in the reflection of an elastic wave at the boundary between a hexagonal crystal and an anisotropic film, *JETP* **80**, No. 2, 317-323.
124. Darinskii, A.N. and Maugin, G.A. (1995) The elastic wave resonance reflection from a thin solid layer in a crystal, *Wave Motion* **23**, 363-385.
125. Alshits, V.I., Lyubimov, V.N. and Radowicz, A. (1996) Resonance excitation of Love waves in sandwich-type structures, *Phys. Solid State* **38**, No. 4, 604-608.
126. Alshits, V.I., Gierulski, W., Lyubimov, V.N. and Radowicz, A. (1997) Resonance excitation of quasi-Rayleigh waves in plates on soft or hard substrates, *Crystallogr. Reports* **42**, No. 1, 20-27.
127. Darinskii, A.N. (1998) Leaky waves and the elastic wave resonance reflection on a crystal-thin solid layer interface. II. Leaky waves given rise to by exceptional bulk waves, *J. Acoust. Soc. Am.* **103**, 1845-1854.

128. Darinskii, A.N. and Maugin, G.A. (1998) Features of elastic wave reflection in the piezoelectric-thin solid layer-piezoelectric structure, *Acta Acustica* **84**, 455-464.
129. Alshits, V.I., Gorkunova, A.S., Lyubimov, V.N. and Nowacki, J.P. (1999) Elastic waves in the anisotropic layer-substrate structure. I. General theory, *Crystallography Reports* **44**, No. 4, 592-602.
130. Alshits, V.I., Gorkunova, A.S., Lyubimov, V.N. and Kotowski, R.K. (1999) Elastic waves in the anisotropic layer-substrate structure. II. Case of transversal isotropy, *Crystallography Reports* **44**, No. 5, 799-805.
131. Alshits, V.I., Gorkunova, A.S., Lyubimov, V.N., Gierulski, W., Radowicz, A. and Kotowski, R.K. (1999) Methods of resonant excitation of surface waves in crystals, in B.T. Maruszewski, W. Muschik and A. Radowicz (eds.), *Trends in Continuum Physics (TRECOP '88)*, World Scientific, Singapore, pp. 28-34.
132. Alshits, V.I., Lyubimov, V.N. and Radowicz A. (2000) Quasi-Rayleigh waves in sandwich structures: dispersion equation, eigenmodes and resonance reflection, *Crystallography Reports* **45**, No. 3, 457-465.
133. Nowacki, J.P., Alshits, V.I. and Lyubimov, V.N. (2000) Magnetoelastic waves in a bicrystalline gap structure under a bias magnetic field, *Int. J. Appl. Electromagnetics and Mechanics* **11**, No. 4, 223-232.
134. Barnett, D.M., Lothe, J., Gavazza, S.D. and Musgrave, M.J.P. (1985) Considerations of the existence of interfacial (Stoneley) waves in bonded anisotropic half-spaces, *Proc. R. Soc. Lond.* A **402**, 153-166.
135. Barnett, D.M., Gavazza, S.D. and Lothe, J. (1988) Slip waves along the interface between two anisotropic elastic half-spaces in sliding contact, *Proc. R. Soc. Lond.* A **415**, 389-419.
136. Abudi, M. and Barnett, D.M. (1990) On the existence of interfacial (Stoneley) waves in bonded piezoelectric half-spaces, *Proc. R. Soc. Lond.* A **429**, 587-611.
137. Alshits, V.I., Barnett, D.M., Darinskii, A.N. and Lothe, J. (1994) On the existence problem for localized acoustic waves on the interface between two piezocrystals, *Wave Motion* **20**, 233-244.
138. Lekhnitskii, S.G. (1947) *Anizotropnye Plastinki*, Nauka, Moscow (Engl. transl. Anisotropic Plates, Gordon & Breach, New York).
139. Newman, E.G. and Mindlin, R.D. (1957) Vibrations of a monoclinic crystal plate, *J. Acoust. Soc. Am.* **29**, 1206-1218.
140. Lyubimov, V.N. (1980) Elastic surface waves of the Bleustein-Gulyaev type in piezoelectric plate, *Sov. Phys. Crystallogr.* **25**, No. 3, 265-268.
141. Lyubimov, V.N. (1980) Lamb surface elastic waves in an anisotropic piezoelectric plate, Sov. Phys. Crystallogr. 25, No. 4, 389-392.
142. Lyubimov, V.N. (1981) Elastic quasi-body surface waves in an anisotropic plate, *Sov. Phys. Crystallogr.* **26**, No. 4, 377-381.
143. Markus, S.A. (1987) Low-frequency approximations for zero-order normal modes in anisotropic plates, *Sov. Phys. Acoust. 33*, 634-636.
144. Alshits, V.I. and Lyubimov, V.N. (1988) Anormalous dispersion of surface elastic waves in an anisotropic plate, *Sov. Phys. Crystallogr.* **33**, No. 2, 163-166.
145. Alshits, V.I. and Lyubimov, V.N. (1988) Exceptional body solutions for elastic waves in a piezoelectric plate, *Sov. Phys. Crystallogr.* **33**, No. 2, 166-168.
146. Shuvalov, A.L. (2000) On the theory of wave propagation in anisotropic plates, *Proc. R. Soc. Lond.* A **456**, 2197-2222.
147. Shuvalov, A.L. (2002) Theory of plane subsonic elastic waves in fluid-loaded anisotropic plates, *Proc. R. Soc. Lond.* A **458**, 1323-1352.
148. Darinskii, A.N. (2000) On the theory of the elastic wave propagation in a crystal

coated with a solid layer. I. Two-component surface waves and simple reflection. II. Surface and leaky waves near a line of exceptional bulk waves, *Proc. R. Soc. Lond.* A **456**, 1897-1929.

149. Shuvalov, A.L. and Every, A. (2002) Characteristic features of the velocity dispersion of surface acoustic waves in anisotropic coated solids, *Ultrasonics* **40**, 939-942.
150. Shuvalov, A.L. and Every, A. (2002) Some properties of surface acoustic waves in anisotropic coated solids, studied by the impedance method, *Wave Motion* **36**, 257-273.

SURFACE WAVES OF NON-RAYLEIGH TYPE

SERGEY V. KUZNETSOV
Institute for Problems in Mechanics
Prosp. Vernadskogo, 101, 119526 Moscow, Russia

Abstract. In a previous publication [1] existence of "forbidden" planes for some transversely isotropic half spaces upon which the genuine Rayleigh waves cannot propagate was established. Now, it is proved that for some cubic crystals and the directions of elastic symmetry there arise exponentially attenuating with depth surface waves of the non-Rayleigh type.

1. Introduction

In our previous paper [1] it was shown that some transversely isotropic media exhibit property of non-existence of the genuine Rayleigh waves. The latter can be defined by the following expression

$$\mathbf{u}(\mathbf{x}) = \sum_{k=1}^{3} C_k \mathbf{m}_k e^{ir(\gamma_k \, \nu \cdot \mathbf{x} + \mathbf{n} \cdot \mathbf{x} - ct)} \tag{1.1}$$

where C_k are complex coefficients determined up to a multiplier by the traction-free boundary conditions; $\mathbf{m}_k$ are complex eigenvectors of the Christoffel equation, which will be introduced further; these eigenvectors correspond to complex roots γ_k of the characteristic polynomial; r is the (real) wave number; ν is an outward normal to the boundary Π_ν of the half-space along which the surface wave propagates; $\mathbf{n} \in \Pi_\nu$ is the unit vector determining direction of propagation of the surface wave, and c is the phase speed. The terms

$$\mathbf{u}_k(\mathbf{x}) \equiv \mathbf{m}_k e^{ir(\gamma_k \nu \cdot \mathbf{x} + \mathbf{n} \cdot \mathbf{x} - ct)} \tag{1.2}$$

are called partial waves.

As was shown in [1], the existence of the "forbidden" planes upon which the genuine Rayleigh wave cannot propagate is due to appearing the Jordan blocks in a specially constructed 6×6 -matrix associated with the Christoffel equation.

R.V. Goldstein and G.A. Maugin (eds.),
Surface Waves in Anisotropic and Laminated Bodies and Defects Detection, 69–78.

The following analysis reveals that the situation regarded in [1] appears to be more complicated. The Jordan blocks in the regarded matrix lead to a qualitative change of the structure of the partial waves (1.2) and, while the genuine Rayleigh wave (1.1) at the situation considered in [1] does not exist, there remains an exponentially attenuating with depth surface wave of the non-Rayleigh type. It should also be noted that some surface waves of the non- Rayleigh type were reported in [3- 5] by applying Stroh's sextic formalism.

2. BASIC NOTATIONS

Equations of motion for an anisotropic elastic medium can be written in the form

$$\mathbf{A}(\partial_x, \partial_t)\mathbf{u} \equiv \mathrm{div}_x\, \mathbf{C} \cdot \cdot \nabla_x \mathbf{u} - \rho \ddot{\mathbf{u}} = 0, \tag{2.1}$$

where $\mathbf{u}$ is the displacement field; ρ is the density of a medium; and $\mathbf{C}$ is the fourth-order elasticity tensor assumed to be positive definite:

$$\underset{\mathbf{A} \in sym(R^3 \otimes R^3), \mathbf{A} \neq 0}{\forall \mathbf{A}} (\mathbf{A} \cdot \cdot \mathbf{C} \cdot \cdot \mathbf{A}) > 0 \tag{2.2}$$

The sign "· " in (2.1), (2.2) and henceforth means the scalar multiplication in the corresponding unitary or Euclidian vector space:

$$\mathbf{a} \cdot \mathbf{b} = \sum_k a_k \overline{b^k} \tag{2.3}$$

Substituting partial waves (1.2) in Eq. (2.1) produces the Christoffel equation:

$$\left[(\gamma_k \nu + \mathbf{n}) \cdot \mathbf{C} \cdot (\mathbf{n} + \gamma_k \nu) \; - \rho c^2 \mathbf{I} \right] \cdot \mathbf{m}_k = 0, \tag{2.4}$$

where $\mathbf{I}$ is the unit diagonal matrix. Equation (2.3) can be written in the equivalent form:

$$\det \left[(\gamma_k \nu + \mathbf{n}) \cdot \mathbf{C} \cdot (\mathbf{n} + \gamma_k \nu) \; - \rho c^2 \mathbf{I} \right] = 0 \tag{2.4'}$$

The left-hand side of Eq. (2.4') represents a polynomial of degree 6 with respect to γ_k .

REMARK 2.1. It can be shown, see [1], that if the phase speed does not exceed the so called lower limiting speed ($c_3^{\lim}$):

$$c < c_3^{\lim}, \tag{2.5}$$

then all the roots of Eq. (2.3) are complex with $\mathrm{Im}(\gamma_k) \neq 0$. The inequality (2.5) ensures that three partial waves (1.2) with $\mathrm{Im}(\gamma_k) < 0$ attenuate with depth in a "lower" half-space at $(\nu \cdot \mathbf{x}) < 0$. Only attenuating with depth partial waves, as being physically reasonable, are regarded.

3. SIX-DIMENSIONAL FORMALISM

Following [1], a more general representation for the partial wave than (1.2), will be considered:

$$\mathbf{v}(x'')\, e^{ir(\mathbf{n}\cdot\mathbf{x}-ct)} \tag{3.1}$$

where $x'' = ir\,\nu\cdot\mathbf{x}$ is the dimensionless complex coordinate, $\mathbf{v}(x'')$ is an unknown vector function, and the exponential multiplier in (3.1) corresponds to propagation of the plane wave front along the direction $\mathbf{n}$ with the phase speed c. Substituting representation (3.1) into Eq. (2.1) yields the following system of ordinary differential equations:

$$\left(\begin{array}{c} (\nu\cdot\mathbf{C}\cdot\nu)\,\partial^2_{x''} + (\nu\cdot\mathbf{C}\cdot\mathbf{n}+\mathbf{n}\cdot\mathbf{C}\cdot\nu)\,\partial_{x''} - \\ (\mathbf{n}\cdot\mathbf{C}\cdot\mathbf{n}-\rho c^2\mathbf{I}) \end{array} \right) \mathbf{v}(x'') = 0 \tag{3.2}$$

Direct analysis of system (3.2) is rather difficult, and reduction to the first-order system can considerably simplify it.

Introduction of a new vector-function $\mathbf{w} = \partial_{x''}\mathbf{v}$ allows us to reduce the second-order system (3.2) in C^3 to the first-order one in C^6 :

$$\partial_{x''}\begin{pmatrix}\mathbf{v}\\ \mathbf{w}\end{pmatrix} = \mathbf{R}_6\cdot\begin{pmatrix}\mathbf{v}\\ \mathbf{w}\end{pmatrix} \tag{3.3}$$

In (3.3) the complex six-dimensional matrix $\mathbf{R}_6$ has the form

$$\mathbf{R}_6 = \begin{pmatrix}\mathbf{0} & \mathbf{I}\\ -\mathbf{M} & -\mathbf{N}\end{pmatrix} \tag{3.4}$$

where three-dimensional matrices $\mathbf{M}$ and $\mathbf{N}$ have the form

$$\begin{array}{l} \mathbf{M} = (\nu\cdot\mathbf{C}\cdot\nu)^{-1}\cdot(\mathbf{n}\cdot\mathbf{C}\cdot\mathbf{n}-\rho c^2\mathbf{I}) \\ \mathbf{N} = (\nu\cdot\mathbf{C}\cdot\nu)^{-1}\cdot(\nu\cdot\mathbf{C}\cdot\mathbf{n}+\mathbf{n}\cdot\mathbf{C}\cdot\nu) \end{array} \tag{3.5}$$

In (3.4) $\mathbf{I}$ stands for the unit (diagonal) matrix in the three-dimensional space. A surjective homomorphism $\Im m : C^6 \to C^3$, such that

$$\Im m(\mathbf{v},\,\mathbf{w}) = \mathbf{v} \tag{3.6}$$

will be needed for the subsequent analysis. The following Proposition takes place [1]:

PROPOSITION 3.1. Let $c \in (0; c_3^{\lim})$:

a) Spectrum of the matrix $\mathbf{R}_6$ coincides with the set of all roots of polynomial (2.4);

b) If γ is a complex eigenvalue and $\mathbf{m} = (\mathbf{m}', \mathbf{m}'')$, $\mathbf{m}', \mathbf{m}'' \in C^3$ is the corresponding six-dimensional eigenvector of the matrix $\mathbf{R}_6$, then $\overline{\gamma}$ is also an eigenvalue with the corresponding eigenvector $\overline{\mathbf{m}} = (\overline{\mathbf{m}', \mathbf{m}''})$;

c) The matrix $\mathbf{R}_6$ admits the following Jordan normal forms

$$\begin{aligned}
\mathbf{J}_6^{(\mathrm{I})} &= \operatorname{diag}\left(\gamma_1, \overline{\gamma}_1, \gamma_2, \overline{\gamma}_2, \gamma_3, \overline{\gamma}_3\right), \\
\mathbf{J}_6^{(\mathrm{II})} &= \operatorname{diag}\left(\begin{pmatrix} \gamma_1 & 1 \\ 0 & \gamma_1 \end{pmatrix}, \begin{pmatrix} \overline{\gamma}_1 & 1 \\ 0 & \overline{\gamma}_1 \end{pmatrix}, \gamma_3, \overline{\gamma}_3\right), \\
\mathbf{J}_6^{(\mathrm{III})} &= \operatorname{diag}\left(\begin{pmatrix} \gamma_1 & 1 & 0 \\ 0 & \gamma_1 & 1 \\ 0 & 0 & \gamma_1 \end{pmatrix}, \begin{pmatrix} \overline{\gamma}_1 & 1 & 0 \\ 0 & \overline{\gamma}_1 & 1 \\ 0 & 0 & \overline{\gamma}_1 \end{pmatrix}\right)
\end{aligned} \tag{3.7}$$

d) According to the Jordan normal forms the following three types of representations for surface waves occur: (i) for the Jordan normal form $\mathbf{J}_6^{(\mathrm{I})}$, the corresponding representation is given by (1.1); (ii) for the Jordan normal form $\mathbf{J}_6^{(\mathrm{II})}$, the representation is as follows:

$$\begin{aligned}
\mathbf{u}(\mathbf{x}) &= (C_1 + ir\, C_2\, \nu \cdot \mathbf{x})\, \mathbf{m}'_1\, e^{ir(\gamma_1\, \nu\cdot\mathbf{x} + \mathbf{n}\cdot\mathbf{x} - ct)} \\
&\quad + C_2\, \mathbf{m}'_2\, e^{ir(\gamma_1\, \nu\cdot\mathbf{x} + \mathbf{n}\cdot\mathbf{x} - ct)} + C_3\, \mathbf{m}'_3\, e^{ir(\gamma_3\, \nu\cdot\mathbf{x} + \mathbf{n}\cdot\mathbf{x} - ct)}
\end{aligned}, \tag{3.8}$$

where $\mathbf{m}'_1 = \Im m(\mathbf{m}_1) \in C^3$, and $\mathbf{m}_1$ is the eigenvector of $\mathbf{R}_6$ corresponding to the eigenvalue γ_1 ; $\mathbf{m}'_2 = \Im m(\mathbf{m}_2) \in C^3$, and $\mathbf{m}_2 \in C^3$ is the generalized eigenvector associated with $\mathbf{m}_1$, and the eigenvector $\mathbf{m}_3 \in C^6$ corresponds to the eigenvalue γ_3 ; (iii) for the Jordan normal form $\mathbf{J}_6^{(\mathrm{III})}$, the representation is as follows:

$$\begin{aligned}
\mathbf{u}(\mathbf{x}) &= (C_1 + ir\, C_2 \nu \cdot \mathbf{x} + \frac{1}{2} C_3\, (ir\, \nu \cdot \mathbf{x})^2) \times \\
&\qquad \times \mathbf{m}'_1\, e^{ir(\gamma_1\, \nu\cdot\mathbf{x} + \mathbf{n}\cdot\mathbf{x} - ct)} + \\
&\quad + (C_2 + ir\, C_3 \nu \cdot \mathbf{x})\, \mathbf{m}'_2\, e^{ir(\gamma_1\, \nu\cdot\mathbf{x} + \mathbf{n}\cdot\mathbf{x} - ct)} + \\
&\qquad + \mathbf{m}'_3\, e^{ir(\gamma_1\, \nu\cdot\mathbf{x} + \mathbf{n}\cdot\mathbf{x} - ct)}
\end{aligned} \tag{3.9}$$

$\mathbf{m}'_1 = \Im m(\mathbf{m}_1) \in C^3$, and $\mathbf{m}_1$ is the eigenvector corresponding to the eigenvalue γ_1 ; and $\mathbf{m}_2$, $\mathbf{m}_3 \in C^6$ are the generalized eigenvectors associated with $\mathbf{m}_1$.

COROLLARY. For any of the Jordan normal forms of the matrix $\mathbf{R}_6$ the three-dimensional components $\mathbf{m}'_k$, $\mathbf{m}''_k$ of the (proper) eigenvector $\mathbf{m}_k$, satisfy the equations

$$\begin{aligned}
\mathbf{m}''_k &= \gamma_k\, \mathbf{m}'_k \\
(\gamma_k^2 \mathbf{I} + \gamma_k \mathbf{N} + \mathbf{M}) \cdot \mathbf{m}'_k &= 0
\end{aligned} \tag{3.10}$$

Proof. When the matrix $\mathbf{R}_6$ has no Jordan blocks, the solution of Eq. (3.3) in view of (3.4) leads to Eqs. (3.10). Thus, the component $\mathbf{m}'_k$ belongs to the kernel space of the matrix $(\gamma_k^2\mathbf{I}+\gamma_k\mathbf{N}+\mathbf{M})$. The analogous situation takes place for any of the proper eigenvectors.

4. CONSTRUCTING THE GENERALIZED EIGENVECTOR

In view of [2] the solution of Eq. (3.3) corresponding to a Jordan block of the second rank can be represented in the form

$$\begin{pmatrix} C_1\left(\mathbf{m}'_1,\ \mathbf{m}''_1\right)+ \\ C_2\left(x''\left(\mathbf{m}'_1,\ \mathbf{m}''_1\right)+\left(\mathbf{m}'_2,\ \mathbf{m}''_2\right)\right) \end{pmatrix} e^{\gamma_1 x''} \tag{4.1}$$

where as before $x'' = ir\,\nu\cdot\mathbf{x}$.

PROPOSITION 4.1. a) The three-dimensional components $\mathbf{m}'_1,\ \mathbf{m}''_1$ of the genuine eigenvector satisfy Eqs. (3.10); b) Components $\mathbf{m}'_2,\ \mathbf{m}''_2$ of the generalized eigenvector satisfy the following equations:

$$\begin{aligned} &(\gamma_1^2\mathbf{I}+\gamma_1\mathbf{N}+\mathbf{M})\cdot\mathbf{m}'_2 = -\left(2\gamma_1\mathbf{I}+\mathbf{N}\right)\cdot\mathbf{m}'_1 \\ &\mathbf{m}''_2 = \mathbf{m}'_1+\gamma_1\mathbf{m}'_2 \end{aligned} \tag{4.2}$$

c) At $c\in(0;c_3^{\lim})$ the matrix $(2\gamma_1\mathbf{I}+\mathbf{N})$ is not degenerate; d) At the same speed interval the vectors $(2\gamma_1\mathbf{I}+\mathbf{N})\cdot\mathbf{m}'_1$ and $\mathbf{m}'_1\cdot(\nu\cdot\mathbf{C}\cdot\nu)$ are orthogonal.

Proof. Conditions a) and b) flow out by direct substituting the solution (4.1) into Eq. (3.3). To prove c) it is sufficient to demonstrate that the matrix

$$\begin{aligned} &(\nu\cdot\mathbf{C}\cdot\nu)\cdot(2\gamma_1\mathbf{I}+\mathbf{N}) = \\ &\quad = 2\gamma_1(\nu\cdot\mathbf{C}\cdot\nu)+(\nu\cdot\mathbf{C}\cdot\mathbf{n}+\mathbf{n}\cdot\mathbf{C}\cdot\nu) \end{aligned} \tag{4.3}$$

is not degenerate. Considering multiplication of the right-hand side of (4.3) by any nonzero conjugate complex vectors $\mathbf{a},\ \bar{\mathbf{a}}\in C^3$ and accounting Remark 2.1, which ensures $\mathrm{Im}(\gamma_1)\neq 0$, we arrive to

$$\begin{aligned} &\mathrm{Im}\left(\mathbf{a}\cdot(2\gamma_1(\nu\cdot\mathbf{C}\cdot\nu)+(\nu\cdot\mathbf{C}\cdot\mathbf{n}+\mathbf{n}\cdot\mathbf{C}\cdot\nu))\cdot\bar{\mathbf{a}}\right) = \\ &\qquad 2\mathrm{Im}(\overline{\gamma_1})(\mathbf{a}\otimes\nu\cdot\cdot\mathbf{C}\cdot\cdot\nu\otimes\bar{\mathbf{a}})\neq 0 \end{aligned} \tag{4.4}$$

In obtaining (4.4) we took into consideration that

$$\mathrm{Im}\left(\mathbf{a}\cdot(\nu\cdot\mathbf{C}\cdot\mathbf{n}+\mathbf{n}\cdot\mathbf{C}\cdot\nu)\cdot\bar{\mathbf{a}}\right)=0,$$

since the matrix

$$(\nu\cdot\mathbf{C}\cdot\mathbf{n}+\mathbf{n}\cdot\mathbf{C}\cdot\nu)$$

is (real) symmetric. The last inequality in (4.4) completes the proof of condition c).

To prove d) Eq. (4.2) can be transformed into equivalent one by multiplying both sides by the nondegenerate matrix $(\nu \cdot \mathbf{C} \cdot \nu)$, this gives

$$\begin{gathered}(\gamma_1^2(\nu \cdot \mathbf{C} \cdot \nu) + \gamma_1(\nu \cdot \mathbf{C} \cdot \mathbf{n} + \mathbf{n} \cdot \mathbf{C} \cdot \nu) + (\mathbf{n} \cdot \mathbf{C} \cdot \mathbf{n} - \rho c^2 \mathbf{I})) \cdot \mathbf{m}_2' = \\ -(2\gamma_1(\nu \cdot \mathbf{C} \cdot \nu) + (\nu \cdot \mathbf{C} \cdot \mathbf{n} + \mathbf{n} \cdot \mathbf{C} \cdot \nu)) \cdot \mathbf{m}_1'\end{gathered} \tag{4.5}$$

Now, the vector $\mathbf{m}_1'$ belongs to the kernel space of the matrix in the left-hand side of Eq. (4.5), this flows out from Proposition 4.1.a. Moreover, the regarded matrix is complex symmetric, hence its left and right eigenvectors coincide. The latter allows us to write for the left- hand side of Eq. (4.5)

$$\mathbf{m}_1' \cdot \begin{pmatrix} \gamma_1^2(\nu \cdot \mathbf{C} \cdot \nu) + \gamma_1(\nu \cdot \mathbf{C} \cdot \mathbf{n} + \mathbf{n} \cdot \mathbf{C} \cdot \nu) \\ +(\mathbf{n} \cdot \mathbf{C} \cdot \mathbf{n} - \rho c^2 \mathbf{I}) \end{pmatrix} \cdot \mathbf{m}_2' = 0 \tag{4.6}$$

Similarly, for the right-hand side of Eq. (4.5)

$$\mathbf{m}_1' \cdot (2\gamma_1(\nu \cdot \mathbf{C} \cdot \nu) + (\nu \cdot \mathbf{C} \cdot \mathbf{n} + \mathbf{n} \cdot \mathbf{C} \cdot \nu)) \cdot \mathbf{m}_1' = 0 \tag{4.7}$$

In view of (3.5), Eq. (4.7) completes the proof.

COROLLARY. In the factor-space $C^3/\mathrm{Ker}\,(\gamma_1^2 \mathbf{I} + \gamma_1 \mathbf{N} + \mathbf{M})$, the vector $\mathbf{m}_2'$ admits the following representation

$$\mathbf{m}_2' = -\left(\gamma_1^2 \mathbf{I} + \gamma_1 \mathbf{N} + \mathbf{M}\right)^{-1} \cdot (2\gamma_1 \mathbf{I} + \mathbf{N}) \cdot \mathbf{m}_1' \tag{4.8}$$

REMARK 4.1. a) At the regarded speed interval $c \in (0; c_3^{\lim})$ the eigenvectors of the complex symmetric matrix appearing in Eq. (4.6) may not form a set of mutually orthogonal vectors in C^3 , in contrast to the mutually orthogonal eigenvectors of any real symmetric matrix. b) For the supersonic Lamb waves propagating with the phase speed exceeding the greatest limiting speed $c_1^{\lim}$, all eigenvalues of the matrix $\mathbf{R}_6$ become real. Presumably, in such a case the condition c) of Proposition 4.1 and the subsequent Corollary can be violated.

5. DISPERSION EQUATION

The traction-free boundary conditions on the surface Π_ν can be written in the form:

$$\mathbf{t}_\nu \equiv \nu \cdot \mathbf{C} \cdot \cdot \nabla \mathbf{u} \Big|_{\mathbf{x} \in \Pi_\nu} = 0 \tag{5.1}$$

Substituting the displacement field into Eq. (5.1) yields

$$\sum_{k=1}^{3} C_k \mathbf{t}_k = 0, \tag{5.2}$$

where $\mathbf{t}_k$ are the partial surface traction.

The following two cases for the partial surface traction fields will be considered:

(i) For the Jordan normal form $\mathbf{J}_6^{(\mathrm{I})}$ and the representation (1.1), the partial surface tractions are of the form

$$\mathbf{t}_k = (\gamma_k \nu \cdot \mathbf{C} \cdot \nu + \nu \cdot \mathbf{C} \cdot \mathbf{n}) \cdot \mathbf{m}'_k \, e^{ir(\mathbf{n}\cdot\mathbf{x}-ct)} \tag{5.3}$$

(ii) For the Jordan normal form $\mathbf{J}_6^{(\mathrm{II})}$ and the representation (3.8), the partial surface tractions are of the form

$$\begin{aligned} \mathbf{t}_1 &= (\gamma_1 \nu \cdot \mathbf{C} \cdot \nu + \nu \cdot \mathbf{C} \cdot \mathbf{n}) \cdot \mathbf{m}'_1 \, e^{ir(\mathbf{n}\cdot\mathbf{x}-ct)} \\ \mathbf{t}_2 &= \begin{pmatrix} \gamma_1 \, (\nu \cdot \mathbf{C} \cdot \nu) \cdot \mathbf{m}'_1 + \\ (\gamma_1 \nu \cdot \mathbf{C} \cdot \nu + \nu \cdot \mathbf{C} \cdot \mathbf{n}) \cdot \mathbf{m}'_2 \end{pmatrix} e^{ir(\mathbf{n}\cdot\mathbf{x}-ct)} \\ \mathbf{t}_3 &= (\gamma_3 \nu \cdot \mathbf{C} \cdot \nu + \nu \cdot \mathbf{C} \cdot \mathbf{n}) \cdot \mathbf{m}'_3 \, e^{ir(\mathbf{n}\cdot\mathbf{x}-ct)} \end{aligned} \tag{5.4}$$

Equations (5.2) can be regarded as linear system with respect to the unknown coefficients C_k . The existence of the nontrivial solution of Eqs. (5.2) is equivalent to vanishing of the following determinant:

$$\mathbf{t}_1 \wedge \mathbf{t}_2 \wedge \mathbf{t}_3 = 0 \tag{5.5}$$

Equation (5.5) provides a necessary and sufficient condition for the existence of the surface wave.

Equation (5.5) is known as the dispersion equation despite the fact, that the phase speed determined by this equation does not depend upon the wave number, or the wave frequency.

6. NON-RAYLEIGH WAVES IN CUBIC CRYSTALS

Let the unit vectors $\mathbf{e}_k$ and $k = 1, 2, 3$ be oriented along the directions of elastic symmetry for a cubic crystal, then the elasticity tensor $\mathbf{C}$ can be represented in the following form:

$$\begin{aligned} \mathbf{C} = c_{11} &\sum_k \mathbf{e}_k \otimes \mathbf{e}_k \otimes \mathbf{e}_k \otimes \mathbf{e}_k + \\ c_{12} &\sum_{k \neq m} \mathbf{e}_k \otimes \mathbf{e}_k \otimes \mathbf{e}_m \otimes \mathbf{e}_m + \\ 4c_{44} &\sum_{k<m} \operatorname{sym}(\mathbf{e}_k \otimes \mathbf{e}_m) \otimes \operatorname{sym}(\mathbf{e}_k \otimes \mathbf{e}_m) , \end{aligned} \;, \tag{6.1}$$

where c_{11}, c_{12} , and c_{44} are the elastic constants, and

$$\operatorname{sym}(\mathbf{e}_k \otimes \mathbf{e}_m) \equiv 1/2 (\mathbf{e}_k \otimes \mathbf{e}_m + \mathbf{e}_m \otimes \mathbf{e}_k) \tag{6.2}$$

The positive definite condition (2.2) applied to the elasticity tensor (6.1) gives

$$c_{11} - c_{12} > 0, c_{11} + 2c_{12} > 0, c_{44} > 0. \tag{6.3}$$

REMARK 6.1. If $c_{11} = c_{12} + 2c_{44}$, then such an elasticity tensor corresponds to isotropic material with Lame's constants $\lambda = c_{11}$ and $\mu = c_{44}$. For isotropic material the positive definite condition (6.3) yields the well known

$$3\lambda + 2\mu > 0, \mu > 0. \tag{6.4}$$

Let the vectors ν, $\mathbf{n}$, and $\mathbf{w} = \nu \times \mathbf{n}$ be directed along the crystallographical axes defined by the vectors $\mathbf{e}_k$ $k = 1, 2, 3$, substitution of the elasticity tensor (5.1) into Eq. (3.5) gives

$$\mathbf{M} = \frac{c_{44} - \rho c^2}{c_{11}} \nu \otimes \nu + \frac{c_{11} - \rho c^2}{c_{44}} \mathbf{n} \otimes \mathbf{n} + \frac{c_{44} - \rho c^2}{c_{44}} \mathbf{w} \otimes \mathbf{w}$$

$$\mathbf{N} = \frac{c_{12} + c_{44}}{c_{11}} \nu \otimes \mathbf{n} + \frac{c_{12} + c_{44}}{c_{44}} \mathbf{n} \otimes \nu \tag{6.5}$$

The following Proposition flows out directly from the analysis of the eigenproblem for the matrix $\mathbf{R}_6$ [1]:

PROPOSITION 6.1. Suppose that the condition (6.3) is satisfied and the phase speed determined by the equation

$$\rho c^2 = \frac{2(c_{12} + c_{44})\sqrt{c_{11}c_{44}(c_{11} + c_{12})(c_{12} + 2c_{44} - c_{11})}}{(c_{11} - c_{44})^2} - \frac{(c_{11} + c_{44})(c_{11} + c_{12})(c_{12} + 2c_{44} - c_{11})}{(c_{11} - c_{44})^2} \tag{6.6}$$

satisfies also the equation

$$\begin{aligned} &c_{11}(c_{11} - c_{44})x^3 - 2c_{11}(c_{11}^2 - c_{12}^2)x^2 + \\ &(c_{11}^2 - c_{12}^2)(c_{11}^2 - c_{12}^2 + 2c_{11}c_{44})x - c_{44}(c_{11}^2 - c_{12}^2)^2 = 0 \end{aligned} \tag{6.7}$$

where $x = \rho c^2$, then

a) At this value of the phase speed the matrix $\mathbf{R}_6$ has the Jordan normal form $\mathbf{J}_6^{(II)}$;

b) The different roots γ_k of the Christoffel equation (2.4?) with $\mathrm{Im}\gamma_k \equiv \alpha_k < 0$ corresponding to Eq. (6.6) are as follows:

$$\begin{aligned} \gamma_1 &= -i\left(1 - \frac{\rho c^2}{c_{44}}\right)^{1/4}\left(1 - \frac{\rho c^2}{c_{11}}\right)^{1/4}, \\ \gamma_3 &= -i\left(1 - \frac{\rho c^2}{c_{44}}\right)^{1/2}, \end{aligned} \tag{6.8}$$

where γ_1 is the multiple root.

c) The complex amplitudes $\mathbf{m}'_k$ in the representation (3.8) have the form:

$$\mathbf{m}'_1 = p\left(\nu\left(1-\frac{\rho c^2}{c_{11}}\right)^{1/4} + i\,\mathbf{n}\left(1-\frac{\rho c^2}{c_{44}}\right)^{1/4}\right),$$
$$\mathbf{m}'_2 = sp\left(\nu c_{44}\left(1-\frac{\rho c^2}{c_{44}}\right)^{1/4} + i\mathbf{n}c_{11}\left(1-\frac{\rho c^2}{c_{11}}\right)^{1/4}\right) \tag{6.9}$$
$$\mathbf{m}'_3 = \mathbf{w},$$

where p is the normalization factor:

$$p = \left(\left(1-\frac{\rho c^2}{c_{11}}\right)^{1/2} + \left(1-\frac{\rho c^2}{c_{44}}\right)^{1/2}\right)^{-1/2} \tag{6.10}$$

and the parameter s is obtained by Eq. (4.8):

$$s = -c_{11}\frac{c_{11}\left(1-\frac{\rho c^2}{c_{11}}\right)^{1/2} + c_{44}\left(1-\frac{\rho c^2}{c_{44}}\right)^{1/2}}{\left(c_{11}\left(1-\frac{\rho c^2}{c_{11}}\right)^{1/2} - c_{44}\left(1-\frac{\rho c^2}{c_{44}}\right)^{1/2}\right)^2} \tag{6.11}$$

d) The dispersion Eq. (5.5) takes the form:

$$\mathbf{t}_1 \times \mathbf{t}_2 \equiv 0, \qquad C_3 = 0 \tag{6.12}$$

REMARK 6.2. Direct analysis reveals that the complex amplitudes $\mathbf{m}'_k$ defined by Eq. (6.8) are not orthogonal (the case $c_{11} = c_{44}$ at which they could be orthogonal, is inconsistent with the positive definite condition and Eqs.(6.6), (6.7)).

Thus, Proposition 6.1 completely characterizes the surface wave propagating on a basal plane of the cubic crystal half space and corresponding to the representation (3.8).

7. ACKNOWLEDGMENT

Author thanks Professor T.C.T.Ting for valuable comments and corrections.

References

1. Kuznetsov S.V. (2002) "Forbidden" planes for Rayleigh waves, *Quart. Appl. Math.* **60**, 87–97.

2. Hartman P. (1964) Ordinary Differential Equations, *Wiley, New York*
3. Ting T.C.T., Barnett D.M. (1997) Classification of surface waves in anisotropic elastic materials, *Wave Motion* **26**, 207–218.
4. Ting T.C.T. (1997) On extraordinary semisimple matrix N(v) for anisotropic elastic materials, *Q. Appl. Math.* **55**, 723–728.
5. Wang Y.M., Ting T.C.T. (1997) The Stroh formalism for anisotropic materials that possess an almost extraordinary degenerate matrix N, *Int. J. Solids Struct.* **34**, 401–413.

NONLINEARITY IN ELASTIC SURFACE WAVES ACTS NONLOCALLY

D. F. PARKER
School of Mathematics, University of Edinburgh,
The King's Buildings, Edinburgh, EH9 3JZ, U.K.

Abstract. Since elastic surface waves are examples of guided waves, nonlinear effects are significant only between linearized modes which have good matching of both phase and group velocities. Within homogeneous half-spaces, linearized modes travelling across the surface in any direction are completely non-dispersive. The phase speed can depend upon direction, but not upon frequency (or wavelength). Consequently, the standard weakly nonlinear theory equates the derivative of the (complex) Fourier transform of the surface displacement to an integral of convolution type, with a kernel which involves various elastic moduli and which takes account of the depth-dependence of the displacement fields within interacting pairs of modes having any two distinct wavenumbers.

The direct formulation of the equation governing the evolution of surface slope involves a quadratically nonlinear, nonlocal operator, incorporating the fact that waveform evolution is influenced by quadratically nonlinear contributions to the stress at all depths. This kernel splits naturally into one entirely local part, a nonlocal part allowing wave profiles to preserve symmetry and one necessarily causing asymmetry. Details are determined for elastic materials of arbitrary anisotropy.

Key words: nonlinear surface waves, elasticity, Rayleigh waves, scale invariance, nonlocality, spectral kernel, spatial kernel, Hilbert transform

1. Introduction

Elastic surface waves (Rayleigh waves) and interface waves (Stoneley waves) have long been known to be surface-guided waves having phase speed determined purely by material moduli. With the growth of interest in the 1960's and 70's in nonlinear wave theories, together with interest in electro-mechanical interactions (Bleustein-Gulyaev waves), it was natural that nonlinear theories of surface-guided waves should be sought. This proved surprisingly challenging. Most weakly nonlinear theories are either long-wave theories (e.g. leading to the Korteweg-deVries equation) or involve interactive perturbations between a fundamental sinusoidal wave and

R.V. Goldstein and G.A. Maugin (eds.),
Surface Waves in Anisotropic and Laminated Bodies and Defects Detection, 79–94.

certain of its harmonics. For elastic and electro-elastic surface and interface waves this is inappropriate, since the phase speed is entirely independent of wavelength, so that a fundamental wavetrain interacts cumulatively with all its harmonics and, indeed, with all other sinusoidal wavetrains.

A theory for the modulation of the Fourier amplitudes of each wavelength within a periodic elastic signal was first devised, using multiple-scale techniques, by Kalyanasundaram [1981] and Kalyansundaram *et al.* [1982]. The generalization to non-periodic waveforms was developed by Lardner [1983], [1986] and to electroelastic materials by Parker and David [1989]. In various of these, numerical computations have shown that sinusoidal waveforms tend to steepen, possibly leading to 'shock' formation. Nevertheless, the possibility of nondistorting periodic waves also exists [Parker and Talbot, 1985; David and Parker, 1990]. Unfortunately, numerical investigation of either phenomenon is limited by the need to truncate series expansions (though, see Mayer [1995] and Panayotaros [2002] for calculations involving improved accuracy). This led Hunter [1989] to devise an alternative formulation, describing the evolution of the surface slope $v(\theta, X)$ directly through the equation

$$\frac{\partial v}{\partial X} + \frac{\partial \mathsf{Q}}{\partial \theta} = 0 \quad , \qquad \mathsf{Q} = \mathsf{Q}[v] \, , \tag{1.1}$$

where X is a (stretched) propagation distance (or time) and θ is a travelling wave coordinate, while $\mathsf{Q}[v]$ is a quadratically nonlinear, but nonlocal, functional of $v(\theta, X)$ taken over all values $\theta \in (-\infty, \infty)$ at fixed X.

The formal simplicity of (1.1) is appealing, since it applies both for isotropic and for anisotropic media and suggests possibilities for qualitative analysis which might resolve questions about the occurrence of wave singularities and the stability of periodic waveforms. The drawback is that identifying $\mathsf{Q}[\cdot]$ in terms of material properties is indirect, involving substantial use of Fourier transforms. The present account aims to extend the methods of Hamilton, Il'insky and Zabolotskaya [1995], [1997] for isotropic materials to anisotropic materials, meanwhile elucidating the rôle of nonlocality within (1.1). It also indicates how dispersive or surface tension effects might be included into (1.1), in order to improve the predictions based on assuming a perfectly uniform elastic (or electroelastic) half-space.

2. The Linearized Displacement Field

Let X_J $(J = 1, 2, 3)$ and $\varepsilon u_i = \varepsilon u_i(X_1, X_2, t)$ $(i = 1, 2, 3)$ denote respectively Lagrangian coordinates and small $(\varepsilon << 1)$ displacement components within an elastic material occupying $X_2 \leq 0$, so that $\varepsilon \partial u_m / \partial X_M \equiv \varepsilon u_{m,M}$ are components of the displacement gradient. The components of the Piola-Kirchhoff stress $\boldsymbol{\tau}$ are derived from the strain energy density $W = W(\boldsymbol{F})$

by $\tau_{Jj} \equiv \partial W/\partial F_{jJ}$, where $F_{jJ} = \delta_{jJ} + \varepsilon u_{j,J}$ are components of deformation gradient. In the limit $\varepsilon \to 0$ this reduces to the familiar linear law $\tau_{Jj} = c_{jJmM}(\varepsilon u_{m,M})$. These generalized plane-strain (independent of X_3) deformations describe surface waves if the material surface $X_2 = 0$ is traction-free ($\tau_{2j} = 0$) and if $|\boldsymbol{u}| \to 0$ as $X_2 \to -\infty$ for all t. The governing equations are then the Euler equations, traction-free boundary condition and decay condition

$$\frac{\partial \tau_{Jj}}{\partial X_J} = \rho\varepsilon \frac{\partial^2 u_j}{\partial t^2} \ , \qquad X_2 \leq 0 \, , \tag{2.1}$$

$$\tau_{2j}(X_1, 0, t) = 0 \ , \qquad |\boldsymbol{u}| \to 0 \ \text{ as } \ X_2 \to -\infty \, . \tag{2.2}$$

Collectively, this system is *scale-invariant* whenever the material is homogeneous (so that $\tau_{Jj} = \partial W/\partial(\varepsilon u_{j,J})$ does not depend explicitly on X_1, X_2 or X_3). Consequently, if $\boldsymbol{u} = \boldsymbol{u}(\boldsymbol{X}, t)$ is one solution to (2.1) and (2.2), then $\boldsymbol{u} = \alpha \boldsymbol{u}(\alpha \boldsymbol{X}, \alpha t)$ defines another solution for *all* choices of $\alpha > 0$. In each of these, the pattern of deformation gradient and strain is similar, but with lengths and time scaled by the factor α. This scale-invariance is directly responsible for the absence of dispersion from linear theories and is central to Hunter's direct formulation (1.1).

In the linear theory for which $\tau_{Jj} = \varepsilon c_{jJmM} u_{m,M}$, travelling wave solutions $\boldsymbol{u} = \boldsymbol{u}(\theta, X_2)$ to (2.1) and (2.2) are sought, with $\theta \equiv X_1 - ct$. The Euler equations obtained from (2.1) as

$$c_{jJmM} u_{m,JM} - \rho \partial^2 u_j / \partial t^2 = 0$$

are satisfied by complex-valued partial waves having the form

$$\boldsymbol{u} = \boldsymbol{a} \exp ik(\theta + sX_2) = \boldsymbol{a} \mathrm{e}^{k\beta X_2} \exp ik(X_1 + \alpha X_2 - ct)$$

provided that the (complex) depth factors $s \equiv \alpha - i\beta$ and polarization vectors $\boldsymbol{a}$ are eigenvalues and eigenvectors of an algebraic problem in which the wavespeed c appears as a parameter. The roots s and $\boldsymbol{a}$ appear as complex conjugate pairs. Let $s = s^{(p)} = \alpha^{(p)} - i\beta^{(p)}$ ($p = 1, 2, 3$) label the three eigenvalues having $\beta^{(p)} = -\mathrm{Im}\, s^{(p)} > 0$ and let $\boldsymbol{a} = \boldsymbol{a}^{(p)}$ be corresponding eigenvectors (arbitrarily normalized). Then, imposing upon the linear combination

$$\mathcal{U}_m(kX_2) = \sum_{p=1}^{3} d_p a_m^{(p)} \mathrm{e}^{iks^{(p)} X_2} \ , \qquad k > 0 \tag{2.3}$$

the traction-free boundary condition (2.2) gives

$$0 = \varepsilon^{-1} \tau_{2j}(X_1, 0, t) = c_{jJmM} u_{m,M} = ik\mathrm{e}^{ik\theta} \sum_p (c_{j2m1} + s^{(p)} c_{j2m2}) a_m^{(p)} d_p$$

which is a further eigenvalue problem which fixes the speed c and determines the (complex) ratios $d_1 : d_2 : d_3$. Use of a normalization $\mathcal{U}_2(0) = 1$ then fixes d_p $(p = 1, 2, 3)$ and determines completely the *depth structure* $\mathcal{U}_m(kX_2)$ for a surface wave sinusoidal in $\theta = X_1 - ct$ and having wavelength $2\pi/k$ $(k > 0)$. Since c is independent of k, the general waveform travelling parallel to $\mathrm{O}X_1$ and having speed c is

$$u_m = \int_{-\infty}^{\infty} C(k)\mathcal{U}_m(kX_2)\mathrm{e}^{ik(X_1-ct)}\mathrm{d}k \equiv \tilde{u}_m(\theta, X_2)\,. \tag{2.4}$$

Here, for $k < 0$, $\mathcal{U}_m(kX_2) = \mathcal{U}_m^*(|k|\,X_2)$, where * denotes the complex conjugate and the complex-valued function $C(k)$ is related to the normal displacement $\tilde{u}_2(\theta, 0)$ of the traction-free surface through the Fourier transform pair

$$\tilde{u}_2(\theta, 0) = \int_{-\infty}^{\infty} C(k)\mathrm{e}^{ik\theta}\mathrm{d}k\,, \quad C(k) = \frac{1}{2\pi}\int_{-\infty}^{\infty} \tilde{u}_2(\theta, 0)\mathrm{e}^{-ik\theta}\mathrm{d}\theta\,. \tag{2.5}$$

When $C(k)$ satisfies $C(-k) = C^*(k)$, the expressions (2.4) and (2.5) for the displacement field and its component $u_2 = \tilde{u}_2(\theta, X_2)$ are real.

The linear theory predicts that (for arbitrary choice $\mathrm{O}X_1$ of propagation direction across the material surface), surface waves of *arbitrary* waveform $\tilde{u}_2(\theta, 0)$ travel *without distortion* while the associated displacements at all depths X_2 are related to $u_2(\theta, 0)$ through (2.4). This general, nondistorting wavefield, rather than a single sinusoidal constituent, must form the basis of the perturbation procedure incorporating nonlinearity. Some of its features are clarified through reference to the familiar isotropic case. Here, all three factors $s^{(p)}$ are pure imaginary. When they are written as $s^{(1)} = s^{(3)} = i\beta_1$, $s^{(2)} = -i\beta_2$, where $\beta_1 = (1 - \rho c^2/\mu)^{1/2}$, $\beta_2 = \{1 - \rho c^2/(\lambda + 2\mu)\}^{1/2}$, the corresponding eigenvectors are $\boldsymbol{a}^{(1)} = (i\beta_1 \;\; 1 \;\; 0)^T$, $\boldsymbol{a}^{(2)} = (i \;\; \beta_2 \;\; 0)^T$, $\boldsymbol{a}^{(3)} = (0 \;\; 0 \;\; 1)^T$. The traction-free condition (2.2) gives $d_3 = 0$ (so that $u_3 \equiv 0$, describing plane strain) and the compatibility condition $4\beta_1\beta_2 = (1 + \beta_1^2)^2$, which is equivalent to the standard *secular equation*

$$(\lambda + 2\mu)\{(\rho c^2)^3 - 8\mu(\rho c^2)^2\} + 8\mu^2(3\lambda + 4\mu)\rho c^2 - 32\mu^3(2\lambda + 3\mu) = 0$$

defining the Rayleigh speed c. The ratio $d_1 : d_2$ is now fixed and the normalization $\mathcal{U}_2(0) = 1$ then yields (for $k > 0$)

$$\begin{aligned}\mathcal{U}_1(kX_2) &= i2\beta_1(1 - \beta_1^4)^{-1}\{(1 + \beta_1^2)\mathrm{e}^{\beta_1 kX_2} - 2\mathrm{e}^{\beta_2 kX_2}\}\,,\\ \mathcal{U}_2(kX_2) &= (1 - \beta_1^2)^{-1}\{2\mathrm{e}^{\beta_1 kX_2} - (1 + \beta_1^2)\mathrm{e}^{\beta_2 kX_2}\}\,.\end{aligned} \tag{2.6}$$

It may be observed that, since $\mathrm{e}^{ix}\mathrm{e}^{-y}$ is an analytic function of $z \equiv x + iy$ then so also is

$$f(x + iy) \equiv 2\int_0^{\infty} C(k)\mathrm{e}^{ikx}\mathrm{e}^{-ky}\mathrm{d}k \equiv \phi(x, y) + i\psi(x, y)\,.$$

Moreover, its real and imaginary parts are conjugate harmonic functions satisfying $\partial\phi/\partial x = \partial\psi/\partial y$, $\partial\phi/\partial y = -\partial\psi/\partial x$. Using $\mathcal{U}_m(kX_2)e^{ik\theta} = \sum_p d_p a_m^{(p)} \exp ik(\theta - i\beta_p X_2)$ in (2.4) then yields

$$\begin{aligned}\tilde{u}_m(\theta, X_2) &= \int_0^\infty C(k)\mathcal{U}_m(kX_2)e^{ik\theta}dk + \text{c.c.} \\ &= \tfrac{1}{2}\sum_p d_p a_m^{(p)} \{\phi(\theta, \beta_p X_2) + i\psi(\theta, \beta_p X_2)\} + \text{c.c.}\end{aligned}$$

so that

$$\begin{aligned}\tilde{u}_1(\theta, X_2) &= \frac{-2\beta_1}{1-\beta_1^2}\psi(\theta, \beta_1 X_2) + \frac{4\beta_1}{1-\beta_1^4}\psi(\theta, \beta_2 X_2)\,, \\ \tilde{u}_2(\theta, X_2) &= \frac{2}{1-\beta_1^2}\phi(\theta, \beta_1 X_2) - \frac{1+\beta_1^2}{1-\beta_1^2}\phi(\theta, \beta_2 X_2)\,.\end{aligned} \tag{2.7}$$

Hence, as shown by Chadwick [1976], displacements within a Rayleigh wave may be represented in terms of a single analytic function but involving a linear combination with two distinct arguments $\theta - i\beta_1 X_2$ and $\theta - i\beta_2 X_2$. A similar result applies for *arbitrary anisotropy*. Expressions (2.7) are generalized to

$$\begin{aligned}\tilde{\boldsymbol{u}} = \sum_{p=1}^{3} \Big\{ &\text{Re}(d_p \boldsymbol{a}^{(p)})\phi(\theta + \alpha_p X_2, \beta_p X_2) \\ &- \text{Im}(d_p \boldsymbol{a}^{(p)})\psi(\theta + \alpha_p X_2, \beta_p X_2)\Big\}\,,\end{aligned} \tag{2.8}$$

so showing that *each* displacement component is some linear combination of the harmonic conjugate pair ϕ and ψ, evaluated at three distinct arguments $(\theta + \alpha_p X_2, \beta_p X_2)$, $p = 1, 2, 3$. Note how, generally, these do not just involve three different scalings β_p normal to the surface but (for $\alpha_p \neq 0$) they involve an advance or delay in X_1, since $\theta + \alpha_p X_2 = X_1 + \alpha_p X_2 - ct$.

At the surface $X_2 = 0$, the normalization of the vector $\mathcal{U}$ gives $\phi(\theta, 0) = \tilde{u}_2(\theta, 0) \equiv u_0(\theta)$ (say), while (2.8) reduces to

$$\tilde{\boldsymbol{u}}(\theta, 0) = \boldsymbol{u}(X_1, 0, t) = \boldsymbol{\gamma}^+\phi(\theta, 0) - \boldsymbol{\gamma}^-\psi(\theta, 0) = \boldsymbol{\gamma}^+ u_0(\theta) - \boldsymbol{\gamma}^- \mathsf{H}u_0(\theta)\,,$$

where $\boldsymbol{\gamma}_+ \equiv \text{Re}\{\sum_p d_p \boldsymbol{a}^{(p)}\}$, $\boldsymbol{\gamma}_- \equiv \text{Im}\{\sum_p d_p \boldsymbol{a}^{(p)}\}$. Thus, all components of the surface displacement are linear combinations of the normal displacement $u_0(\theta)$ and its Hilbert transform, defined by

$$\mathsf{H}u_0(\theta) = \frac{1}{\pi} \int_{-\infty}^{\infty} \frac{u_0(\theta')}{\theta' - \theta} d\theta'$$

in which the bar denotes the Cauchy principal part integral. Thus, throughout the linearized wavefield, all components of displacement, displacement gradient and stress are nonlocal functionals of $u_0(\theta) = u_2(X_1, 0, t)$. At the surface, they are simply expressed in terms of (derivatives of) $u_0(\theta)$ and its Hilbert transform, but at general location the more complicated formula (2.8) must be used.

3. Treatment of Quadratic Nonlinearity

The linearized theory is perturbed by terms arising from the $\mathrm{O}(\varepsilon^2)$ term in

$$\tau_{Jj} = \varepsilon c_{jJmM}\, u_{m,M} + \varepsilon^2 c_{jJmMnN}\, u_{m,M}\, u_{n,N} + \dots . \tag{3.1}$$

Modifications to the displacement field (2.4) are written as functions of $\theta \equiv X_1 - ct$, X_2 and the gradual evolution variable $X \equiv \varepsilon X_1$, in the form

$$\boldsymbol{u}(X_1, X_2, t) = \tilde{\boldsymbol{u}}(\theta, X_2, X) + \varepsilon \boldsymbol{w}(\theta, X_2, X) + \dots,$$
$$\text{where} \quad \tilde{u}_m \equiv \int_{-\infty}^{\infty} C(k, X)\mathcal{U}_m(kX_2)\mathrm{e}^{ik\theta}\mathrm{d}k . \tag{3.2}$$

Notice that this is *not* a multiple scales scheme. The three variables X_1, X_2 and t are merely replaced by θ, X_2 and X, exploiting the fact that in the limit $\varepsilon = 0$ requisite solutions depend *solely* on θ and X_2. Details of the analysis have been presented elsewhere [Parker, 1990; Parker and Hunter, 1998], but follow from replacement of derivatives according to $\partial/\partial X_1 = \partial/\partial\theta + \varepsilon\partial/\partial X$ and $\partial/\partial t = -c\partial/\partial\theta$, so yielding from (2.1)

$$c_{jJmM}\, w_{m,JM} - \rho c^2\, w_{m,11} = -(c_{j1mM} + c_{jMm1})\tilde{u}_{m,MX}$$
$$\qquad - c_{jJmMnN}\partial(\tilde{u}_{m,M}\tilde{u}_{n,N})/\partial X_J \equiv -P_j(\theta, X) ,$$

and from (2.2) the boundary condition on $X_2 = 0$:

$$c_{jJmM} w_{m,M} = -c_{j2m1}\tilde{u}_{m,X} - c_{j2mMnN}\tilde{u}_{m,M}\tilde{u}_{n,N} \equiv -Q_j(\theta, X_2, X) .$$

Here, the indices J and M range only over the values $1, 2$, with $\partial/\partial\theta$ denoted by $_{,1}$ as in $\tilde{u}_{j,1} \equiv \partial\tilde{u}_j/\partial\theta$.

Coupled with the decay condition $|\boldsymbol{w}| \to 0$ as $X_2 \to -\infty$, this defines for $w_m(\theta, X_2, X)$ an inhomogeneous version of the linear boundary-value problem defining $\tilde{u}_m$, in which X appears only as a parameter. The condition ensuring solvability for this system (the Fredholm alternative) is first deduced by taking Fourier transforms $\bar{\boldsymbol{w}}(k, X_2, X) = \mathcal{F}\{\boldsymbol{w}(\theta, X_2, X)\} \equiv (2\pi)^{-1}\int_{-\infty}^{\infty} \boldsymbol{w}\mathrm{e}^{-ik\theta}\mathrm{d}\theta$, so yielding a system for which the compatibility condition is (for $k > 0$)

$$\int_{-\infty}^{0} \mathcal{U}_j^*(kX_2)\bar{P}_j(k, X_2, X)\mathrm{d}X_2 = \Big[\mathcal{U}_j^*(kX_2)\bar{Q}_j(k, X)\Big]_{X_2=0} , \tag{3.3}$$

where $\bar{P}_j = \mathcal{F}\{P_j\}$, $\bar{Q}_j = \mathcal{F}\{Q_j\}$. In constructing these Fourier transforms, the identities

$$\mathcal{F}\{\tilde{u}_m\} = \mathcal{U}_m(kX_2)C(k,X)\,,$$
$$\mathcal{F}\{\tilde{u}_{m,M}\} = \begin{cases} D_M\mathcal{U}_m(kX_2)ikC(k,X) & k>0 \\ D_M^*\mathcal{U}_m^*(|k|X_2)i|k|C^*(|k|,X) & k<0 \end{cases} \tag{3.4}$$

where $D_M \equiv \delta_{M1} - i\delta_{M2}\mathrm{d}/\mathrm{d}(|k|X_2)$, are used to obtain a convolution product

$$\mathcal{F}\{c_{jJmMnN}\tilde{u}_{m,M}\tilde{u}_{n,N}\} = \int_{-\infty}^{\infty} K_{jJ}(\mu,l,X_2)\mu l C(\mu,X)C(l,X)\,\mathrm{d}l \tag{3.5}$$

with $\mu + l = k > 0$. The kernel K_{jJ} in (3.5) is defined by

$$K_{jJ}(\mu,l,X_2) = \begin{cases} c_{jJmMnN}D_M^*\mathcal{U}_m^*(-\mu X_2)D_N\mathcal{U}_n(lX_2) & \mu<0<l \\ -c_{jJmMnN}D_M\mathcal{U}_m(\mu X_2)D_N\mathcal{U}_n(lX_2) & 0<\mu\,,\, l<k \\ c_{jJmMnN}D_M\mathcal{U}_m(\mu X_2)D_N^*\mathcal{U}_n^*(-lX_2) & l<0<\mu\,. \end{cases} \tag{3.6}$$

Notice that this kernel involves knowledge of the displacement $\mathcal{U}_m(kX_2)$ at *all depths* $X_2 < 0$ within a mode having wavenumber k.

Inserting (3.6) and the X-derivative of $\mathcal{F}\{\tilde{u}_{m,M}\}$ from (3.4) into (3.3) yields the equation

$$iJ\frac{\partial C(k,X)}{\partial X} + \int_{-\infty}^{\infty} \Gamma(k-l,l)(k-l)lC(k-l,X)C(l,X)\,\mathrm{d}l = 0\,. \tag{3.7}$$

In this *evolution equation* for the complex amplitude $C(k,X)$ in (3.2) showing quadratic interaction between all wavenumbers, the real constant J represents the modal energy while the kernel $\Gamma(k-l,l)$ involves integration with respect to X_2, arising from (3.3). Notice how, in the three-wave interactions, the contributions from $0<\mu\,,\, l<k$ with $\mu+l=k$ correspond to all the 'sum frequency' interactions, while those from $\mu<0$ or $l<0$ with $0<\mu+l=k$ correspond to all the 'difference frequency' interactions (e.g. $|l|-|\mu| = k > 0$). The kernel is symmetric and homogeneous of degree zero in its arguments. Indeed it may be written as

$$\Gamma(\mu,l) = -iJ\Lambda(\mu,l) = -iJ|\nu|\tilde{\Lambda}(\nu,\mu,l)\,, \qquad \text{where } \nu+\mu+l=0 \tag{3.8}$$

and $\tilde{\Lambda}(\nu,\mu,l)$ is symmetric under *all interchanges* of its arguments

$$\begin{aligned}\tilde{\Lambda}(\nu,\mu,l) &= \tilde{\Lambda}(\mu,\nu,l) = \tilde{\Lambda}(l,\nu,\mu) = \tilde{\Lambda}^*(-\nu,-\mu,-l) \\ &= \alpha\tilde{\Lambda}(\alpha\nu,\alpha\mu,\alpha l) \qquad \forall\,\alpha>0\end{aligned}$$

and homogeneous of degree -1. It is given explicitly for $\mu > 0$, $l > 0$ (so that $\nu < 0$) by

$$\tilde{\Lambda}(\nu, \mu, l) = c_{jJmMnN} \sum_{p,q,r} b_{jJ}^{(p)*} b_{mM}^{(q)} b_{nN}^{(r)} \int_{-\infty}^{0} \mathrm{e}^{i[\nu s^{(p)*} + \mu s^{(q)} + l s^{(r)}] X_2} \mathrm{d}X_2$$
$$= \sum_{p,q,r} \Delta_{pqr} \frac{-i}{\nu s^{(p)*} + \mu s^{(q)} + l s^{(r)}} \tag{3.9}$$

where

$$b_{mM}^{(q)} = d_q a_m^{(q)} (\delta_{M1} + i s^{(q)} \delta_{M2}), \quad \Delta_{pqr} = c_{jJmMnN} b_{jJ}^{(p)*} b_{mM}^{(q)} b_{nN}^{(r)}.$$

It is clear that although the spectral evolution equation (3.7) has the same form for homogeneous materials of any anisotropy and for propagation in arbitrary direction across a traction-free half-space, the details of the kernel $\Lambda(\mu, l) = iJ^{-1}\Gamma(\mu, l)$ depend upon the elastic moduli c_{jJmM} and c_{jJmMnN} in a very complicated way. To unravel how the various contributions affect waveform evolution was the stimulus for the reformulation as an evolution equation (1.1) directly for the displacements, rather than for their Fourier transforms.

4. The Direct Formulation

Hunter [1989] realized that *scale-invariance* is the key feature of Rayleigh waves and so introduced, as a primary dependent variable, a scale-invariant quantity, such as the surface slope

$$v(\theta, X) \equiv \tilde{u}_{2,1}(\theta, 0, X) \equiv \int_{-\infty}^{\infty} ikC(k, X)\mathrm{e}^{ik\theta} \mathrm{d}k\,. \tag{4.1}$$

He then showed that $v(\theta, X)$ satisfies an equation of the form (1.1), with the non-local functional $\mathsf{Q}[v]$ related to $\Lambda(\mu, l)$ *via* a double Fourier transform.

The derivation uses $\mathcal{F}\{v(\theta, X)\} = ikC(k, X) \equiv U(k, X)$, so that inverting (3.7) and using $J^{-1}\Gamma(k-l, l) = -i\Lambda(k-l, l)$ leads to

$$\frac{\partial v}{\partial X} + \int_{-\infty}^{\infty} ik\mathrm{e}^{ik\theta} \int_{-\infty}^{\infty} \Lambda(k-l, l) U(k-l, X) U(l, X)\, \mathrm{d}l\, \mathrm{d}k = 0\,.$$

This has the 'divergence form' shown in (1.1) provided that the 'nonlocal flux' $\mathsf{Q}[v]$ has Fourier transform

$$\mathcal{F}\{\mathsf{Q}\} = \int_{-\infty}^{\infty} \Lambda(k-l, l) U(k-l, X) U(l, X)\, \mathrm{d}l\,. \tag{4.2}$$

Inverting this relation gives

$$\begin{aligned}
\mathsf{Q}[v] &= \int_{-\infty}^{\infty} \mathrm{e}^{ik\theta}\,\mathrm{d}k \int_{-\infty}^{\infty} \Lambda(k-l,l)U(k-l,X)U(l,X)\,\mathrm{d}l \\
&= \int_{-\infty}^{\infty}\int_{-\infty}^{\infty} \mathrm{e}^{ik\theta}\frac{\Lambda(k-l,l)}{(2\pi)^2} \\
&\qquad \times \int_{-\infty}^{\infty}\int_{-\infty}^{\infty} \mathrm{e}^{-i(k-l)y}\mathrm{e}^{-ilz}v(y,X)v(z,X)\,\mathrm{d}y\,\mathrm{d}z\,\mathrm{d}l\,\mathrm{d}k \\
&= \int_{-\infty}^{\infty}\int_{-\infty}^{\infty} v(y,X)v(z,X) \\
&\qquad \times \left\{\frac{1}{(2\pi)^2}\int_{-\infty}^{\infty}\int_{-\infty}^{\infty} \mathrm{e}^{i(k-l)(\theta-y)}\mathrm{e}^{il(\theta-z)}\Lambda(k-l,l)\,\mathrm{d}l\,\mathrm{d}k\right\}\mathrm{d}y\,\mathrm{d}z \\
&= \int_{-\infty}^{\infty}\int_{-\infty}^{\infty} K(\theta-y,\theta-z)v(y,X)v(z,X)\,\mathrm{d}y\,\mathrm{d}z
\end{aligned}$$

or, equivalently,

$$\mathsf{Q}[v](\theta,X) = \int_{-\infty}^{\infty}\int_{-\infty}^{\infty} K(y,z)v(\theta-y,X)v(\theta-z,X)\,\mathrm{d}y\,\mathrm{d}z\,, \tag{4.3}$$

in which the kernel is

$$K(y,z) \equiv \frac{1}{(2\pi)^2}\int_{-\infty}^{\infty}\int_{-\infty}^{\infty} \Lambda(\mu,l)\mathrm{e}^{i(\mu y+lz)}\mathrm{d}\mu\,\mathrm{d}l\,. \tag{4.4}$$

Here the *spatial kernel* $K(y,z)$ in (4.3) records the influence on the slope v at (θ,X) of the slopes at all pairs of points distant y and z from θ (at fixed X), arising because the linearized deformation gradient at each point in $X_2 < 0$ is itself representable as a single integral of $v(\theta-y,X)$ over all y. The spatial kernel forms a (double) Fourier transform pair with the *spectral kernel* $\Lambda(\mu,l)$ through (4.4) and

$$\Lambda(\mu,l) \equiv \int_{-\infty}^{\infty}\int_{-\infty}^{\infty} K(y,z)\mathrm{e}^{-i(\mu y+lz)}\mathrm{d}y\,\mathrm{d}z\,. \tag{4.5}$$

The two kernels have interrelated properties associated with symmetry, with the reality of v and with scale invariance. They are

$K(y,z) = K(z,y)$	$\Lambda(\mu,l) = \Lambda(l,\mu)$	Symmetry
$K(y,z) = K^*(y,z)$	$\Lambda(-\mu,-l) = \Lambda^*(l,\mu)$	Reality
$K(\alpha y,\alpha z) = \alpha^{-2}K(y,z)$	$\Lambda(\alpha\mu,\alpha l) = \Lambda(\mu,l)$	Scale invariance

where $\alpha > 0$. The reality and scale invariance properties simplify whenever (as for isotropic materials) all factors $s^{(p)}$ are pure imaginary, but, in general, the spectral kernel $\Lambda(\mu,l) = (\mu+l)\tilde{\Lambda}(-\mu-l,\mu,l)$ is complex, so that care must be taken over the signs of μ, l and $\nu = -\mu - l$. In $0 < \mu\,,\, 0 < l$ it is given, using (3.9), as

$$\tilde{\Lambda}(-\mu - l, \mu, l) = \sum_{p,q,r} \frac{-i\Delta_{pqr}}{\mu(s^{(q)} - s^{(p)*}) + l(s^{(r)} - s^{(p)*})} \,.$$

5. Locality versus nonlocality

It is well known that, unless $f(v)$ is linear, the first-order conservation law

$$v_X + (f(v))_{,\theta} = 0 \tag{5.1}$$

predicts 'shock formation' from any non-monotonic initial conditions $v(\theta, 0)$. If $f(v)$ is quadratic, the Fourier transform of v is governed by an equation of the form (1.1) and (4.3). This suggests isolating from $\mathsf{Q}\,[v]$ its local part, in order to investigate how the remaining nonlocal part might inhibit the onset of singular behaviour. This separation was accomplished, for real $\Lambda(\mu, l)$, by Hamilton *et al.* [1995] and is here extended to the general case. The remaining nonlocal contributions are also treated in generality. However, a warning about the splitting into local and nonlocal contributions is in order.

Hunter [1989] showed how to solve the equation

$$v_X + (\mathsf{Q}[v])_{,\theta} = 0 \tag{5.2}$$

when

$$\mathsf{Q}\,[v] = \tfrac{1}{2}\alpha v^2 + \tfrac{1}{2}\beta \mathsf{H}\left[\mathsf{H}[v^2]\right] = \tfrac{1}{2}\alpha v^2 + \tfrac{1}{2}\beta\mathsf{H}[v^2] - \tfrac{1}{2}\beta v\mathsf{H}[v]\,. \tag{5.3}$$

This sub-class of equations of the type (1.1), (4.3) is related to the 'diffusionless Burgers equation'

$$a_X + a a_\theta = 0 \tag{5.4}$$

(i.e. $\mathsf{Q}\,[a] = \frac{1}{2}a^2$ is a local nonlinearity) through

$$v(\theta, X) \equiv \alpha a(\theta, X) - \beta\mathsf{H}[a](\theta, X) \qquad (\alpha,\, \beta \text{ real constants})$$

or, equivalently, through

$$a(\theta, X) \equiv (\alpha^2 + \beta^2)^{-1}\{\alpha v(\theta, X) + \beta\mathsf{H}[v](\theta, X)\}\,.$$

Since typical solutions $a(\theta, X)$ form singularities at finite X, solutions $u(\theta, X)$ to (5.2) and (5.3) will form singularities at finite X. Even though expression (5.3) includes nonlocal contributions, the shock-forming tendency inherent in (5.4) is not suppressed. Solutions will still form singularities, but of modified type (even for $\alpha = 0$, so that in (5.3) Q becomes entirely nonlocal).

6. Calculation of the Spatial Kernel

Evaluation of $K(y,z)$ from (4.4), (3.8) and (3.9) follows Hamilton *et al.* [1995] by introducing polar variables (λ, η) in the l, μ plane and (s, ξ) in the y, z plane through

$$l = \lambda \cos \eta\,, \quad \mu = \lambda \sin \eta \quad ; \qquad z = s\cos\xi\,, \quad y = s\sin\xi$$

so that $\Lambda(\mu, l) \equiv G(\eta)$ and

$$4\pi^2 K(y,z) = \lim_{\delta\to 0+} \int_0^{2\pi} G(\eta) \int_0^{\infty} \lambda e^{is\lambda\cos(\eta-\xi)} \mathrm{d}\lambda\, \mathrm{d}\eta\,. \tag{6.1}$$

Here $G(\eta)$ is 2π-periodic, with $G^*(\eta+\pi) = G(\eta) = G(\eta+2\pi)$ (reality). Also, the λ-integration is readily performed to yield the closed-form expression $\{s\cos(\eta-\xi)+i\delta\}^{-2}$. Splitting this and $G(\eta)$ into real and imaginary parts, with $\mathrm{Re}\, G(\eta) \equiv G^+(\eta)$, $\mathrm{Im}\, G(\eta) \equiv G^-(\eta)$, then yields

$$\begin{aligned} 4\pi^2 K(y,z) = &\lim_{\delta\to 0+} \int_0^{2\pi} G^+(\eta) \frac{s^2\cos^2(\eta-\xi) - \delta^2}{[s^2\cos^2(\eta-\xi)+\delta^2]^2} \mathrm{d}\eta \\ &+ \lim_{\delta\to 0+} \int_0^{2\pi} G^-(\eta) \frac{2s\delta\cos(\eta-\xi)}{[s^2\cos^2(\eta-\xi)+\delta^2]^2} \mathrm{d}\eta\,. \end{aligned} \tag{6.2}$$

Observe that the second integral disappears when $G(\eta)$ is real (as for isotropic materials). However, extending the technique of Hamilton *et al.* [1995] by writing the multipliers of $G^\pm(\eta)$ as

$$\frac{s^2\cos^2(\eta-\xi) - \delta^2}{[s^2\cos^2(\eta-\xi)+\delta^2]^2} = \frac{1}{s^2+\delta^2} \left\{ \frac{-\delta^2}{s^2\cos^2(\eta-\xi)+\delta^2} + \frac{\partial \Theta^+}{\partial \eta} \right\}$$
$$\frac{2s\delta\cos(\eta-\xi)}{[s^2\cos^2(\eta-\xi)+\delta^2]^2} = \frac{1}{s^2+\delta^2} \frac{\partial \Theta^-}{\partial \eta}$$

with $\Theta^\pm(\eta-\xi;\varepsilon)$ being single-valued, 2π-periodic functions of $\eta-\xi$ and with $\varepsilon \equiv \delta/s$, then allows $K(y,z)$ to be analysed as $K = K_0 + K^+ + K^-$, where

$$4\pi^2 K_0(y,z) \equiv \lim_{\delta\to 0+} \frac{1}{s^2+\delta^2} \int_0^{2\pi} \frac{-\delta^2}{s^2\cos^2(\eta-\xi)+\delta^2} G^+(\eta)\, \mathrm{d}\eta\,. \tag{6.3}$$

In (6.3), it may be seen that, for all $s \neq 0$, the integrand vanishes as $\delta \to 0+$ so that $K_0(y,z) = 0$ for $y^2+z^2 \neq 0$. However, as shown by

Hamilton *et al.* [1995], integration over the whole y, z plane gives

$$\begin{aligned}4\pi^2 K_0(y,z) &= -\lim_{\delta\to 0+}\int_0^{2\pi}\frac{\delta^2\,\mathrm{d}s}{s^2+\delta^2}\int_0^{2\pi}G^+(\eta)\mathrm{d}\eta\int_0^{2\pi}\frac{s\,\mathrm{d}\xi}{s^2\cos^2(\eta-\xi)+\delta^2}\\ &= -\lim_{\delta\to 0+}\int_0^{2\pi}\frac{\delta^2}{s^2+\delta^2}2\pi\langle G^+(\eta)\rangle\frac{2\pi s}{\delta\sqrt{s^2+\delta^2}}\mathrm{d}s\\ &= -4\pi^2\langle G^+(\eta)\rangle\,,\end{aligned}$$

where $\langle G^+(\eta)\rangle$ denotes the mean value of $\operatorname{Re} G(\eta)$. Consequently, $K_0(y,z)$ is singular and may be written as

$$K_0(y,z) \equiv -\langle G^+(\eta)\rangle\,\delta(y)\,\delta(z)\,. \tag{6.4}$$

This shows that the contribution $K_0(y,z)$ to $K(y,z)$ is *purely local.* Within the flux $\mathsf{Q}[v]$ it gives a contribution

$$\begin{aligned}\mathsf{Q}_0[v] &\equiv \int_{-\infty}^{\infty}\int_{-\infty}^{\infty}K_0(y,z)v(\theta-y,X)v(\theta-z,X)\,\mathrm{d}y\,\mathrm{d}z\\ &= -\langle G^+\rangle\,v^2(\theta,X)\end{aligned} \tag{6.5}$$

just like the 'shock-forming' term $\frac{1}{2}a^2$ in the diffusionless Burgers equation (5.4).

The corresponding formula

$$4\pi^2 K^{\pm}(y,z) \equiv \lim_{\delta\to 0+}\frac{1}{s^2+\delta^2}\int_0^{2\pi}\frac{\partial\Theta^{\pm}}{\partial\eta}G^{\pm}(\eta)\,\mathrm{d}\eta \tag{6.6}$$

shows that both of $K^{\pm}(y,z)$ are bounded as $s\to 0$. In evaluating each, the symmetries $G^{\pm}(\eta+\pi)=\pm G^{\pm}(\eta)$ may be used, to reduce the range of integration to $-\pi/4<\eta<3\pi/4$ (i.e. to $\mu+l>0$). In this interval, write $S\equiv\tan\eta=\mu/l$, so that

$$\begin{aligned}\frac{\partial\Theta^+}{\partial\eta} &= \frac{(T^2+\varepsilon^2T^2-\varepsilon^2)S^2+2(1+2\varepsilon^2)TS+1+\varepsilon^2-\varepsilon^2T^2}{\{(T^2+\varepsilon^2T^2+\varepsilon^2)S^2+2TS+1+\varepsilon^2+\varepsilon^2T^2\}^2}\sec^2\xi\frac{\partial S}{\partial\eta}\,,\\ \frac{\partial\Theta^-}{\partial\eta} &= \frac{2\varepsilon(1+\varepsilon^2)(1+TS)\sec\eta}{\{(T^2+\varepsilon^2T^2+\varepsilon^2)S^2+2TS+1+\varepsilon^2+\varepsilon^2T^2\}^2}\sec^3\xi\frac{\partial S}{\partial\eta}\,,\end{aligned}$$

where $T\equiv\tan\xi=y/z$, $\varepsilon=\delta/s$ and $\sec\eta=\sqrt{1+S^2}$ for $-\pi/4<\eta<\pi/2$ (i.e. $-1<S$) while $\sec\eta=-\sqrt{1+S^2}$ for $\pi/2<\eta<3\pi/4$ (i.e. $S<-1$). The functions $G^{\pm}(\eta)$ are then the real and imaginary parts of $G(\eta)$ given (using in (3.9) the symmetries of $\tilde{\Lambda}(\nu, \mu, l)$) by

$$G(\eta) = \begin{cases} \sum_{p,q,r} \Delta^*_{pqr} \dfrac{S+1}{\gamma_{pq}(S+1) - \gamma_{pr} S} & \text{for } -1 < S < 0 \\ \sum_{p,q,r} \Delta_{pqr} \dfrac{S+1}{\gamma_{qp} S + \gamma_{rp}} & \text{for } 0 < S < \infty \\ \sum_{p,q,r} \Delta^*_{pqr} \dfrac{S+1}{\gamma_{pr}(S+1) - \gamma_{pq}} & \text{for } S < -1\,, \end{cases}$$

where $\gamma_{rp} = i(s^{(r)} - s^{(p)*}) = \gamma^*_{pr}$. In the first quadrant $0 < \eta < \pi/2$ (i.e. for $0 < S < \infty$), the contribution K_1^+ to K^+ is

$$K_1^+ = \frac{1}{2\pi^2} \mathrm{Re} \sum_{p,q,r} \Delta_{pqr} \lim_{\delta \to 0+} \frac{I_{pqr}}{s^2 + \delta^2}\,, \text{ for } I_{pqr} = \int_0^\infty \frac{S+1}{\gamma_{qp} S + \gamma_{rp}} \frac{\partial \Theta^+}{\partial S} \mathrm{d}S\,.$$

After considerable use of partial fractions, the limiting process gives

$$\lim_{\delta \to 0+} \frac{I_{pqr}}{s^2 + \delta^2} = \frac{\gamma_{qp} - \gamma_{rp}}{(\gamma_{qp} z - \gamma_{rp} y)^2} (\log \gamma_{qp} y - \log \gamma_{rp} z) + \frac{\gamma_{rp} z^2 + \gamma_{qp} y^2}{\gamma_{qp} \gamma_{rp} y z (y^2 + z^2)} - \frac{\gamma_{qp} - \gamma_{rp}}{\gamma_{qp} \gamma_{rp}} \frac{\gamma_{qp} z + \gamma_{rp} y}{(\gamma_{qp} z - \gamma_{rp} y)(y^2 + z^2)}\,.$$

Analogous calculations in the two inter-related intervals $-1 < S < 0$ and $S < -1$ (i.e. $-1 < S^{-1} < 0$) combine to give (after appreciable cancellation) the kernel

$$K^+(y, z) = \frac{1}{\pi^2} \mathrm{Symm}\, \mathrm{Re} \sum_{p,q,r} \Delta_{pqr} \gamma_{qp} L^+_{pqr}(y, z)\,, \tag{6.7}$$

where

$$L^+_{pqr}(y, z) = \frac{\log(\gamma_{qp} y) - \log(\gamma_{rp} z)}{(\gamma_{rp} y - \gamma_{qp} z)^2} - \frac{\log[\gamma_{rp}(y - z)] - \log(\gamma_{qp} z)}{[\gamma_{rp}(y - z) + \gamma_{qp} z]^2} + \frac{\gamma_{rp}(y + z) - \gamma_{qp} y}{\gamma_{qp} \gamma_{rp} [\gamma_{rp}(y + z) - \gamma_{qp} z](y^2 + z^2)} \tag{6.8}$$

and where Symm denotes the part symmetric under interchange of y and z. In (6.7) and (6.8) it may be noted that $\gamma_{qp} = \beta^{(p)} + \beta^{(q)} + i(\alpha^{(q)} - \alpha^{(p)})$ has a positive real part. For all materials having reflectional symmetry in planes $X_1 =$ constant, all the factors $s^{(p)}$ are purely imaginary (i.e. $\alpha^{(p)} = 0$, $p = 1, 2, 3$) so that $G(\eta)$ is real. In these cases, the total kernel is given by $K(y, z) = K_0(y, z) + K^+(y, z)$ and, moreover, γ_{rp} is real for all pairs (r, p), so that all the logarithms within (6.8) have real arguments (and so do not involve the arctan function).

Evaluation of $K^-(y,z)$ in each of $-1 < S < 0$, $0 > S$ and $S < -1$ involves many integrals of the type

$$I \equiv \lim_{\varepsilon\to 0+} \int_a^b \frac{2\varepsilon(z+yS)(S+1)(1+S^2)^{1/2}}{(\delta_1 S+\delta_2)\sigma^2(S-S_+)^2(S-S_-)^2}\mathrm{d}S\,,$$

where $S_\pm = -z/y \pm i\varepsilon\sigma^{-1}\sqrt{1+\varepsilon^2 s^2}$ are the roots of

$$\sigma S^2 + 2yzS + \nu = (yS+z)^2 + \varepsilon^2 s^2(1+S^2) = 0$$

so that $\sigma = y^2+\varepsilon^2 s^2$, $\nu = z^2+\varepsilon^2 s^2$ with $s = (y^2+z^2)^{1/2}$. Since the integrand is $\mathrm{O}(\varepsilon S^{-2})$ except near $S = -z/y$ (each of $S = S_\pm$ is a double pole, while $S = -z/y$ is a zero), the limit ($\varepsilon \to 0$) of the integral arises purely as a contribution from the vicinity of $S = -z/y$. Moreover, the integral vanishes unless $-z/y \in (a,b)$.

Making the substitution $S = -z/y + \varepsilon u$ leads to

$$I \equiv \lim_{\varepsilon\to 0+} \int_{-\infty}^{\infty} \frac{(y-z+\varepsilon yu)\{s^2 - 2\varepsilon zyu + \varepsilon^2 y^2 u^2\}^{1/2} u}{(\delta_2 y - \delta_1 z + \varepsilon\delta_1 yu)[y^4u^2 + s^4 - 2\varepsilon s^2 yzu + \varepsilon^2 s^2 y^2 u^2]}\mathrm{d}u\,.$$

If $J(y,z,u,\varepsilon)$ denotes the new integrand, then $J_0(y,z,u) \equiv J(y,z,u,0)$ is an odd function of u, so that I is finite. Expanding J as $J = J_0(y,z,u) + \varepsilon J_1(y,z,u) + \ldots$ gives $I = 2y^3|y|\int_{-\infty}^{\infty} J_1(y,z,u)\,\mathrm{d}u$. It is found that

$$J_1(y,z,u) = \frac{(\delta_2-\delta_1)y^2 s u^2}{(\delta_2 y - \delta_1 z)^2(y^4u^2+s^4)^2} - \frac{yz(y-z)s^{-1}u^2}{(\delta_2 y - \delta_1 z)(y^4u^2+s^4)^2} + \frac{4yz(y-z)s^3u^2}{(\delta_2 y - \delta_1 z)(y^4u^2+s^4)^3}\,,$$

so that the substitution $u = (s/y)^2\tan\phi$ readily yields

$$\int_{-\infty}^{\infty} J_1(y,z,u)\,\mathrm{d}u = \frac{(\delta_2-\delta_1)y^{-4}s^{-1}}{(\delta_2 y - \delta_1 z)^2}\int_{-\pi/2}^{\pi/2}\sin^2\phi\,\mathrm{d}\phi + \frac{z(y-z)y^{-5}s^{-3}}{(\delta_2 y - \delta_1 z)}\int_{-\pi/2}^{\pi/2}(4\cos^2\phi - 1)\sin^2\phi\,\mathrm{d}\phi$$

which then gives $I = \pi s^{-1}(\delta_2 y - \delta_1 z)^{-2}\operatorname{sgn} y$.

The three cases $yz < 0$, $z/y > 1$ and $y/z > 1$ now require separate consideration. For $yz < 0$ so that $S = -z/y \in (0,\infty)$, it is appropriate to choose $\delta_1 = \gamma_{qp}$ and $\delta_2 = \gamma_{rp}$, so yielding

$$K^-(y,z) = \frac{\operatorname{sgn} y}{2\pi}\sum_{p,q,r}\operatorname{Im}\left\{\frac{\Delta_{pqr}(\gamma_{rp}-\gamma_{qp})}{[\gamma_{rp}y - \gamma_{qp}z]^2}\right\}, \qquad yz < 0.$$

For $z/y > 1$, the value $S = -z/y$ lies in $(-\infty, -1)$ (where $\sec\eta < 0$), so that the choice $\delta_1 = \gamma_{pr}$, $\delta_2 = \gamma_{pr} - \gamma_{pq}$ gives

$$K^-(y,z) = \frac{\operatorname{sgn} y}{2\pi} \sum_{p,q,r} \operatorname{Im}\left\{ \frac{\Delta^*_{pqr}\gamma_{pq}}{[\gamma_{pq}y + \gamma_{pr}(z-y)]^2} \right\}, \qquad 1 < z/y.$$

Similarly, taking $\delta_1 = \gamma_{pr} - \gamma_{pq}$, $\delta_2 = -\gamma_{pq}$ for $y/z > 1$ gives

$$K^-(y,z) = \frac{\operatorname{sgn} y}{2\pi} \sum_{p,q,r} \operatorname{Im}\left\{ \frac{\Delta^*_{pqr}\gamma_{pr}}{[\gamma_{pr}z + \gamma_{pq}(y-z)]^2} \right\}, \qquad 1 < y/z.$$

Collectively, these may be rearranged (using $\Delta_{prq} = \Delta_{pqr}$) as

$$K^-(y,z) = \begin{cases} \dfrac{\operatorname{sgn} y}{2\pi} \displaystyle\sum_{p,q,r} \operatorname{Im}\left\{ \frac{\Delta_{pqr}(\gamma_{qp} - \gamma_{rp})}{[\gamma_{rp}z - \gamma_{qp}y]^2} \right\} & \text{for } yz < 0\,, \\ \dfrac{-\operatorname{sgn} y}{2\pi} \displaystyle\sum_{p,q,r} \operatorname{Im}\left\{ \frac{\Delta_{pqr}\gamma_{qp}}{[\gamma_{qp}y + \gamma_{rp}(z-y)]^2} \right\} & \text{for } \dfrac{z-y}{y} > 0, \\ \dfrac{-\operatorname{sgn} y}{2\pi} \displaystyle\sum_{p,q,r} \operatorname{Im}\left\{ \frac{\Delta_{pqr}\gamma_{qp}}{[\gamma_{qp}z + \gamma_{rp}(y-z)]^2} \right\} & \text{for } \dfrac{y-z}{z} > 0. \end{cases} \tag{6.9}$$

Observe that, in (6.9), all the identities (taking $\alpha > 0$)

$$K^-(z,y) = K^-(y,z) = -K^-(-y,-z)\,, \qquad K^-(\alpha y, \alpha z) = \alpha^{-2}K^-(y,z)$$

are satisfied. Thus, not only does K^- satisfy symmetry, reality and scale-invariance, but also it describes the part of $K(y,z)$ which violates forward-backward symmetry. Indeed, since $K^+(-y,-z) = K^+(y,z)$, it follows that $K^-(y,z) = \frac{1}{2}\{K(y,z) - K(-y,-z)\}$, which is non-zero only in anisotropic media appropriately oriented relative to the OX_J axes.

7. Summary and Conclusions

To summarize, expressions (6.4), (6.7), (6.8) and (6.9) determine the kernel $K(y,z) = K_0(y,z) + K^+(y,z) + K^-(y,z)$ so defining the flux (4.3) for use in the direct evolution equation (1.1). Expressions (6.7) and (6.8) for $K^+(y,z)$ generalize those in Hamilton *et al.* [1995; 1997] for isotropic materials. The present treatment allows for arbitrary anisotropy, so requiring expression (6.9) for $K^-(y,z)$ also.

Two final remarks are appropriate. Although linear theory (see §2) allows all components of deformation gradient at $X_2 = 0$ to be inter-related *via* Hilbert transforms, using these solely within the quadratically nonlinear traction boundary condition (as in Gusev *et al.*, 1997) is incorrect, since it takes no account of quadratic nonlinearity everywhere in $X_2 < 0$. In practice, no material is exactly homogeneous and nondissipative. Additional physical effects should be included. Material dispersion and dissipation give rise to linear terms in the spectral treatment leading to (3.7), and so add linear terms within the *nonlocal kernel.* Fine-scale effects such as surface tension, which become increasingly important as singularities in $v(\theta, X)$ form, are more naturally included into (1.1) as *local* terms involving higher derivatives with respect to θ.

References

Chadwick, P. Surface and Interfacial Waves of Arbitrary Form in Isotropic Elastic Media. *Journal of Elasticity*, 6:73–80, 1976.

David, E. A. and D. F. Parker. Nondistorting Waveforms of Electroelastic Surface Waves. *Wave Motion* 12:315–327, 1990.

Gusev, V., W. Lauriks and J. Thoen. Theory for the Time-evolution of the Nonlinear Rayleigh Waves in an Isotropic Solid *Physical Review B*, 55(15):9344–9347, 1997.

Hamilton, M. F., Yu. A. Il'insky and E. A. Zabolotskaya. Local and Nonlocal Nonlinearity in Rayleigh Waves. *Journal of the Acoustical Society of America*, 97:882–890, 1995.

Hamilton, M. F., Yu. A. Il'insky and E. A. Zabolotskaya. General Theory for the Spectral Evolution of Nonlinear Rayleigh Waves. *Journal of the Acoustical Society of America*, 102:1402–1417, 1997.

Hunter, J. K. Nonlinear Surface Waves. In W. B. Lindquist, editor, *Current Progress in Hyperbolic Problems and Computations, Contemporary Mathematics*, 100:185-202. American Mathematical Society, Providence, R.I., 1989.

Kalyanasundaram, N. Nonlinear Surface Acoustic Waves on an Isotropic Solid. *International Journal of Engineering Science*, 19:279–286, 1981.

Kalyanasundaram, N., R. Ravindran and P. Prasad. Coupled Amplitude Theory of Nonlinear Surface Acoustic Waves. *Journal of the Acoustical Society of America*, 72:488–493, 1982.

Lardner, R. W. Nonlinear Surface Acoustic Waves on an Elastic Solid. *International Journal of Engineering Science*, 21:1331–1342, 1983.

Lardner, R. W. Nonlinear Surface Acoustic Waves on an Elastic Solid of General Anisotropy. *Jounal of Elasticity*, 16:63–73, 1986.

Mayer, A. P. Surface Acoustic Waves in Nonlinear Elastic Media. *Physics Reports*, 256:237–366, 1995.

Panayotaros, P. An Expansion Method for Non-linear Rayleigh Waves. *Wave Motion*, 36:1–21, 2002.

Parker, D. F. Nonlinear Surface Acoustic Waves on Homogeneous Media. In A.D. Boardman, M. Bertolotti and T. Twardowski, editors, *Nonlinear Waves in Solid State Physics*, pages 163 – 193. Plenum, New York, 1990.

Parker, D. F. and E. A. David. Nonlinear Piezoelectric Surface Waves. *International Journal of Engineering Science*, 27:565–581, 1989.

Parker, D. F. and J. K. Hunter. Scale Invariant Elastic Surface Waves. *Supplemento ai Rendiconti del Circolo Matematico di Palermo*, 57:381–392, 1998.

Parker, D. F. and F. M. Talbot. Analysis and Computation for Nonlinear Elastic Surface Waves of Permanent form. *Journal of Elasticity*, 15:389–426, 1985.

EXPLICIT SECULAR EQUATIONS FOR SURFACE WAVES IN AN ANISOTROPIC ELASTIC HALF-SPACE FROM RAYLEIGH TO TODAY

T. C. T. TING
Stanford University, Division of Mechanics and Computation, Durand 262, Stanford, CA, 94305, U.S.A.

Key words: nonlinear surface waves, elasticity, Rayleigh waves, scale invariance, nonlocality, spectral kernel, spatial kernel, Hilbert transform

1. Introduction

An explicit secular equation for surface waves propagating in an elastic half-space $x_2 \geq 0$, in the direction of the x_1 -axis, was first obtained by Rayleigh (1885) for isotropic materials. Explicit secular equations were derived by Stoneley (1949) and Alshits and Lothe (1978) for transversely isotropic (hexagonal) materials, by Stoneley (1955) for cubic materials, by Sveklo (1948) and Stoneley (1963) for orthotropic materials, and by Currie (1979), Destrade (2001) and Ting (2002a,b,c) for monoclinic materials with the symmetry plane at x_3 =0. For monoclinic materials with the symmetry plane at x_1 =0 or x_2 =0, explicit secular equations were presented by Ting (2002a,b, 2003). Explicit secular equations for general anisotropic materials were obtained by Taziev (1989) and Ting (2002b, 2003). The secular equations mentioned above employed different derivations. In most cases, the same derivation for a general anisotropic material and for a special anisotropic material has to be carried out separately. We show here that all derivations can be presented using the Stroh (1962) formalism or its modified version (Ting, 2002a,b,c). We also show how the derivations can be improved or made more general. While numerical schemes are available for computing the surface wave speed, an explicit secular equation allows us to analyze the dependence of the surface wave speed on the elastic constants. For instance, for the special case of monoclinic materials with the symmetry

R.V. Goldstein and G.A. Maugin (eds.),
Surface Waves in Anisotropic and Laminated Bodies and Defects Detection, 95–116.

plane at x_3 =0, the secular equation is independent of the reduced elastic compliances s'_{16} and s'_{26} when (a) s'_{16} =0, (b) s'_{12} =0 and s'_{16} =$2s'_{26}$, or (c) s'_{16} =s'_{26} and $s'_{12}s'_{66}+s'(1,2) = {s'_{16}}^2$. It is shown that, when s'_{12} =0 (of which (b) is a special case), the secular equation reduces to a quadratic equation so that an exact expression of the surface wave speed is obtained. Other special materials such as (c) are presented for which an exact expression of the surface wave speed can be obtained.

2. Rayleigh Equation

In a fixed rectangular coordinate system x_i (i=1,2,3) let u_i and σ_{ij} be the displacement and stress in an anisotropic elastic material. The stress-strain law and the equation of motion are

$$\sigma_{ij} = C_{ijks}u_{k,s}, \tag{2.1}$$

$$C_{ijks}u_{k,sj} = \rho \ddot{u}_i, \tag{2.2}$$

in which repeated indices imply summation, ρ is the mass density, the comma denotes differentiation with x_i, the dot stands for differentiation with time t, and C_{ijks} are the elastic stiffnesses that are assumed to possess the full symmetry. For a surface wave propagating in the half-space $x_2 \geq 0$ with a steady wave speed v in the direction of the x_1 -axis, the problem is to find a solution to (2.1) and (2.2) that satisfies the conditions,

$$u_i = 0 \text{ at } x_2 = \infty \quad \text{and} \quad \sigma_{i2} = 0 \text{ at } x_2 = 0. \tag{2.3}$$

An *explicit secular equation* is an equation that contains the elastic constants C_{ijks} (or the reduced elastic compliances $s'_{\alpha\beta}$), the mass density ρ and the wave speed v only.

When the material is isotropic, Rayleigh (1885) obtained an explicit secular equation which is often written in the literature as

$$(2 - X/C_{66})^2 = 4\sqrt{1 - X/C_{11}}\sqrt{1 - X/C_{66}}, \tag{2.4}$$

where $C_{\alpha\beta}$ is the contracted notation of C_{ijks} and

$$X = \rho v^2. \tag{2.5}$$

Although (2.4) can be identified with equation (23) in the original paper of Rayleigh (1885), the more explicit secular equation is the one listed between equations (23) and (24) of his paper. That equation is not numbered. It is in the form in which both sides of (2.4) are squared. He employed different notations. What is not mentioned often in the literature is that, after deriving the secular equation in the first half of his paper, Rayleigh studied numerically in the remaining half of his paper the special cases of Poisson ratio $\nu = 1/2$ (incompressible) and $\nu = 1/4$. It was pointed out later by Lamb (1904) that, when the Poisson ratio $\nu = 1/4$, (2.4) has the exact root

$$X/C_{66} = 2(1 - 1/\sqrt{3}). \tag{2.6}$$

Another exact expression of X is when $\nu = 0$ (Eringen and Suhubi, 1975). It is

$$X/C_{66} = 3 - \sqrt{5}. \tag{2.7}$$

Rahman and Barber (1995) presented exact expressions of the roots X of (2.4) for any Poisson ratio. We show in Section 10 that (2.6), (2.7) are special cases of (10.7), (10.2) for certain monoclinic materials with the symmetry plane at x_3 =0. Since Rayleigh's paper, several different approaches have been employed to derive explicit secular equations for surface waves in a material more general than isotropic materials. They can all be connected to the Stroh (1962) formalism and its modified version (Ting, 2002a,b,c). Hence the Stroh formalism is presented next.

3. The Stroh Formalism

When the surface wave is propagating in the half-space $x_2 \geq 0$ with a steady wave speed v in the direction of the x_1 -axis, a solution for the displacement vector u of (2.2) can be written as (Stroh, 1962; see also Ting, 1996, Chapter 12)

$$\mathbf{u} = \mathbf{a}e^{ikz}, \quad z = x_1 - vt + px_2, \tag{3.1}$$

in which k is the real wave number, and p and $\mathbf{a}$ satisfy the eigenrelation

$$\Gamma\mathbf{a} = \mathbf{0}, \tag{3.2}$$

$$\Gamma = \mathbf{Q} - X\mathbf{I} + p(\mathbf{R} + \mathbf{R}^T) + p^2\mathbf{T}. \tag{3.3}$$

In the above X is defined in (2.5), the superscript T stands for the transpose, $\mathbf{I}$ is the unit matrix, and $\mathbf{Q}$, $\mathbf{R}$, $\mathbf{T}$ are 3×3 matrices whose elements are

$$Q_{ik} = C_{i1k1}, R_{ik} = C_{i1k2}, \; T_{ik} = C_{i2k2}. \tag{3.4}$$

The matrices $\mathbf{Q}$ and $\mathbf{T}$ are symmetric and positive definite. In the contracted notation $C_{\alpha\beta}$, the symmetric 3×3 matrix Γ in (3.3) has the expression

$$\Gamma = \begin{bmatrix} C_{11}+2pC_{16}+p^2C_{66}-X & C_{16}+p(C_{12}+C_{66})+p^2C_{26} & C_{15}+p(C_{14}+C_{56})+p^2C_{46} \\ C_{16}+p(C_{12}+C_{66})+p^2C_{26} & C_{66}+2pC_{26}+p^2C_{22}-X & C_{56}+p(C_{46}+C_{25})+p^2C_{24} \\ C_{15}+p(C_{14}+C_{56})+p^2C_{46} & C_{56}+p(C_{46}+C_{25})+p^2C_{24} & C_{55}+2pC_{45}+p^2C_{44}-X \end{bmatrix} \tag{3.5}$$

Introducing the new vector **b** defined by

$$\mathbf{b} = (\mathbf{R}^T + p\mathbf{T})\mathbf{a} = -[p^{-1}(\mathbf{Q} - X\mathbf{I}) + \mathbf{R}]\mathbf{a} \tag{3.6}$$

in which the second equality follows from (3.2), the stress determined from (2.1) can be written as

$$\sigma_{i1} = -\phi_{i,2} - \rho v \dot{u}_i, \ \sigma_{i2} = \phi_{i,1}. \tag{3.7}$$

The ϕ_i $(i = 1, 2, 3)$ are the components of the stress function vector

$$\boldsymbol{\phi} = \mathbf{b}e^{ikz}. \tag{3.8}$$

The vanishing of the determinant of Γ in (3.5) yields a sextic equation in p. In terms of X, it is a cubic equation. There are six eigenvalues p_α and six Stroh eigenvectors $\mathbf{a}_\alpha$ and $\mathbf{b}_\alpha$ $(\alpha = 1, 2, \ldots, 6)$. When p_α are complex they consist of three pairs of complex conjugates. If p_1, p_2, p_3 are the eigenvalues with a positive imaginary part, the remaining three eigenvalues are the complex conjugates of p_1, p_2, p_3. When p_1, p_2, p_3 are distinct, the general solution obtained from superposing three solutions of (3.1) and (3.8) associated with p_1, p_2, p_3 can be written as

$$\mathbf{u} = \mathbf{A}\left\langle e^{ikz_\alpha}\right\rangle \mathbf{q}, \quad \boldsymbol{\phi} = \mathbf{B}\left\langle e^{ikz_\alpha}\right\rangle \mathbf{q}, \tag{3.9}$$

where **q** is an arbitrary constant vector and

$$\mathbf{A} = [\mathbf{a}_1, \mathbf{a}_2, \mathbf{a}_3], \quad \mathbf{B} = [\mathbf{b}_1, \mathbf{b}_2, \mathbf{b}_3], \tag{3.10a}$$

$$\left\langle e^{ikz_\alpha}\right\rangle = diag\left[\, e^{ikz_1}, \ e^{ikz_2}, \ e^{ikz_3}\,\right], \tag{3.10b}$$

$$z_\alpha = x_1 - vt + p_\alpha x_2. \tag{3.10c}$$

Since the imaginary parts of p_1, p_2, p_3 are positive, $(3.9)_1$ assures us that $u \to 0$ as $x_2 \to \infty$. The surface traction at x_2 =0 vanishes if $\boldsymbol{\phi} = \mathbf{0}$ at x_2 =0, i.e.,

$$\mathbf{B}\mathbf{q} = \mathbf{0}. \tag{3.11}$$

This has a nontrivial solution for **q** when the determinant of **B** vanishes,

$$|\mathbf{B}| = 0. \tag{3.12}$$

This is the secular equation for v or for X. When the material is isotropic, the columns of **B** can be computed explicitly in terms of X and $C_{\alpha\beta}$ so that (3.12) leads to (2.4).

In the case p_1, p_2, p_3 are not distinct, the system is degenerate. The derivation in (3.9)-(3.12) is not valid. The modification needed when the system is degenerate can be found in (Wang and Ting, 1997; Ting, 1997; Ting and Barnett, 1997).

4. A Modified Stroh Formalism

The two equations in (3.6) can be written in a standard eigenrelation as (Ingebrigtsen and Tonning 1969; Barnett and Lothe 1973; Chadwick and Smith 1977)

$$\mathbf{N}\xi = p\xi, \tag{4.1}$$

$$\mathbf{N} = \begin{bmatrix} \mathbf{N}_1 & \mathbf{N}_2 \\ \mathbf{N}_3 + X\mathbf{I} & \mathbf{N}_1^T \end{bmatrix}, \quad \xi = \begin{bmatrix} \mathbf{a} \\ \mathbf{b} \end{bmatrix}, \tag{4.2}$$

$$\mathbf{N}_2 = \mathbf{T}^{-1}, \quad \mathbf{N}_1 = -\mathbf{N}_2\mathbf{R}^T, \quad \mathbf{N}_3 = \mathbf{R}\mathbf{N}_2\mathbf{R}^T - \mathbf{Q}. \tag{4.3}$$

It was shown in (Ting, 1988) that $\mathbf{N_1}$, $\mathbf{N_2}$, $\mathbf{N_3}$ have the structures

$$-\mathbf{N}_1 = \begin{bmatrix} r_6 & 1 & s_6 \\ r_2 & 0 & s_2 \\ r_4 & 0 & s_4 \end{bmatrix}, \ \mathbf{N}_2 = \begin{bmatrix} n_{66} & n_{26} & n_{46} \\ n_{26} & n_{22} & n_{24} \\ n_{46} & n_{24} & n_{44} \end{bmatrix}, \ -\mathbf{N}_3 = \begin{bmatrix} \eta & 0 & -\kappa \\ 0 & 0 & 0 \\ -\kappa & 0 & \mu \end{bmatrix}. \tag{4.4}$$

An explicit expression of the elements of $\mathbf{N_1}$, $\mathbf{N_2}$, $\mathbf{N_3}$ was given in (Ting, 1988) in terms of the reduced elastic compliances $s'_{\alpha\beta}$ as (see also Ting, 1996, p.167)

$$\begin{aligned} &\mu = s'_{11}/\Delta, \ \eta = s'_{55}/\Delta, \ \kappa = s'_{15}/\Delta, \ n_{\alpha\beta} = s'(\alpha,1,5|\beta,1,5)/\Delta, \\ &r_\alpha = s'(1,5|5,\alpha)/\Delta, \ s_\alpha = s'(1,5|\alpha,1)/\Delta, \ \Delta = s'(1,5) > 0. \end{aligned} \tag{4.5}$$

In the above, $s'(n_1,\ldots,n_k|m_1,\ldots,m_k)$ is the $k \times k$ minor of the matrix $s'_{\alpha\beta}$, the elements of which belong to the rows of $s'_{\alpha\beta}$, numbered $n_1,\ldots,n_k$ and columns numbered $m_1,\ldots,m_k$, $1 \le k \le 6$. A principal minor is $s'(n_1,\ldots,n_k|n_1,\ldots,n_k)$, which is written as $s'(n_1,\ldots,n_k)$ for simplicity. Removing the third row and the third column of $s'_{\alpha\beta}$ that contain only zero elements, the 5×5 matrix is positive definite. Hence $\Delta > 0$. μ is the shear modulus when the material is isotropic. An explicit expression of $\mathbf{N_1}, \mathbf{N_2}, \mathbf{N_3}$ in terms of the elastic stiffnesses $C_{\alpha\beta}$ is given in (Barnett and Chadwick, 1990).

Equation (4.1) consists of two equations,

$$(\mathbf{N}_1 - p\mathbf{I})\mathbf{a} + \mathbf{N}_2\mathbf{b} = \mathbf{0}, \quad (\mathbf{N}_3 + X\mathbf{I})\mathbf{a} + (\mathbf{N}_1^T - p\mathbf{I})\mathbf{b} = \mathbf{0}. \tag{4.6}$$

Assuming that the determinant $|\mathbf{N}_3 + X\mathbf{I}| \neq 0$, $(4.6)_2$ can be solved for $\mathbf{a}$ and $(4.6)_1$ gives (Ting, 2002a,c)

$$\hat{\Gamma}\mathbf{b} = \mathbf{0}, \tag{4.7}$$

$$\hat{\Gamma} = \hat{\mathbf{Q}} + p(\hat{\mathbf{R}} + \hat{\mathbf{R}}^T) + p^2\hat{\mathbf{T}}, \tag{4.8}$$

where

$$\hat{\mathbf{T}} = -(\mathbf{N}_3 + X\mathbf{I})^{-1}, \quad \hat{\mathbf{R}} = -\mathbf{N}_1\hat{\mathbf{T}}, \quad \hat{\mathbf{Q}} = \mathbf{N}_1\hat{\mathbf{T}}\mathbf{N}_1^T + \mathbf{N}_2. \tag{4.9}$$

The matrices $\hat{\mathbf{T}}$ and $\hat{\mathbf{Q}}$ are symmetric. Equation (4.7) provides a direct computation of the eigenvector $\mathbf{b}$. The vector $\mathbf{a}$ obtained from $(4.6)_2$ is, using (4.9) and (4.7),

$$\mathbf{a} = -(\hat{\mathbf{R}}^T + p\hat{\mathbf{T}})\mathbf{b} = (p^{-1}\hat{\mathbf{Q}} + \hat{\mathbf{R}})\mathbf{b}. \tag{4.10}$$

Equations (4.7)-(4.10) and (3.2), (3.3), (4.3), (3.6) are *dual* to each other. The similarities between the two formalisms are striking.

With the use of (4.5) in $(4.4)_3$, it can be shown that

$$|\mathbf{N}_3 + X\mathbf{I}| = wX/\Delta, \tag{4.11}$$

$$w = 1 - (s'_{11} + s'_{55})X + s'(1,5)X^2 = (1 - s'_{11}X)(1 - s'_{55}X) - (s'_{15}X)^2. \tag{4.12}$$

Hence (4.7)-(4.10) are not valid when $X = 0$ or $w = 0$. $w = 0$ means that

$$X^{-1} = \tfrac{1}{2}(s'_{11} + s'_{55}) \pm \tfrac{1}{2}\sqrt{(s'_{11} - s'_{55})^2 + (2s'_{15})^2} > 0. \tag{4.13}$$

After a minor adjustment, (4.7)-(4.10) remain valid for $X = 0$ and $w = 0$ (Ting, 2002c). It is clear from (4.11) that $X = 0$ and the two X's in (4.13) are the eigenvalues of $-\mathbf{N}_3$, which is positive semi-definite (Ting, 1988). Hence the two X's in (4.13) are positive.

5. Bounds on Surface Wave Speed

At the free surface $x_2 = 0$ $(3.9)_1$ gives

$$\mathbf{u}(x_1, 0, t) = \mathbf{a}_R e^{ik(x_1 - vt)}, \quad \mathbf{a}_R = \mathbf{A}\mathbf{q}, \tag{5.1}$$

where $\mathbf{a}_R$ is the polarization vector of the surface waves at the free boundary. A linear superposition of three equations obtained from $(4.6)_2$ with p=p_1, p_2, p_3 leads to

$$(\mathbf{N}_3 + X\mathbf{I})\mathbf{A}\mathbf{q} + \mathbf{N}_1^T\mathbf{B}\mathbf{q} = \mathbf{B} < p_* > \mathbf{q}. \tag{5.2}$$

In view of (3.11) and $(5.1)_2$ we have

$$(\mathbf{N}_3 + X\mathbf{I})\mathbf{a}_R = \mathbf{B} < p_* > \mathbf{q}. \tag{5.3}$$

Pre-multiply by $\bar{\mathbf{q}}^T\bar{\mathbf{A}}^T$ and use (3.11), $(5.1)_2$ and $(6.1b)_4$ in Section 6, (5.3) gives

$$\bar{\mathbf{a}}_R^T(\mathbf{N}_3 + X\mathbf{I})\mathbf{a}_R = 0. \tag{5.4}$$

Hence X is bounded by the largest and smallest eigenvalues of $-\mathbf{N}_3$ (Ting, 1996, p. 472). This result is valid regardless of whether the surface wave is subsonic or not. The largest eigenvalue of $-\mathbf{N}_3$ is the larger of the two X's shown in (4.13).

6. Secular Equations in Terms of the Barnett-Lothe Tensors

The 6-vector $\xi = [\mathbf{a}, \mathbf{b}]$ shown in (4.1) is a right eigenvector of the 6×6 matrix N. The left eigenvector of $\mathbf{N}$ can be shown to be $[\mathbf{b}, \mathbf{a}]$ (Chadwick and Smith, 1977). The left and right eigenvectors associated with different eigenvalues p are orthogonal to each other. The vectors $\mathbf{a}_1, \mathbf{a}_2, \mathbf{a}_3$ and $\mathbf{b}_1, \mathbf{b}_2, \mathbf{b}_3$ are unique up to an arbitrary constant multiplier. They can be normalized so that the following *orthogonality* relation holds,

$$\begin{bmatrix} \mathbf{B}^T & \mathbf{A}^T \\ \bar{\mathbf{B}}^T & \bar{\mathbf{A}}^T \end{bmatrix} \begin{bmatrix} \mathbf{A} & \bar{\mathbf{A}} \\ \mathbf{B} & \bar{\mathbf{B}} \end{bmatrix} = \begin{bmatrix} \mathbf{I} & \mathbf{0} \\ \mathbf{0} & \mathbf{I} \end{bmatrix}, \tag{6.1a}$$

or

$$\begin{aligned} \mathbf{B}^T\mathbf{A} + \mathbf{A}^T\mathbf{B} = \mathbf{I} = \bar{\mathbf{B}}^T\bar{\mathbf{A}} + \bar{\mathbf{A}}^T\bar{\mathbf{B}}, \\ \mathbf{B}^T\bar{\mathbf{A}} + \mathbf{A}^T\bar{\mathbf{B}} = \mathbf{0} = \bar{\mathbf{B}}^T\mathbf{A} + \bar{\mathbf{A}}^T\mathbf{B}. \end{aligned} \tag{6.1b}$$

The over bar denotes the complex conjugate. The two 6×6 matrices in (6.1a) are the inverse of each other so that their products commute, i.e.,

$$\begin{bmatrix} \mathbf{A} & \bar{\mathbf{A}} \\ \mathbf{B} & \bar{\mathbf{B}} \end{bmatrix} \begin{bmatrix} \mathbf{B}^T & \mathbf{A}^T \\ \bar{\mathbf{B}}^T & \bar{\mathbf{A}}^T \end{bmatrix} = \begin{bmatrix} \mathbf{I} & \mathbf{0} \\ \mathbf{0} & \mathbf{I} \end{bmatrix}, \tag{6.2a}$$

or

$$\begin{aligned} \mathbf{A}\mathbf{B}^T + \bar{\mathbf{A}}\bar{\mathbf{B}}^T = \mathbf{I} = \mathbf{B}\mathbf{A}^T + \bar{\mathbf{B}}\bar{\mathbf{A}}^T, \\ \mathbf{A}\mathbf{A}^T + \bar{\mathbf{A}}\bar{\mathbf{A}}^T = \mathbf{0} = \mathbf{B}\mathbf{B}^T + \bar{\mathbf{B}}\bar{\mathbf{B}}^T. \end{aligned} \tag{6.2b}$$

These are the *closure* relations (Stroh, 1962). They tell us that $(2\mathbf{A}\mathbf{B}^T - \mathbf{I})$, $\mathbf{A}\mathbf{A}^T$ and $\mathbf{B}\mathbf{B}^T$ are purely imaginary. Hence the three Barnett-Lothe tensors

$$\mathbf{S} = i(2\mathbf{A}\mathbf{B}^T - \mathbf{I}), \quad \mathbf{H} = 2i\mathbf{A}\mathbf{A}^T, \quad \mathbf{L} = -2i\mathbf{B}\mathbf{B}^T, \tag{6.3}$$

are real. From $(6.3)_3$, the secular equation (3.12) can be replaced by

$$|\mathbf{L}| = 8i|\mathbf{B}|^2 = 0. \tag{6.4}$$

An alternate secular equation derived by Chadwick and Smith (1982) is

$$|\mathbf{I} \pm i\mathbf{S}| = 1 + \tfrac{1}{2}tr[\mathbf{S}^2] = 0. \tag{6.5}$$

Defining the impedance tensor $\mathbf{M}$ as

$$\mathbf{M} = -i\mathbf{B}\mathbf{A}^{-1} = -i(\mathbf{A}\mathbf{B}^T)^T(\mathbf{A}\mathbf{A}^T)^{-1} = \mathbf{H}^{-1}(\mathbf{I} + i\mathbf{S}), \tag{6.6}$$

another secular equation (Barnett and Lothe, 1985) is

$$|\mathbf{M}| = 0. \tag{6.7}$$

Instead of (6.3), the three Barnett-Lothe tensors $\mathbf{S}$, $\mathbf{H}$, $\mathbf{L}$ can be computed using an integral formalism without finding the eigenvalues p and the eigenvectors $\mathbf{a}$ and $\mathbf{b}$ (Barnett and Lothe, 1973). Unfortunately, an explicit expression of $\mathbf{S}, \mathbf{H}, \mathbf{L}$ is available only for orthotropic materials (Dongye and Ting, 1989; Chadwick and Wilson, 1992; see also Ting, 1996, p. 480). Hence the secular equations (6.4), (6.5) and (6.7) are not explicit for materials more general than orthotropic materials.

Equation $(6.4)_1$ suggests that (Ting, 1996, p. 469)

$$|\mathbf{B}| = \pm\frac{1-i}{4}|\mathbf{L}|^{1/2}. \tag{6.8}$$

Hence the real and imaginary parts of $|\mathbf{B}|$ are identical but opposite in sign. If the columns of $\mathbf{B}$ are not normalized, the real and imaginary parts of $|\mathbf{B}|$ are linearly dependent. Thus (3.12) gives only one secular equation, not two as Synge (1956) suggested.

7. Secular Equations Derived from $|\mathbf{B}| = \mathbf{0}$

Using (4.4), the $\hat{\mathbf{T}}$, $\hat{\mathbf{R}}$, $\hat{\mathbf{Q}}$ in (4.9) can be written as (Ting, 2002c)

$$\hat{\mathbf{T}} = w^{-1}\begin{bmatrix} e_{11} & 0 & e_{15} \\ 0 & -wX^{-1} & 0 \\ e_{15} & 0 & e_{55} \end{bmatrix}, \quad \hat{\mathbf{R}} = -w^{-1}\begin{bmatrix} e_{16} & wX^{-1} & e_{56} \\ e_{12} & 0 & e_{52} \\ e_{14} & 0 & e_{54} \end{bmatrix}$$

$$\hat{\mathbf{Q}} = w^{-1}\begin{bmatrix} e_{66} - wX^{-1} & e_{26} & e_{46} \\ e_{26} & e_{22} & e_{24} \\ e_{46} & e_{24} & e_{44} \end{bmatrix}. \tag{7.1}$$

In (7.1),

$$e_{11} = s'_{11} - s'(1,5)X, \quad e_{15} = s'_{15}, \quad e_{55} = s'_{55} - s'(1,5)X, \tag{7.2a}$$

for the matrix $\hat{\mathbf{T}}$

$$e_{1\beta} = s'_{1\beta} - s'(1,5|\beta,5)X, \quad e_{5\beta} = s'_{5\beta} - s'(1,5|1,\beta)X, \tag{7.2b}$$

for the matrix $\hat{\mathbf{R}}$, and

$$e_{\alpha\beta} = s'_{\alpha\beta} - [s'(1,\alpha|1,\beta) + s'(\alpha,5|\beta,5)]X + s'(\alpha,1,5|\beta,1,5)X^2, \tag{7.2c}$$

for the matrix $\hat{\mathbf{Q}}$. Noticing that (7.2a,b) are special cases of (7.2c) which shows that $e_{\alpha\beta} = e_{\beta\alpha}$, (7.2c) is a universal expression of $e_{\alpha\beta}$ for the matrices $\hat{\mathbf{T}}$, $\hat{\mathbf{R}}$, $\hat{\mathbf{Q}}$. It should be pointed out that $e_{\alpha\beta} = s'_{\alpha\beta}$ and $w = 1$ when $X = 0$ (elastostatics).

When $w \neq 0$ the eigenrelation (4.7) remains valid if $\hat{\Gamma}$ is multiplied by the factor w. With (7.1), the $\hat{\Gamma}$ in (4.8) has the expression

$$\hat{\Gamma} = \begin{bmatrix} e_{66} - wX^{-1} - 2e_{16}p + e_{11}p^2 & e_{26} - (wX^{-1} + e_{12})p & e_{46} - (e_{14} + e_{56})p + e_{15}p^2 \\ e_{26} - (wX^{-1} + e_{12})p & e_{22} - wX^{-1}p^2 & e_{24} - e_{25}p \\ e_{46} - (e_{14} + e_{56})p + e_{15}p^2 & e_{24} - e_{25}p & e_{44} - 2e_{45}p + e_{55}p^2 \end{bmatrix}. \quad (7.3)$$

For a monoclinic material with the symmetry plane at $x_3 = 0$ the inplane and antiplane deformations are uncoupled. Consider the inplane deformation. The displacement $\mathbf{u}$ and the stress function $\boldsymbol{\phi}$ are 2-vectors. The third row and the third column of the matrix $\hat{\Gamma}$ in (7.3) can be deleted so that

$$\hat{\Gamma} = \begin{bmatrix} \hat{\Gamma}_{11}(p) & \hat{\Gamma}_{12}(p) \\ \hat{\Gamma}_{12}(p) & \hat{\Gamma}_{22}(p) \end{bmatrix}. \quad (7.4)$$

Since $\left|\hat{\Gamma}\right| = \hat{\Gamma}_{11}\hat{\Gamma}_{22} - \hat{\Gamma}_{12}\hat{\Gamma}_{12} = 0$, it is easily shown that the eigenrelation (4.7) for the vector $\mathbf{b}$ is satisfied by

$$\mathbf{b} = \begin{bmatrix} \hat{\Gamma}_{22}(p) \\ -\hat{\Gamma}_{12}(p) \end{bmatrix} \text{ or } \mathbf{b} = \begin{bmatrix} -\hat{\Gamma}_{12}(p) \\ \hat{\Gamma}_{11}(p) \end{bmatrix}. \quad (7.5)$$

If we take $(7.5)_1$ as the eigenvector for $\mathbf{b}_1$ and $\mathbf{b}_2$, the matrix $\mathbf{B}$ is

$$\mathbf{B} = \begin{bmatrix} \hat{\Gamma}_{22}(p_1) & \hat{\Gamma}_{22}(p_2) \\ -\hat{\Gamma}_{12}(p_1) & -\hat{\Gamma}_{12}(p_2) \end{bmatrix}. \quad (7.6)$$

The vanishing of the determinant of $\mathbf{B}$ gives the secular equation

$$(p_1 + p_2)e_{26}wX^{-1} - (wX^{-1} + e_{12})(p_1 p_2 wX^{-1} + e_{22}) = 0, \quad (7.7)$$

where we have dropped the common factor $(p_1 - p_2)$. Similarly, if we take $(7.5)_2$ as the eigenvector for $\mathbf{b}_1$ and $\mathbf{b}_2$, we obtain

$$[(p_1 + p_2)e_{11} - 2e_{16}]e_{26} - (wX^{-1} + e_{12})[p_1 p_2 e_{11} + wX^{-1} - e_{66}] = 0. \quad (7.8)$$

Subtraction of (7.8) multiplied by wX^2 from (7.7) multiplied by $e_{11}X^3$ leads to the explicit secular equation

$$(w + e_{12}X)[w(w - e_{66}X) - e_{11}e_{22}X^2] + 2we_{16}e_{26}X^2 = 0. \quad (7.9)$$

Using (7.2), this can be written in full as (Ting, 2002c)

$$[1-(s'_{11}-s'_{12})X]\,\{(1-s'_{11}X)^2(1-s'_{66}X)-s'_{11}[s'_{22}-s'(1,2)X]X^2\}$$
$$-\,s'_{16}(1-s'_{11}X)[(s'_{16}-2s'_{26})(1-s'_{11}X)-s'_{12}s'_{16}X]X^2=0. \tag{7.10}$$

This is a quartic equation in X. It is more explicit than the ones obtained in (Currie, 1979; Destrade 2001; Ting, 2002a,b). In the special case of orthotropic materials, $s'_{16}=s'_{26}=0$ so that, by inspection, (7.10) simplifies to a cubic equation

$$(1-s'_{11}X)^2(1-s'_{66}X)-s'_{11}[s'_{22}-s'(1,2)X]X^2=0, \tag{7.11}$$

because $[1-(s'_{11}-s'_{12})X]\neq 0$. The secular equation for an orthotropic material has been obtained by Sveklo (1948) and Stoneley (1963) in terms of the elastic stiffnesses $C_{\alpha\beta}$. Gilinskii and Shcheglov (1981) studied the case in which only the plane x_3 =0 coincides with a symmetry plane of the orthotropic material. However, they assumed that C_{11} =C_{22}.

Let $\hat{\Gamma}^*$ be the adjoint of $\hat{\Gamma}$ so that

$$\hat{\Gamma}\hat{\Gamma}^*=\left|\hat{\Gamma}\right|\mathbf{I}=\mathbf{0}. \tag{7.12}$$

The element $\hat{\Gamma}^*_{ij}$ is the co-factor of $\hat{\Gamma}_{ij}$ in $\left|\hat{\Gamma}\right|$. Comparison with (4.7) suggests that $\mathbf{b}$ is proportional to any column of $\hat{\Gamma}^*$. For monoclinic materials with the symmetry plane at x_1 =0, use of the first column of $\hat{\Gamma}^*$ as the vector $\mathbf{b}$ leads to a secular equation that contains Γ_I, Γ_{II}, Γ_{III} where

$$\Gamma_I=p_1+p_2+p_3,\quad \Gamma_{II}=p_1p_2+p_2p_3+p_3p_1,\quad \Gamma_{III}=p_1p_2p_3. \tag{7.13}$$

The Γ_I, Γ_{II}, Γ_{III} can be eliminated from the sextic equation for p obtained from $\left|\hat{\Gamma}\right|=0$. This procedure yields an explicit secular equation (Ting, 2002a). The same procedure applies to monoclinic materials with the symmetry plane at $x_2=0$ except that the second column of $\hat{\Gamma}^*$ is used as the vector $\mathbf{b}$.

For a general anisotropic material, the elements of the 3×3 adjoint matrix $\hat{\Gamma}^*$ deduced from $\hat{\Gamma}$ in (7.3) are polynomials in p of degree not higher than four. They can be written as

$$\hat{\Gamma}^*=wX^{-1}\begin{bmatrix} t_kp^k & g_kp^k & h_kp^k \\ g_kp^k & m_kp^k & n_kp^k \\ h_kp^k & n_kp^k & e_kp^k \end{bmatrix},\quad g_4=n_4=0, \tag{7.14}$$

where the repeated k implies summation with k from $k=0$ to $k=4$. The $t_k, g_k, h_k, m_k, n_k, e_k$ are the coefficients of the polynomials. If we ignore the factor wX^{-1} and use the first column of $\hat{\Gamma}^*$ for the vectors $\mathbf{b}_1, \mathbf{b}_2, \mathbf{b}_3$, the matrix $\mathbf{B}$ is

$$\mathbf{B} = \begin{bmatrix} t_k p_1^k & t_k p_2^k & t_k p_3^k \\ g_k p_1^k & g_k p_2^k & g_k p_3^k \\ h_k p_1^k & h_k p_2^k & h_k p_3^k \end{bmatrix}. \tag{7.15}$$

The determinant $|\mathbf{B}|$ computed from (7.15) has a lengthy expression, but the result can be cast neatly in the form of the determinant of a 5×5 matrix $\mathbf{K}$ as (Ting, 2002b)

$$|\mathbf{B}| = Y\,|\mathbf{K}|, \tag{7.16}$$

$$\mathbf{K} = \begin{bmatrix} \Gamma_{III} & 0 & t_0 & g_0 & h_0 \\ -\Gamma_{II} & \Gamma_{III} & t_1 & g_1 & h_1 \\ \Gamma_{I} & -\Gamma_{II} & t_2 & g_2 & h_2 \\ -1 & \Gamma_{I} & t_3 & g_3 & h_3 \\ 0 & -1 & t_4 & 0 & h_4 \end{bmatrix}, \tag{7.17}$$

where

$$Y = (p_1 - p_2)(p_2 - p_3)(p_3 - p_1). \tag{7.18}$$

Likewise one can also use the second and the third columns of $\hat{\Gamma}^*$ in (7.14) for the vectors $\mathbf{b}_1, \mathbf{b}_2, \mathbf{b}_3$ and compute the determinant $|\mathbf{B}|$. Using the argument that five of the six vectors $t_k, g_k, h_k, m_k, n_k, e_k$ must be linearly dependent, Ting (2002b) obtained the secular equation

$$\begin{vmatrix} \delta_{k1} & \delta_{k2} & \delta_{k3} & \delta_{k4} & \delta_{k5} & \delta_{k6} \\ t_0 & g_0 & m_0 & h_0 & n_0 & e_0 \\ t_1 & g_1 & m_1 & h_1 & n_1 & e_1 \\ t_2 & g_2 & m_2 & h_2 & n_2 & e_2 \\ t_3 & g_3 & m_3 & h_3 & n_3 & e_3 \\ t_4 & 0 & m_4 & h_4 & 0 & e_4 \end{vmatrix} = 0, \tag{7.19}$$

$k = 1, 2, \ldots 6$, where $\delta_{\alpha\beta}$ is the Kronecker delta. This is equivalent to six secular equations but not all of them provides a valid X.

For monoclinic materials with the symmetry plane at $x_3 = 0$, (7.19) reduces to

$$\begin{vmatrix} t_2 & g_2 & m_2 \\ t_3 & g_3 & m_3 \\ t_4 & 0 & m_4 \end{vmatrix} = 0. \tag{7.20}$$

Substitution of the expressions for t_k, g_k, m_k in (7.20) leads to the secular equation (7.10). In fact the vanishing of the determinant of a 3×3 matrix taken from any three rows of the columns t_k, g_k, m_k leads to (7.10) (Ting,

2002b). For monoclinic materials with the symmetry plane at x_1 =0 or $x_2 = 0$, (7.19) simplifies to

$$\begin{vmatrix} t_0 & m_0 & e_0 \\ t_2 & m_2 & e_2 \\ t_4 & m_4 & e_4 \end{vmatrix} = 0. \tag{7.21}$$

The $t_k, g_k, h_k, m_k, n_k, e_k$ can be computed in terms of $e_{\alpha\beta}$ and 2×2 minors of $e_{\alpha\beta}$. It was shown in (Ting, 2002c) that any 2×2 minor of $e_{\alpha\beta}$ is a product of w and a polynomial in X of degree no more than two. One can then show that the t_k, g_k, h_k, m_k, n_k, e_k are polynomial in X of degree no more than three for $k = 0$, no more than two for $k = 1$ and 2, and no more than one for $k = 3$ and 4. Thus the secular equation computed from (7.19) is a polynomial in X of degree no more than nine. For monoclinic materials with the symmetry plane at $x_3 = 0$ shown in (7.20), it is of degree no more than four, while for monoclinic materials with the symmetry plane at x_1 =0 or $x_2 = 0$ given in (7.21) it is of degree no more than six. Using a different derivation, Taylor (1981) has shown that his (not an explicit) secular equation for a general anisotropic material is a polynomial in X of degree no more than 27 (see also Taziev, 1989, Ting, 2003), while for monoclinic materials with the symmetry plane at $x_1 = 0$ or $x_2 = 0$ it is of degree no more than twelve. The paper by Taylor is 36 pages long. It was submitted for publication in 1976 and appeared in print five years later in 1981, possibly a Guinness record for the longest time that took a paper to publish.

8. Secular Equations Derived from the Polarization Vector

Instead of driving secular equations from $|\mathbf{B}| = 0$ that requires computation of the eigenvectors $\mathbf{b}$, Currie (1979) and Taziev (1989) derived secular equations from the governing equation for the polarization vector $\mathbf{a}_R$ defined in $(5.1)_2$. Before we present their derivations using the formalism in Section 4, we generalize the eigenrelation (4.1) as

$$\mathbf{N}^n \xi = p^n \xi, \tag{8.1}$$

where n is any positive or negative integer. Let

$$\mathbf{N}^n = \begin{bmatrix} \mathbf{N}_1^{(n)} & \mathbf{N}_2^{(n)} \\ \mathbf{K}^{(n)} & {\mathbf{N}_1}^{(n)T} \end{bmatrix} \tag{8.2}$$

in which

$$\mathbf{N}_1^{(1)} = \mathbf{N}_1, \quad \mathbf{N}_2^{(1)} = \mathbf{N}_2, \quad \mathbf{K}^{(1)} = \mathbf{N}_3 + X\mathbf{I}. \tag{8.3}$$

For $n = -1$, it can be shown (Ting, 1996, p. 451) that

$$\mathbf{N}_1^{(-1)} = \mathbf{N}_2^{(-1)}\mathbf{R},\ \mathbf{N}_2^{(-1)} = -(\mathbf{Q} - X\mathbf{I})^{-1},\ \mathbf{K}^{(-1)} = \mathbf{T} + \mathbf{R}^T\mathbf{N}_2^{(-1)}\mathbf{R}. \quad (8.4)$$

The 3×3 symmetric matrices $\mathbf{K}^{(n)}$ are of particular interest here. For $n = 2$ and -2 we have

$$\mathbf{K}^{(2)} = \mathbf{K}^{(1)}\mathbf{N}_1 + \mathbf{N}_1^T\mathbf{K}^{(1)}, \quad (8.5)$$

$$\mathbf{K}^{(-2)} = \mathbf{K}^{(-1)}\mathbf{N}_1^{(-1)} + \mathbf{N}_1^{(-1)^T}\mathbf{K}^{(-1)}. \quad (8.6)$$

Equation (8.1) consists of two equations of which the second one is

$$\mathbf{K}^{(n)}\mathbf{a} + \mathbf{N}_1^{(n)^T}\mathbf{b} = p^n\mathbf{b}. \quad (8.7)$$

Following the derivation of (5.3) and (5.4) from $(4.6)_2$, we obtain from (8.7)

$$\mathbf{K}^{(n)}\mathbf{a}_R = \mathbf{B} < p_*^n > \mathbf{q}, \quad (8.8)$$

$$\bar{\mathbf{a}}_R^T\mathbf{K}^{(n)}\mathbf{a}_R = 0. \quad (8.9)$$

Equations (8.8) and (8.9) are the generalization of (5.3) and (5.4). When $n = 1$ and $X = 0$, (8.9) recovers the identity given by Stroh (1958). The identities (3.23), $(3.24)_{1,2}$ in (Currie, 1979) are $n = 1, 2, -2$, respectively, in (8.9).

When the vector $\mathbf{b}$ in $(3.6)_{1,2}$ is inserted into the columns of $\mathbf{B}$ in (3.11) we obtain

$$\mathbf{R}^T\mathbf{a}_R + \mathbf{TA} < p_* > \mathbf{q} = \mathbf{0}, \quad (8.10a)$$

$$(\mathbf{Q} - X\mathbf{I})\mathbf{A} < p_*^{-1} > \mathbf{q} + \mathbf{R}\mathbf{a}_R = \mathbf{0}. \quad (8.10b)$$

The three equations $(5.1)_2$ and (8.10a,b) can be solved for $q_1\mathbf{a}_1$, $q_2\mathbf{a}_2$,$q_3\mathbf{a}_3$ in terms of $\mathbf{a}_R$. Satisfaction of the eigenrelation (4.7) gives three equations for $\mathbf{a}_R$ but only two deduced from the three are independent. They are

$$[\Gamma_{III}\mathbf{K}^{(-1)} + \Gamma_I\mathbf{K}^{(1)} - \mathbf{K}^{(2)}]\mathbf{a}_R = \mathbf{0}, \quad (8.11a)$$

$$[\Gamma_{III}\mathbf{K}^{(-2)} - \Gamma_{II}\mathbf{K}^{(-1)} - \mathbf{K}^{(1)}]\mathbf{a}_R = \mathbf{0}. \quad (8.11b)$$

These are (3.16) and (3.17) in (Currie, 1979).

We will derive (8.11a,b) using an alternate approach that seems more straightforward. The three equations (3.11) and (8.8) for $n = 1$ and -1 can be solved for $q_1\mathbf{b}_1, q_2\mathbf{b}_2, q_3\mathbf{b}_3$. The solution for $q_1\mathbf{b}_1$ is

$$q_1\mathbf{b}_1 = -Y^{-1}(p_2 - p_3)[\Gamma_{III}\mathbf{K}^{(-1)} + p_1\mathbf{K}^{(1)}]\mathbf{a}_R, \quad (8.12)$$

while that for $q_2\mathbf{b}_2, q_3\mathbf{b}_3$ are obtained from (8.12) by a cyclic permutation of the subscripts 1,2,3. It can then be shown that

$$\mathbf{B} < p_*^2 > \mathbf{q} = (\Gamma_{III}\mathbf{K}^{(-1)} + \Gamma_I\mathbf{K}^{(1)})\mathbf{a}_R, \quad (8.13a)$$

$$\mathbf{B} < p_*^{-2} > \mathbf{q} = \Gamma_{III}^{-1}(\Gamma_{II}\mathbf{K}^{(-1)} + \mathbf{K}^{(1)})\mathbf{a}_R. \tag{8.13b}$$

For $n = 2$ and -2, (8.8) with the use of (8.13) leads to (8.11).

In the special case of monoclinic materials with the symmetry plane at $x_3 = 0$, $\mathbf{b}_1$ and $\mathbf{b}_2$ are two-vectors while $\mathbf{b}_3$ is not needed. The $\mathbf{K}^{(n)}$ are 2×2 matrices. Equation (8.8) for $n = 1$ and (3.11) can be solved for $q_1\mathbf{b}_1$, $q_2\mathbf{b}_2$ as

$$q_1\mathbf{b}_1 = -q_2\mathbf{b}_2 = (p_1 - p_2)^{-1}\mathbf{K}^{(1)}\mathbf{a}_R. \tag{8.14}$$

Hence

$$\mathbf{B} < p_*^2 > \mathbf{q} = \tilde{\Gamma}_I\mathbf{K}^{(1)}\mathbf{a}_R, \quad \mathbf{B} < p_*^{-1} > \mathbf{q} = -\tilde{\Gamma}_{II}^{-1}\mathbf{K}^{(1)}\mathbf{a}_R, \tag{8.15}$$

$$\tilde{\Gamma}_I = p_1 + p_2, \quad \tilde{\Gamma}_{II} = p_1 p_2. \tag{8.16}$$

For $n = 2$ and -1, (8.8) with the use of (8.15) gives

$$[\tilde{\Gamma}_I\mathbf{K}^{(1)} - \mathbf{K}^{(2)}]\mathbf{a}_R = \mathbf{0}, \quad [\tilde{\Gamma}_{II}\mathbf{K}^{(-1)} + \mathbf{K}^{(1)}]\mathbf{a}_R = \mathbf{0}. \tag{8.17}$$

Currie (1979) presented (8.17) from (8.11) by arguing that it can be formally deduced by setting $p_3 = 0$ and restricting $\mathbf{K}^{(n)}$ to 2×2 matrices. With the statement "it follows from (8.17) that", he presented without a proof,

$$[\mathbf{K}^{(2)}(\mathbf{K}^{(1)})^{-1}\mathbf{K}^{(-1)} - \mathbf{K}^{(-1)}(\mathbf{K}^{(1)})^{-1}\mathbf{K}^{(2)}]\mathbf{a}_R = \mathbf{0}. \tag{8.18}$$

Equation (8.18) is obtained when $(8.17)_1$ multiplied by $[\tilde{\Gamma}_{II}\mathbf{K}^{(-1)}(\mathbf{K}^{(1)})^{-1} + \mathbf{I}]$ is added to (8.17)2 multiplied by $[\mathbf{K}^{(2)}(\mathbf{K}^{(1)})^{-1} - \tilde{\Gamma}_I\mathbf{I}]$. The matrix inside the brackets in (8.18) is a 2×2 skew-symmetric matrix so that the two off-diagonal elements are equal but opposite in sign. The vanishing of the offdiagonal elements is the secular equation

$$[\mathbf{K}^{(2)}(\mathbf{K}^{(1)})^{-1}\mathbf{K}^{(-1)} - \mathbf{K}^{(-1)}(\mathbf{K}^{(1)})^{-1}\mathbf{K}^{(2)}]_{12} = 0. \tag{8.19}$$

Currie (1979) expressed the left hand side of (8.19) in the form of a 3×3 determinant. He did not present an explicit expression because he employed the elastic stiffnesses which are more complicated in computing $\mathbf{K}^{(1)} = \mathbf{N}_3 + X\mathbf{I}$. With the use of the reduced elastic compliance, (8.19) does recover the secular equation (7.10).

Although the secular equation (8.19) by Currie (1979) is not as explicit as that of Destrade (2001) and Ting (2002c), he is the first one who obtained an explicit secular equation for monoclinic materials with the symmetry plane at $x_3 = 0$. Unaware of the work by Currie (1979), Destrade (2001) derived the secular equation for monoclinic materials with the symmetry plane at $x_3 = 0$ employing the method of first integrals to be discussed in the next section. It is interesting to note that Currie and Destrade are, respectively, the first and the last Ph. D. students of Professor Michael Hayes.

For a general anisotropic material Currie (1979) derived from (8.11) a secular equation in (5.12) of his paper. He discovered later that it was an identity, and presented a revised secular equation in (12) of (Taylor and Currie, 1981). However, their secular equation involves the axial vectors of five skew-symmetric matrices. Since the axial vector of a skew-symmetric matrix is unique up to an arbitrary multiplicative parameter, it is an open question if their secular equation is valid. The same comment applies to their secular equations for monoclinic materials with the symmetry plane at $x_1 = 0$ or $x_2 = 0$. There is also an inconsistency with the secular equation derived by Taziev (1989). In (8.11), Currie needs only four $\mathbf{K}^{(n)}$, $n = 1, 2, -1, -2$. As shown below, Taziev needs five matrices $\mathbf{K}^{(n)}$, which can be $n = 1, 2, 3, 4, 5$ or $n = 1, 2, 3, -1, -2$.

Taziev (1989) employed a different approach. Let

$$\mathbf{a}_R^T = [\, 1, \ \alpha, \ \beta \,] \tag{8.20}$$

where α and β are complex. Equation (8.9) gives

$$\begin{gathered} K_{11}^{(n)} + K_{22}^{(n)}\alpha\bar{\alpha} + K_{33}^{(n)}\beta\bar{\beta} + K_{12}^{(n)}(\alpha + \bar{\alpha}) + \\ + K_{13}^{(n)}(\beta + \bar{\beta}) + K_{23}^{(n)}(\alpha\bar{\beta} + \beta\bar{\alpha}) = 0. \end{gathered} \tag{8.21}$$

Setting $n = 1, 2, 3, 4, 5$, (8.21) provides five equations that can be solved for, say,

$$\alpha + \bar{\alpha} = f_1, \quad \alpha\bar{\alpha} = f_2, \quad \beta + \bar{\beta} = f_3, \quad \beta\bar{\beta} = f_4, \quad \alpha\bar{\beta} + \beta\bar{\alpha} = f_5. \tag{8.22}$$

When α, $\bar{\alpha}$,β, $\bar{\beta}$ are computed from the first four equations, the fifth of (8.22) is

$$f_1^2 f_4 + f_3^2 f_2 + f_5^2 - f_1 f_3 f_5 - 4 f_2 f_4 = 0. \tag{8.23}$$

This is the secular equation obtained by Taziev (1989).

The computation of the matrices $\mathbf{K}^{(n)}$ becomes more complicated for the higher orders $n = 4$ and 5. Instead of choosing $n = 1, 2, 3, 4, 5$, we could choose $n = 1, 2, 3, -1, -2$. This would provide a *new* secular equation (Ting, 2003).

The approach employed by Taziev (1989) has in fact been used independently by Ting (1996, pp. 472 and 482) for orthotropic materials. In this case, the matrices $\mathbf{K}^{(n)}$ for $n = 1$ and -1 are 2×2 diagonal matrices so that (8.21) for $n = 1$ and -1 are

$$K_{11}^{(1)} + K_{22}^{(1)}\alpha\bar{\alpha} = 0, \quad K_{11}^{(-1)} + K_{22}^{(-1)}\alpha\bar{\alpha} = 0. \tag{8.24}$$

Elimination of $\alpha\bar{\alpha}$ yields

$$K_{11}^{(1)} K_{22}^{(-1)} - K_{22}^{(1)} K_{11}^{(-1)} = 0. \tag{8.25}$$

This leads to the secular equation in (7.11). Taziev (1989) wrote (8.23) as a product of three factors when the material is orthotropic. The vanishing of one of the three factors is the secular equation. He presented it in terms of the elastic stiffnesses.

9. The Method of First Integrals

The method of first integrals has been employed in other branches of mathematical physics, but Mozhaev (1995) appears to be the first one to use it for surface waves.

With the solution for the displacement $\mathbf{u}$ given in $(3.9)_1$ the differential equation (2.2) takes the form

$$(\mathbf{Q} - X\mathbf{I})\mathbf{u} - i(\mathbf{R} + \mathbf{R}^T)\mathbf{u}' - \mathbf{T}\mathbf{u}'' = \mathbf{0}, \tag{9.1}$$

where the prime denotes differentiation with (kx_2). Indeed, substitution of $(3.9)_1$ into (9.1) yields the eigenrelation (3.2). The vanishing of the surface traction at $x_2 = 0$ means that, inserting $(3.6)_1$ into (3.11),

$$\mathbf{R}^T\mathbf{u}_R - i\mathbf{T}\mathbf{u}'_R = \mathbf{0}, \tag{9.2}$$

where $\mathbf{u}_R$ is the displacement at the free surface $x_2 = 0$. When (9.1) is post-multiplied by $i\bar{\mathbf{u}}^T$ and $\bar{\mathbf{u}}'^T$, and integrated from $(kx_2) = 0$ to ∞, we have after taking the real parts,

$$(\mathbf{Q} - X\mathbf{I})(\mathbf{u}, i\bar{\mathbf{u}}) + (\mathbf{R} + \mathbf{R}^T)(\mathbf{u}', \bar{\mathbf{u}}) - \mathbf{T}(\mathbf{u}'', i\bar{\mathbf{u}}) = \mathbf{0}, \tag{9.3a}$$

$$(\mathbf{Q} - X\mathbf{I})(\mathbf{u}, \bar{\mathbf{u}}') - (\mathbf{R} + \mathbf{R}^T)(i\mathbf{u}', \bar{\mathbf{u}}') - \mathbf{T}(\mathbf{u}'', \bar{\mathbf{u}}') = \mathbf{0}, \tag{9.3b}$$

where

$$(\mathbf{g}, \mathbf{h}) = \int_0^\infty (\mathbf{g}\mathbf{h}^T + \bar{\mathbf{g}}\bar{\mathbf{h}}^T) d(kx_2) \tag{9.4}$$

is a 3×3 matrix. It is easily shown that the first integrals $(\mathbf{u}, i\bar{\mathbf{u}})$ and $(i\mathbf{u}', \bar{\mathbf{u}}')$ are skew-symmetric and $(\mathbf{u}', \bar{\mathbf{u}}) = (\mathbf{u}, \bar{\mathbf{u}}')^T$. Upon integration by parts we obtain

$$(\mathbf{u}, \bar{\mathbf{u}}') = -2\mathbf{W} - (\mathbf{u}, \bar{\mathbf{u}}')^T, \quad 2\mathbf{W} = \mathbf{u}_R\bar{\mathbf{u}}_R^T + \bar{\mathbf{u}}_R\mathbf{u}_R^T. \tag{9.5}$$

Hence $(\mathbf{u}, \bar{\mathbf{u}}')$ can be represented by a skew-symmetric matrix minus the 3×3 symmetric matrix $\mathbf{W}$. Equation (9.2) can be written as, using (4.3),

$$\mathbf{u}'_R = i\mathbf{N}_1\mathbf{u}_R, \tag{9.6}$$

so that

$$-i(\mathbf{u}'_R\bar{\mathbf{u}}_R^T - \bar{\mathbf{u}}'_R\mathbf{u}_R^T) = 2\mathbf{N}_1\mathbf{W}, \quad \tfrac{1}{2}(\mathbf{u}'_R\bar{\mathbf{u}}'^T_R + \bar{\mathbf{u}}'_R\mathbf{u}'^T_R) = \mathbf{N}_1\mathbf{W}\mathbf{N}_1^T. \tag{9.7}$$

Computation of $(\mathbf{u}'', i\bar{\mathbf{u}})$ shows that it is $-(i\mathbf{u}', \bar{\mathbf{u}}')$ plus the matrix shown on the left of $(9.7)_1$. As to the first integral $(\mathbf{u}'', \bar{\mathbf{u}}')$ it is a skew symmetric matrix minus the symmetric matrix shown on the left of $(9.7)_2$. The net result is that the six first integrals in (9.3) are represented by four 3×3 skewsymmetric matrices and the symmetric matrix $\mathbf{W}$. Thus there are a total of 18 unknowns in the six first integrals. Equation (9.3) consists of 18 scalar equations which can be re-written as a product of 18×18 matrix and an 18×1 column matrix whose elements are the 18 unknowns. The vanishing of the determinant of the 18×18 matrix is the secular equation obtained by Mozhaev (1995).

Using $(4.3)_2$ and observing that the trace of the product of a symmetric matrix and a skew-symmetric matrix vanishes, it can be shown that the trace of (9.3a) vanishes (Destrade, 2002). Hence three of the nine equations in (9.3a) are linearly dependent. This means that the 18×18 determinant vanishes identically so that the secular equation obtained by Mozhaev (1995) is a trivial identity. However, in the beginning of his paper he derived the secular equation for orthotropic materials using the first integrals, and obtained the correct secular equation. It shows that while the derivation for a general anisotropic material may not lead to a secular equation, the derivation for a special anisotropic material can lead to a correct secular equation.

Instead of employing the differential equation (9.1) for $\mathbf{u}$, we could use the differential equation for the stress function vector $\boldsymbol{\phi}$. The differential equation for $\boldsymbol{\phi}$ can be inferred from (4.7) as

$$\hat{\mathbf{Q}}\boldsymbol{\phi} - i(\hat{\mathbf{R}} + \hat{\mathbf{R}}^T)\boldsymbol{\phi}' - \hat{\mathbf{T}}\boldsymbol{\phi}'' = \mathbf{0}. \tag{9.8}$$

Following the derivation of (9.3) from (9.1) we have

$$\hat{\mathbf{Q}}(\boldsymbol{\phi}, i\bar{\boldsymbol{\phi}}) + (\hat{\mathbf{R}} + \hat{\mathbf{R}}^T)(\boldsymbol{\phi}', \bar{\boldsymbol{\phi}}) - \hat{\mathbf{T}}(\boldsymbol{\phi}'', i\bar{\boldsymbol{\phi}}) = \mathbf{0}, \tag{9.9a}$$

$$\hat{\mathbf{Q}}(\boldsymbol{\phi}, \bar{\boldsymbol{\phi}}') - (\hat{\mathbf{R}} + \hat{\mathbf{R}}^T)(i\boldsymbol{\phi}', \bar{\boldsymbol{\phi}}') - \hat{\mathbf{T}}(\boldsymbol{\phi}'', \bar{\boldsymbol{\phi}}') = \mathbf{0}. \tag{9.9b}$$

The first integrals in (9.9) are simpler to evaluate because $\boldsymbol{\phi} = 0$ at $x_2 = 0$. Again, the six first integrals in (9.9) are represented by four 3×3 skew-symmetric matrices and a symmetric matrix $\hat{\mathbf{W}}$,

$$2\hat{\mathbf{W}} = \boldsymbol{\phi}'_R \bar{\boldsymbol{\phi}}'{}_R{}^T + \bar{\boldsymbol{\phi}}'{}_R \boldsymbol{\phi}'{}_R{}^T. \tag{9.10}$$

Thus (9.9) can also be written as a product of 18×18 matrix and an 18×1 column matrix. Since the matrix $\hat{\mathbf{W}}$ here appears only in $(\boldsymbol{\phi}'', \bar{\boldsymbol{\phi}}')$ in (9.9b), (9.9a) represents nine scalar equations in nine unknowns of the three skewsymmetric matrices. However, the trace of (9.9a) vanishes so that the 18×18 determinant vanishes identically. Again, it fails to provide a secular equation for a general anisotropic material. Nevertheless, as in the case of first integrals for the displacement, the derivation can provide a secular equation for monoclinic materials with the symmetry plane at $x_3 = 0$.

For monoclinic materials with the symmetry plane at $x_3 = 0$, the three first integrals in (9.9a) are 2×2 skew-symmetric matrices so that each one of them contains only one unknown. Equation (9.9a) consists of four scalar equations but two of them are identical. The three independent equations can be written as a product of a 3×3 matrix and a 3×1 column matrix whose elements are the three unknowns. The vanishing of the determinant of the 3×3 matrix is the secular equation derived by Destrade (2001).

10. Monoclinic Materials with the Symmetry Plane at $x_3 = 0$

The explicit secular equation for monoclinic materials with the symmetry plane at $x_3 = 0$ is shown in (7.10). When the material is orthotropic, $s'_{16} = s'_{26} = 0$ so that (7.10) reduces to (7.11). It should be noted that, in reducing (7.10) to (7.11), s'_{26} need not vanish. Thus the secular equation for monoclinic materials with the symmetry plane at $x_3 = 0$ is identical to that for orthotropic materials when $s'_{16} = 0$.

The following are special cases for which (7.10) or (7.11) has an exact expression for the root X.

Case (i). When

$$s'_{12} = 0 \text{ or } C(1,6|2,6) = 0, \tag{10.1}$$

(7.10) is a product of $(1 - s'_{11}X)^2$ and a quadratic equation in X. The smaller root of the quadratic equation is

$$X^{-1} = \tfrac{1}{2}(s'_{11} + s'_{66}) + \tfrac{1}{2}\sqrt{(s'_{11} - s'_{66})^2 + 4s'_{11}s'_{22} + 4s'_{16}(s'_{16} - 2s'_{26})}. \tag{10.2}$$

In the special case of isotropic materials, $s'_{12} = 0$ means that Poisson ratio $\nu = 0$, and (10.2) reduces to (2.7).

Case (ii). When

$$s'_{12} = s'_{16} = 0 \text{ or } C_{12} = C_{16} = 0, \tag{10.3}$$

(10.2) simplifies to

$$X^{-1} = \tfrac{1}{2}(s'_{11} + s'_{66}) + \tfrac{1}{2}\sqrt{(s'_{11} - s'_{66})^2 + 4s'_{11}s'_{22}}. \tag{10.4}$$

Case (iii). If

$$s'_{12} = 0 \text{ and } s'_{16} = 2s'_{26}, \tag{10.5a}$$

or

$$C(1,6|2,6) = 0 \text{ and } C_{16}C(2,6) = 2C_{26}C(1,6), \tag{10.5b}$$

(10.2) also reduces to (10.4).

Case (iv). Let

$$s'_{16} = s'_{26} \text{ and } s'_{12}s'_{66} + s'(1,2) = {s'_{16}}^2. \qquad (10.6)$$

The secular equation (7.10) is a product of $[(1 - s'_{11}X)^2 - (s'_{12}X)^2]$ and a quadratic equation. The vanishing of the quadratic equation gives

$$X^{-1} = \tfrac{1}{2}(s'_{11} - s'_{12} + s'_{66}) + \tfrac{1}{2}\sqrt{(s'_{11} - s'_{12} - s'_{66})^2 - 4s'_{12}s'_{66}}. \qquad (10.7)$$

When the material is isotropic, (10.6) leads to $\nu = 1/4$ and (10.7) recovers (2.6). It should be noted that s'_{16} and s'_{26} do not appear in (10.4) and (10.7). Hence the exact roots in (10.4) and (10.7) apply to orthotropic materials as well.

Case (v). An exact expression of X was obtained by Mozhaev (1995) who assumed that the material is orthotropic and $C_{12} = C_{66}$. The condition $C_{12} = C_{66}$ means

$$s'_{12}s'_{66} + s'(1,2) = 0, \qquad (10.8)$$

which is a special case of (10.6) when $s'_{16} = s'_{26} = 0$. Since (10.7) does not involve s'_{16} and s'_{26}, it remains valid here. We rewrite (10.7) using (10.8) as

$$X^{-1} = \tfrac{1}{2}(s'_{11} - s'_{12} + s'_{66}) + \tfrac{1}{2}\sqrt{(s'_{11} - s'_{12} - s'_{66})^2 + 4s'(1,2)}. \qquad (10.9)$$

Condition (10.8) implies that $s'_{12} < 0$ so that (10.9) is not valid when $s'_{12} = 0$. However, when $s'_{12} = 0$, (10.9) recovers (10.4) which is valid for $s'_{12} = 0$. This is a paradox.

When the material is incompressible (Destrade et al., 2002),

$$s'_{1\alpha} + s'_{2\alpha} = 0, \quad \alpha = 1, 2, 4, 5, 6. \qquad (10.10)$$

The secular equation (7.10) reduces to

$$1 - (1 - s'_{11}X)[(4s'_{11} + s'_{66}) + s'(1,6)(3 - 2s'_{11}X)X]X = 0. \qquad (10.11)$$

This has a simpler expression than the ones in (Destrade et al., 2002; Ting, 2002c). It should be noted that the five special cases discussed above do not apply to incompressible materials. This is so because, when $s'_{12} = 0$, (10.10) demands that $s'_{11} = s'_{22} = 0$. It violates the positive semi-definiteness of the strain energy density (Destrade et al., 2002). The condition $(10.6)_1$ and (10.10) for $\alpha = 6$ imply that $s'_{16} = s'_{26} = 0$. Thus (10.6) reduces to (10.8) for incompressible materials. However, it is easily seen from (10.10) that $s'(1,2) = 0$ so that (10.8) demands $s'_{12} = 0$ or $s'_{66} = 0$. In either case the positive semi-definiteness of the strain energy density is violated.

11. Concluding Remarks

In order to find the displacement and stress, a complete solution for surface waves requires not only the computation of the surface wave speed X, but also the eigenvalues p and the eigenvectors $\mathbf{a}$ and $\mathbf{b}$. Explicit expressions of the eigenvectors $\mathbf{a}$ and $\mathbf{b}$ in terms of p and X have been presented in (Ting, 2002c). What remained is to find p and X.

In the special case of monoclinic materials with the symmetry plane at $x_3 = 0$, the algebraic equations for p and X are both quartic equations. It is shown in (Ting, 2002a) that there is no need to solve the quartic equation in p. When X satisfies the secular equation, the eigenvalues p can be found explicitly in terms of X. For the special monoclinic materials with the symmetry plane at $x_3 = 0$ discussed in Section 10, the exact expression of X is available. Therefore, for these materials, exact explicit solutions to the surface waves can be obtained. They will be useful as test examples for a numerical scheme.

We saw that, in some cases, the derivation of the secular equation for a general anisotropic material leads to a trivial identity, but the same derivation for a special material does yield a correct secular equation. This shows that the derivation of a secular equation for a special material has to be considered separately from that for a general anisotropic material.

Although Head (1979a,b) has shown that the sextic equation for the eigenvalues p is insoluble analytically in the sense of Galois, it does not preclude the possibility of finding an explicit secular equation as demonstrated by several secular equations in this paper. Thus an explicit expression of p is not necessary for obtaining an explicit secular equation. Another example is the quartic equation in p for monoclinic materials with the symmetry plane at $x_3 = 0$. Although, in theory, the roots p of the quartic equation can be obtained in radicals, the expression would be too complicated. Nevertheless, in conjunction with the quartic secular equation in X, a simple analytical solution of p in terms of X was given in (Ting, 2002a). Hence, at least for monoclinic materials with the symmetry plane at $x_3 = 0$, only one quartic equation in X needs to be solved.

References

Alshits, V. I. and Lothe, J. (1978) On surface waves in hexagonal crystals, *Crystallographiya* **23**, 901.

Barnett, D. M. and Chadwick, P. (1990) The existence of one-component surface waves and exceptional subsequent transonic states of types 2, 4, and E1 in anisotropic elastic media, in J. J. Wu, J. J., T. C. T. Ting and D. M. Barnett (Eds.), *Modern Theory of Anisotropic Elasticity and Applications*, SIAM, Philadelphia, SIAM Proceedings Series, pp.199-214.

Barnett, D. M. and Lothe, J. (1973) Synthesis of the sextic and the integral formalism for dislocations, Greens function and surface wave (Rayleigh wave) solutions in anisotropic elastic solids, *Phys. Norv.* **7**, 13-19.

Barnett, D. M. and Lothe, J. (1985) Free surface (Rayleigh) waves in anisotropic elastic half-spaces: The surface impedance methods, *Proc. Roy. S. Lond.* A**402**, 135- 152.

Chadwick, P. and Smith, G. D. (1977) Foundations of the theory of surface waves in anisotropic elastic materials, *Adv. Appl. Mech.* **17**, 307-376.

Chadwick, P. and Smith, G. D. (1982) Surface waves in cubic elastic materials, in H.G. Hopkins and M. J. Sewell (eds.), *The Rodney Hill 60th Anniversary Volume*, Pergamon, Oxford, 47-100.

Chadwick, P. and Wilson, N. J. (1992) The behaviour of elastic surface waves polarized in a plane of material symmetry III. Orthorhombic and cubic media, *Proc.R. Soc.Lond.*, A **438**, 225-247.

Currie, P. K. (1979) The secular equation for Rayleigh waves on elastic crystal, *Q. J. Appl. Math. Mech.* **32**, 163-173.

Destrade, M. (2001) The explicit secular equation for surface waves in monoclinic elastic crystals, *J. Acous. S. America* **109**, 1398-1402.

Destrade, M. (2002) Private communications.

Destrade, M., Martin, P. A. and Ting, T. C. T. (2002) The incompressible limit in linear anisotropic elasticity, with applications to surface waves and elastostatics, *J. Mech. Phys. Solids* **50**, 1453-1468.

Dongye, Changsong and Ting, T. C. T. (1989) Explicit expressions of Barnett-Lothe tensors and their associated tensors for orthotropic materials, *Q. Appl. Math.* **47**, 723-734.

Eringen, A. C. and Suhubi, E. S. (1975) *Elastodynamics: Linear Theory, Vol. 2,* Academic Press, New York.

Gilinskii, I. A. and Shcheglov, I. M. (1981) Generalized elastic Rayleigh waves in crystals, *Sov. Phys. Crystallogr.* **26**(2), 143-147.

Head, A. K. (1979a) The Galois unsolvability of the sextic equation of anisotropic elasticity, *J. Elasticity* **9**, 9-20.

Head, A. K. (1979b) The monodromic Galois groups of the sextic equation of anisotropic elasticity, *J. Elasticity* **9**, 321-324.

Ingebrigtsen, K. A. and Tonning, A. (1969) Elastic surface waves in crystal, *Phys. Rev.* **184**, 942-951.

Lamb, H. (1904) Propagation of tremors over the surface of an elastic solid, *Phil. Trans.* **A203**, 1-42.

Mozhaev, V. G. (1995) Some new ideas in the theory of surface acoustic waves in anisotropic media, in D. F. Parker and A. H., England (eds.), *IUTAM Symposium on Anisotropy, Inhomogeneity and Nonlinearity in Solid Mechanics,* Kluwer Academic Pub, Dordrecht, The Netherlands, 455-462.

Rahman, M. and Barber, J. R. (1995) Exact expressions for the roots of the secular equation for Rayleigh waves, *J. Appl. Mech.* **62**, 250-252.

Rayleigh, Lord (1885) On waves propagated along the plane surface of an elastic solid, *Proc. London Math. Soc.* **17**, 4-11.

Stoneley, R. (1949) The seismological implications of aeolotropy in continental structure, *Monthly Notices, Roy. Astronomical S., Geohpys. Suppl.* **5**, 343-353.

Stoneley, R. (1955) The propagation of surface elastic waves in a cubic crystal, *Proc. Roy. S. Lond.* **A232**, 447-458.

Stoneley, R. (1963) The propagation of surface waves in an elastic medium with orthorhombic symmetry, *Geophys. J., Roy. Astronomical S.* **8**, 176-186.

Stroh, A. N. (1958) Dislocations and cracks in anisotropic elasticity, *Phil. Mag.* **3**, 625-646.
Stroh, A. N. (1962) Steady state problems in anisotropic elasticity, *J. Math. Phys.* **41**, 77-103.
Sveklo, V. A. (1948) Plane waves and Rayleigh waves in anisotropic media, *Doklady Akademii Nauk SSSR,* **59**, 871-874.
Synge, J. L. (1956) Elastic waves in anisotorpic media, *J. Math. Phys.* **35**, 323-334.
Taylor, D. B. (1981) Surface waves in anisotropic media: the secular equation and its numerical solution, *Proc. Roy. S. Lond.* **A376**, 265-300.
Taylor, D. B. and Currie, P. K. (1981) The secular equation for Rayleigh waves on elastic crystals II: Corrections and additions, *Q. J. Mech. Appl. Math.* **34**, 231-234.
Taziev, R. M. (1989) Dispersion relation for acoustic waves in an anisotropic elastic half-space, *Sov. Phys. Acous.* **35**(5), 535-538.
Ting, T. C. T. (1988) Some identities and the structure of $\mathbf{N}_i$ in the Stroh formalism of anisotropic elasticity, *Q. Appl. Math.* **46**, 109-120.
Ting, T. C. T. (1996) *Anisotropic Elasticity: Theory and Applications,* Oxford University Press, New York.
Ting, T. C. T. (1997) Surface waves in anisotropic elastic materials for which the matrix $N(v)$ is extraordinary degenerate, degenerate, or semisimple, *Proc. Roy. S. Lond.* **A453**, 449-472.
Ting, T. C. T. (2002a) Explicit secular equations for surface waves in monoclinic materials with the symmetry plane at $x_1 = 0, x_2 = 0$ or $x_3 = 0$, *Proc. Roy. S. London,* **A458**, 1017-1031.
Ting, T. C. T. (2002b) An explicit secular equation for surface waves in an elastic material of general anisotropy, *Q. J. Mech. Appl. Math.* **55**, 297-311.
Ting, T. C. T. (2002c) A unified formalism for elastostatics or steady state motion of compressible or incompressible anisotropic elastic materials, The J. D. Achenbach Symposium volume, *Int. J. solids Structures,* **39**/21-22, 5427-5445.
Ting, T. C. T. (2003) The polarization vector and secular equation for surface waves in an anisotropic elastic half-space, The Bruno Boley symposium volume, *Int. J. Solids Strcutures*, in press.
Ting, T. C. T. and Barnett, D. M. (1997) Classifications of surface waves in anisotropic elastic materials, *Wave Motion*, **26**, 207-218.
Wang, Y. M. and Ting, T. C. T. (1997) The Stroh formalism for anisotropic materials that possess an almost extraordinary degenerate matrix **N**, *Int. J. Solids Structures*, **34**, 401-413.

Part II

Bending and edge waves in plates and shells

“NONGEOMETRICAL PHENOMENA” IN PROPAGATION OF ELASTIC SURFACE WAVES

V. M. BABICH and A. P. KISELEV
St.Petersburg Department of the
Steklov Mathematical Institute
Fontanka 27, St.Petersburg 191011, Russia

Abstract. A review of some theoretical research on surface waves in elastic structures done by the St.Petersburg team during 40 years is presented. By “nongeometrical phenomena” we mean those which can not be simply understood on the basis of plane wave theory but are quantitatively described by sophisticated versions of the ray method. We discuss the following phenomena: the so-called Berry phase in propagation of Rayleigh waves over a curved surface of inhomogeneous body, anomalies of polarization (as compared to the corresponding plane waves), conversion of Love and Rayleigh modes and propagation of narrow beams of surface waves. Ray theory which generalises plane wave solutions to waves with non-plane fronts is the basic tool.

Key words: nonlinear surface waves, elasticity, Rayleigh waves, scale invariance, nonlocality, spectral kernel, spatial kernel, Hilbert transform

1. Introduction

We present a review of some research on the theory of surface waves in elastic structures done during 40 years by the St.Petersburg team. The team included V. M. Babich, N. Ya. Rusakova (Kirpichnikova), P. V. Krauklis, B. A. Chikhachev, T. B. Yanovskaya, V. E. Nomofilov, A. P. Kiselev, V. V. Lukyanov, M. V. Perel and A. V. Aref’ev, and other researchers from St.Petersburg Department of the Steklov Mathematical Institute and St.Petersburg State University.

By “nongeometrical phenomena” in wave propagation we mean those which cannot be easily understood on the basis of plane wave theory but are quantitatively described by application of a more or less sophisticated version of the ray method. This nomenclature is borrowed from Babich and Kiselev [1]. The phenomenon we start with is the “nongeometrical” phase known now as the Berry phase. It appeared first in a short note by Babich

R.V. Goldstein and G.A. Maugin (eds.),
Surface Waves in Anisotropic and Laminated Bodies and Defects Detection, 119–129.

[2] and then in an extended presentation by Babich and Rusakova [3] as a result of an asymptotic calculation when the ray method was applied to propagation of Rayleigh waves over a surface of 3D inhomogeneous elastic body. This theory was generalised to Love and Rayleigh waves in a laterally inhomogeneous layered elastic structures and later to piezoelectric bodies. Gaussian beams of surface waves were constructed. Love and Rayleigh waves in laterally inhomogeneous layered structures were shown to have anomalous polarizations (Love-type component for Rayleigh waves and vice versa) described by higher-order ray theory. We discuss also strong interaction of Love and Rayleigh modes in slightly inhomogeneous and slightly anisotropic layered structures.

All the considerations are based on versions of the ray theory. This review does not cover whispering galleries (see e. g. Babich, Krauklis and Molotkov [4] and Krauklis and Tsepelev [5] and references therein) which require a different approach.

2. Plane body waves and the ray method

2.1. EQUATIONS OF LINEAR ELASTODYNAMICS

We consider linear theory of propagation of surface waves in elastic bodies where a time-harmonic complex displacement $\mathbf{u}(\mathbf{r})e^{-i\omega t}$ is described by

$$\frac{\partial \sigma_{km}}{\partial x_k} + \rho(\mathbf{r})\omega^2 u_m = 0, \quad m = 1, 2, 3. \tag{2.1}$$

Here ω is the angular frequency, $\mathbf{r}$ is the position vector characterised by its Cartesian coordinates $x_1 \equiv x,\ x_2 \equiv y,\ x_3 \equiv z,\ \rho$ is the volume density and $\sigma_{km} = \sigma_{km}(\mathbf{u})$ is the linear stress tensor. In the case of isotropic medium

$$\sigma_{km} = \rho v_S^2 \left(\frac{\partial u_k}{\partial x_m} + \frac{\partial u_m}{\partial x_k} \right) + \rho(v_P^2 - 2v_S^2)\delta_{km} \nabla \cdot \mathbf{u}, \tag{2.2}$$

with the $v_P = v_P(\mathbf{r})$ and $v_S = v_S(\mathbf{r})$ velocities of P and S waves, respectively, $0 < v_S < v_P$.

In homogeneous medium where v_P, v_S and ρ are independent on $\mathbf{r}$ the best known solutions of (2.1) are plane waves.

2.2. PLANE WAVES

To be precise, we start with homogeneous plane waves which are solutions of the electrodynamic equations (2.1) of the form

$$\mathbf{u} = e^{i\omega\tau(\mathbf{r})}\mathbf{U} \quad \text{with} \quad \tau = \tau_1 x + \tau_2 y + \tau_3 z, \tag{2.3}$$

where the amplitude vector $\mathbf{U}$ is constant, i. e. independent $\mathbf{r}$, and τ_j, $j = 1, 2, 3$ are also constants. The eikonal τ necessarily satisfies the eikonal equation

$$(\nabla\tau)^2 = \frac{1}{v^2} \quad \text{where either} \quad v = v_P \quad \text{or} \quad v = v_S. \tag{2.4}$$

The direction of propagation is parallel to $\nabla\tau$. We will use notations $\mathbf{U}^{P,S}$ and $\tau_{P,S}$ to distinguish P and S waves. Polarization, or direction of displacement in P waves is longitudinal and in S waves it is transverse, $\mathbf{U}^P \| \nabla\tau_P$ and $\mathbf{U}^S \perp \nabla\tau_S$.

2.3. RAY METHOD

Analysis shows that the notion of wave is an asymptotic generalisation of plane waves. In the end of 1950-s it was extended to waves with non-plane wave fronts in homogeneous medium and to the case of inhomogeneous medium. We discuss here only the time-harmonic case. The small parameters in the theory describing non-plane waves are

$$\frac{\omega R_{1,2}}{c} \gg 1, \tag{2.5}$$

where $R_{1,2}$ are the main curvatures of the wave front and c is the wave speed. In case of smoothly inhomogeneous media we assume in addition that

$$\frac{|\nabla\nu|}{\nu\omega c} \ll 1, \tag{2.6}$$

with ν standing for any of material parameters v_S, v_S and ρ. Further ω will be considered as a formal large parameter and we write "$\omega \to \infty$" in the above sense. Perturbation theory with respect to large ω which starts from plane waves is known as ray theory or ray method. Historical comments on the ray theory for body waves can be found e. g. in Babich and Kiselev [1].

Displacement vector describing a time-harmonic wave is presented by

$$\mathbf{u} = \exp(i\omega\tau)\left[\mathbf{U}^0 + \frac{\mathbf{U}^1}{-i\omega} + ...\right], \qquad \text{"}\omega \to \infty\text{"}, \tag{2.7}$$

with non-constant amplitudes $\mathbf{U}^j = \mathbf{U}^j(\mathbf{r})$ and the eikonal $\tau = \tau(\mathbf{r})$ not necessarily linear in x, y and z. As well known

i) τ satisfies the eikonal equation (2.4),

ii) the amplitude $\mathbf{U}^0(\mathbf{r})$ is ruled by the energy balance equation, and

iii) the zero-order polarization described by the direction of $\mathbf{U}^0(\mathbf{r})$ is "locally plane" i. e. it is the same as that of the corresponding plane wave at $\mathbf{r}$.

Higher order amplitudes starting from $\mathbf{U}^1(\mathbf{r})$ normally have anomalously polarized or "additional" components. In particular, P wave has a transverse component, see e. g. Babich and Kiselev [1].

Rays appear in the course of solution of the eikonal equation (2.4). They are curves along which travel times are minimal. We set

$$\tau_{P,S} = \int \frac{ds}{v_{P,S}}, \tag{2.8}$$

where the integrals are taken along the corresponding rays and ds stands for the infinitesimal arc length. The expressions (2.8) are travel times of P and S waves, respectively.

Coordinates the most convenient for description of propagation of each wave are orthogonal ray coordinates $(\tau, \alpha_1, \alpha_2)$ where (α_1, α_2) characterise orthogonal frames on the surfaces $\tau = const$ which are wave fronts.

3. Rayleigh waves on curved surface and the Berry phase

Classical Rayleigh wave in a traction-free homogeneous elastic half space $-\infty < x < \infty$, $-\infty < y < \infty$, $0 < z < \infty$ is a solution of (2.1) such that it decreases with depth z, and the faster the larger is ω. We will describe generalisation the Rayleigh wave to the case of non-plane surface C of inhomogeneous elastic body presented by Babich and Rusakova (Kirpichnikova) [3].

3.1. PLANE WAVES WITH COMPLEX EIKONALS AND THE CLASSICAL RAYLEIGH WAVE

For definiteness we consider propagation along the x-axis taking $\tau_2 = 0$ in (2.3). Let $\tau_1 = 1/c$, $c > 0$. In the case of $c < v_S$ τ_3 is purely imaginary. We assume that $Im\ \tau_3 > 0$. The eikonals for P and S waves become

$$\tau_P = \frac{x}{c} + iq_P z, \quad \tau_S = \frac{x}{c} + iq_S z \quad \text{with} \quad q_{P,S} = \sqrt{\frac{1}{c^2} - \frac{1}{v_{P,S}^2}} > 0. \tag{3.9}$$

Such solutions propagate along the x-axis with the velocity c and exponentially decay as $z \to \infty$, or equivalently, $\omega \to \infty$.

Rayleigh introduced a sum of such solutions with the same c,

$$\mathbf{u} = e^{i\omega\tau_P}\mathbf{U}^P + e^{i\omega\tau_S}\mathbf{U}^S = e^{i\omega x/c}\left[e^{-\omega q_P z}\mathbf{U}^P + e^{-\omega q_S z}\mathbf{U}^S\right]. \tag{3.10}$$

Inserting (3.10) into the traction-free surface condition at the plane $z = 0$,

$$\sigma_{3k}\big|_{z=0} = 0, \quad k = 1, 2, 3, \tag{3.11}$$

yielded the unique value for the velocity c, $0 < c < v_S$, and gave also a relation between the amplitude vectors $\mathbf{U}^P$ and $\mathbf{U}^S$. The result can be presented in the form

$$\mathbf{u} = \chi \left[l_P \nabla \tau_P e^{i\omega\tau_P} + l_S \mathbf{m} e^{i\omega\tau_S} \right], \tag{3.12}$$

with $l_P = 1/v_S^2 - 2/c^2$ and $l_S = 2i\sqrt{1/c^2 - 1/v_P^2}\Big/ c$, $\mathbf{m}$ is a complex vector in the plane defined by the normal to the free surface and $\nabla\tau_S$ and such that $\mathbf{m} \cdot \nabla\tau_S \equiv m_k \partial\tau_S/\partial x_k = 0$ and $\mathbf{m} \cdot \mathbf{m} = 1/v_S^2$, and χ is an arbitrary constant factor. This is the classical surface Rayleigh wave.

It is important to note that the phase function $\tau = x/c$ in (3.10) satisfies the "surface" eikonal equation with respect to the lateral variables

$$(\nabla_C \tau)^2 = 1/c^2, \tag{3.13}$$

where $(\nabla_C \tau)^2 = (\partial\tau/\partial x)^2 + (\partial\tau/\partial y)^2$.

3.2. RAYLEIGH WAVE ON A CURVED SURFACE OF INHOMOGENEOUS ELASTIC BODY

Now we consider an inhomogeneous elastic body with a curved surface C and apply a kind of the ray method. It is natural to assume that in vicinity of each point $\mathbf{M}$ on C the solution is similar to the classical Rayleigh solution propagating with the "local" Rayleigh velocity $c(\mathbf{M})$. We expect that it propagates along surface rays which are surface curves minimising travel time for the velocity c.

We introduce the ray coordinates (τ, α) on C in such a way that α is constant along each ray and $\tau = const$ is the equation of a surface wave front. Similarly to (3.13) we have $(\nabla_C \tau)^2 = 1/c^2(\mathbf{M})$ where ∇_C is now the "surface" gradient on C. Convenient coordinates inside the body are (τ, α, n), where $n > 0$ is the inner normal to C.

We seek the expression for the surface wave as a sum of two complex ray solutions, for P and S waves of which the eikonals $\tau_P(\mathbf{r})$ and $\tau_S(\mathbf{r})$ are real on C and coincide there with the travel time of the Rayleigh wave:

$$(\nabla\tau_{P,S}(\mathbf{r}))^2 = \frac{1}{v_{P,S}^2(\mathbf{r})}, \qquad \tau_{P,S}\big|_{\mathbf{r}\in C} = \tau. \tag{3.14}$$

The derivatives of $\frac{\partial\tau_P}{\partial n}$ and $\frac{\partial\tau_S}{\partial n}$ will be purely imaginary. We take their imaginary parts positive to obtain wave fields decaying with depth. Near the surface approximate solutions of (3.14) are

$$\tau_{P,S} = \tau + i q_{P,S} n \tag{3.15}$$

which reduces to (3.9) in the case of a homogeneous half space.

The expression for the wave field in the form of a ray analog of (3.10),

$$\mathbf{u} = e^{i\omega\tau_P}\left[\mathbf{U}^{P0} + \frac{\mathbf{U}^{P1}}{-i\omega} + ...\right] + e^{i\omega\tau_S}\left[\mathbf{U}^{S0} + \frac{\mathbf{U}^{S1}}{-i\omega} + ...\right], \tag{3.16}$$

must be inserted now into the free surface conditions $\sigma_{km}n_m = 0$ where $\mathbf{n}$ is a normal to C.

Terribly long calculation described by Babich and Rusakova (Kirpichnikova) [3] established the expected energy balance along surface rays. In the zero-order approximation the energy was conserved in a narrow "tube" $\alpha_0 \leq \alpha \leq \alpha + d\alpha$, $0 \leq n \leq n_0$ where n_0 is a small positive number. This provided the dependencies of $\mathbf{U}^{P0}$ and $\mathbf{U}^{S0}$ on the geometry of the surface and on the material parameters.

Calculation demonstrated also a quite unexpected phenomenon known now as the Berry phase. The resulting formula for the zero-order displacement was of the form (3.12) where now $\chi = \chi(\tau, \alpha)$ was found as a solution of an ordinary differential equation along rays. It had the structure

$$\chi = \chi^0\sqrt{\frac{j^0}{j}}\exp\left(\int_0^\tau B_1 d\tau\right)\exp\left(i\int_0^\tau\left[\frac{B_2 b_{\tau\tau}}{c^2} + \frac{B_3 b_{\alpha\alpha}}{j^2} + B_4\right]d\tau\right) \tag{3.17}$$

Here $j = j(\alpha, \tau)$ was the geometrical spreading of surface rays[1], $\chi^0 = \chi^0(\alpha) = \chi(\alpha, 0)$ was an arbitrary smooth function describing distribution of the intensity along the initial wave front line $\tau = 0$, $j^0 = j(\alpha, 0)$, $b_{\tau\tau}$ and $b_{\alpha\alpha}$ were the coefficients of the Gauss second quadratic form of C. Explicit expressions for B_j in the general case are extraordinarily long and their detailed calculation by Babich [6], Krauklis [7] and Babich and Kirpichnikova [8] have taken around 30 years. Here we just describe their structures.

B_1, B_2, B_3 and B_4 are all real. B_1 is a linear combination of $\partial v_{P,S}/\partial\tau$, $\partial c/\partial\tau$, and $\partial\rho/\partial\tau$ where the coefficients are algebraic functions of v_P, v_S, c and ρ. The integral $\int B_1 d\tau$ can be calculated in explicit form which gives

$$\exp\left(\int_0^\tau B_1 d\tau\right) = \sqrt{\frac{d(0,\alpha)}{d(\tau,\alpha)}}, \tag{3.18}$$

with

$$d(\tau,\alpha) = \frac{\rho}{c\sqrt{1/c^2 - 1/v_S^2}}\left[\frac{6}{c^2}\left(\frac{1}{v_S^2} - \frac{1}{v_P^2}\right) + \frac{4}{v_P^2 v_S^2} - \frac{6}{v_S^4} + \frac{c^2}{v_S^6}\right], \tag{3.19}$$

[1] Geometrical spreading of surface rays can be defined as follows. Let $\mathbf{r} = \mathbf{r}(\tau, \alpha)$ be the equation of the surface C. Then $j = |\partial\mathbf{r}/\partial\alpha|$.

$d > 0$. Further, B_2 and B_3 are algebraic functions of v_P, v_S and c, and B_4 is linear function of $\partial v_P/\partial n$, $\partial v_S/\partial n$ and $\partial \rho/\partial n$ of which the coefficients are algebraic in v_P, v_S, c and ρ. Nowadays, in the wake of the paper by Berry [9] the integral in the last exponent in (3.17) is known the Berry phase.

One can not find the above formulas in Karal and Keller [10].

3.3. RELATED TOPICS

Gregory [11] considered later simple particular cases of propagation of Rayleigh waves over a circular cylinder, circular cylindric bore and a sphere. Interesting discussion of physical implementations of the theory was given by Mozhaev [12]. The theory was generalised to a curved surface of inhomogeneous arbitrarily anisotropic elastic body by Nomofilov [13, 14] and Mozhaev [12] and to Stoneley waves by Krauklis [7] and Nomofilov [15].

4. Waves in layered structures with weak lateral inhomogeneity

4.1. LATERALLY HOMOGENEOUS STRUCTURE

We consider a traction-free isotropic half space $z > 0$. It is laterally homogeneous when $v_P = v_P(z)$, $v_S = v_S(z)$ and $\rho = \rho(z)$. We assume that a solution of (1) satisfying (3.11) exists in the form of a generalised plane wave

$$\mathbf{u} = e^{i\omega x/c}\left[u(z)\mathbf{e}_1 + v(z)\mathbf{e}_2 + w(z)\mathbf{e}_3\right], \tag{4.20}$$

where $\mathbf{e}_j$ are unit Cartesian coordinate vectors, the phase velocity c is real and the displacement $\mathbf{u}$ decays as $z \to \infty$. Smooth dependencies of material parameters on z and also welded or slipping contacts for some values of depth $z = z_1, \ldots, z = z_N$ are allowed.

The problem splits into a scalar problem for $v(z)$ with $u \equiv w \equiv 0$ for Love modes and a vector problem for $u(z)$ and $w(z)$ with $v \equiv 0$ for Rayleigh modes.

4.2. STRUCTURES WITH WEAK LATERAL INHOMOGENEITY

In case of material parameters slowly depending on x and y a kind of ray theory was developed by Woodhouse [16] and by Babich, Chikhachev and Yanovskaya [17] to describe propagation of modes of the both types. "Topography" was allowed that was smooth variation of depths of layers. Expressions for zero-order amplitudes were found and their relationship with the energy balance were discussed. Here Berry phases do not arise.

Later this theory was generalised to piezoelectric (Babich and Lukyanov [18]) and to viscoelastic (Aref'ev and Kiselev [19, 20]) almost layered structures.

4.3. STRONGLY CURVED LAYERED STRUCTURES

When the whole structure is strongly curved we are coming to a generalisation of the theory discussed in the Section 3. This was done first by Babich and Chikhachev [21] who considered both Love and Rayleigh modes. They presented expressions for the Berry phase factors. These results were generalised to piezoelectricity by Babich and Lukyanov [22] who employed a general approach to such problems developed by Tromp and Dahlen [23].

4.4. POLARIZATION ANOMALIES

First-order corrections provided by the ray theory for amplitudes of either body or surface waves describe "anomalous polarization" or "additional" components. It means that Love modes have also "Rayleigh-type" polarized displacement and vice versa.

Simple expression for the "additional" component of a Rayleigh wave with a non-plane wave front in a laterally homogeneous structure was given by Kirpichnikova and Kiselev [24]. The anomalous polarization in this case is the effect of a non-homogeneous distribution of its amplitude. We consider laterally homogeneous traction-free half space with arbitrary depth dependencies $v_P = v_P(z)$, $v_S = v_S(z)$, $\rho = \rho(z)$, possibly with welded or slipping horizontal contacts . Let c be the phase velocity of one of Rayleigh plane wave modes and $u(z)$ and $w(z)$ be its longitudinal and vertical displacements, respectively, see (4.20). The most simple is the case of a wave with a circular wave front when $\tau = r/c$ where $r = \sqrt{x^2 + y^2}$ is the polar radius. We denote the directivity of this wave by $f(\alpha)$ where $\alpha = \arcsin(y/r)$ stands for the polar angle. The zero-order ray theory easily describes the standard Rayleigh-type polarization i. e. it gives displacements along $\mathbf{e}_3$ and $\mathbf{e}_r$, where $\mathbf{e}_r$ is the unit vector along the polar radius. We present an asymptotic expression for the wave field which includes also the Love-type displacement along $\mathbf{e}_\alpha$, where $\mathbf{e}_\alpha$ is the unit coordinate vector of α. It is

$$\mathbf{u}^0 \approx \frac{e^{i\omega r/c}}{\sqrt{r}} \left\{ [(u(z)\mathbf{e}_r + w(z)\mathbf{e}_3)]\, f(\alpha) - \frac{cu(z)}{-i\omega r} \frac{\partial f(\alpha)}{\partial \alpha} \mathbf{e}_\alpha \right\}, \quad \frac{\omega r}{c} \gg 1. \quad (4.21)$$

This was derived considering the first-order correction for the standard ray formulas.

Further discussion of "additional" components of Love and Rayleigh waves with curved wave fronts in half space and in elastic structures with weak lateral dependencies can be found in Kirpichnikova and Kiselev [24,

25, 26]. Effects of Volterra–Maxwell–Boltzmann viscoelasticity on polarization anomalies were described by Aref'ev and Kiselev [19, 20].

No discussion of polarization anomalies for strongly curved structures is available.

4.5. CONVERSION OF MODES

When the material in a layered structure is both smoothly laterally inhomogeneous and slightly anisotropic a strong interaction between Love and Rayleigh modes can occur. This effect was asymptotically described in Kiselev and Perel [27] under the assumption that the phase curves of these two modes are not tangent. Their approach employs a kind of a boundary-layer technique (see Babič and Kirpičnikova [28] among others) which naturally arises from the ray method. The theory of conversion is in agreement with the concept presented e. g. by Babich and Kiselev [1] who claimed that the ray method always prompts an approach to description of wave fields inside small boundary-layers where its standard form fails.

It is found that when anisotropy dominates over lateral inhomogeneity the mode conversion is almost complete. In the opposite case it follows that the conversion is small.

5. Narrow beams

Ray theory of Rayleigh waves allows description of such exotic wave fields as "Gaussian beams". The procedure can be started with taking the surface eikonal τ in a form of a complex function with non-negative imaginary part and such that $Im\ \tau = 0$ only at a fixed surface ray. The solution would have the standard Rayleigh–type decay with depth and a Gaussian–type decay with respect to the lateral distance from that fixed surface ray. Derivation of Gaussian beams of Rayleigh waves on the surface of a smoothly inhomogeneous body by Kirpichnikova [29] as well as its generalisation to Love and Rayleigh waves in an almost planar layered structure, by Yomogida [30] available at the moment are based on a more complicated "parabolic-equation techniques".

Gaussian-beam solution were not extended yet to strongly curved structures.

6. Conclusion

We have demonstrated that ray theory of surface waves which is a perturbation of plane-wave solutions provides sufficiently new and unexpected effects. We mentioned some yet unsolved related problems in the hope of stimulating further research.

7. Acknowledgments

Preparation of this paper was supported by RFBR grants 02-01-00260 and 00-01-00485.

References

1. Babich, V.M. and Kiselev, A.P. (1989) Non-geometrical waves – are there any ? An asymptotic description of some "nongeometrical" phenomena in seismic wave propagation, *Geophys. J. Intern.* **99**, 412–420.
2. Babich, V.M. (1961) Propagation of Rayleigh waves along the surface of a homogeneous elastic body of arbitrary shape, *Doklady Akad. Nauk SSSR* **137**, 1263–1266 (in Russian).
3. Babich, V.M. and Rusakova (Kirpichnikopva), N.Ya. (1963) The propagation of Rayleigh waves over the surface of a non-homogeneous elastic body with an arbitrary form, *U.S.S.R. Comput. Math. Math. Phys.* **2**, 719–735.
4. Babich, V.M., Krauklis, V.P. and Molotkov, L.A. (1984) Dynamical problems of geoacustics, *Sov. Phys. Acoustics* **30**, 409–410.
5. Krauklis, P.V. and Tsepelev, N.V. (1987) On question of distant propagation of P_n and S_n waves in the oceanic lithosphere, *Izv. Acad. Sci. USSR. Phys. Solid Earth* **23**, 217–220.
6. Babich, V.M. (1967) Conservation of energy in the propagation of nonstationary waves, *Vestnik Leningrad. Univ. Ser. Mat. Mech.* **22**(7), 38–42 (in Russian).
7. Krauklis, P.V. (1971) Estimation of the intensity of Rayleigh and Stonely surface waves on an inhomogeneous path, *Seminars in Mathematics in Steklov Math. Institute* **15**, 63–66.
8. Babich, V.M. and Kirpichnikova, N.Ya. (1990) On the question of Rayleigh waves propagating along the surface of an inhomogeneous elastic body, *J. Sov. Math.* **50**, 1693–1695.
9. Berry, M.V. (1984) Quantal factors accompanying adiabatic changes, *Proc. Roy. Soc. London*, **392**(1802), 45–57.
10. Karal, F.G. and Keller, J.B. (1964) Geometrical theory of surface-wave excitation and propagation, *J. Acoust. Soc. Amer.* **36**, 32–40.
11. Gregory, R.D. (1971) The propagation of Rayleigh waves over curved surfaces at high frequency, *Proc. Camb. Phil. Soc.* **70**, 103–121.
12. Mozhaev, V.G. (1984) Application of the perturbation method for calculating the characteristics of surface-waves in anisotropic and isotropic solids with curved boundaries, *Sov. Phys. Acoustics* **30**, 394–400.
13. Nomofilov, V.E. (1979) Quasistationary Rayleigh waves on the surface of an inhomogeneous, anisotropic elastic body, *Sov. Phys. Doklady* **24**, 609–611.
14. Nomofilov, V.E. (1982) Propagation of quasi-stationary Rayleigh waves in a nonuniform anisotropic elastic medium, *J. Sov. Math.* **19**, 1466–1475.
15. Nomofilov, V.E. (1982) Quasistationary Stoneley waves, *J. Sov. Math.* **20**, 1860–1869.
16. Woodhouse, J.H. (1974) Surface waves in laterally varying structure, *Geophys.J. Roy. Astr. Soc.* **37**, 461–490.
17. Babich, V.M., Chikhachev, B.A. and Yanovskaya, T.B. (1976) Surface waves in a vertically inhomogeneous elastic half-space with a weak horizontal inhomogeneity, *Izv. Ac. Sci. USSR. Phys. Solid Earth* **12**, 242–245.

18. Babich, V.M. and Lukyanov, V.V. (1997) Propagation of waves in a piezoelectric layer with a weak horizontal inhomogeneity, *Wave Motion* **26**, 379–397.
19. Aref'ev, A.V. and Kiselev, A.P. (1998) Viscoeleastic Rayleigh waves in a layered structure with weak lateral inhomogeneity, *J. Math. Sci.* **91**, 2701–2705.
20. Aref'ev, A.V. and Kiselev, A.P. (1999) Viscoeleastic Love waves in a layered structure with weak lateral inhomogeneity, *J. Math. Sci.* **96**, 3289–3291.
21. Babič, V. M. and Čihačev B. A. (1980) Propagation of Love and Rayleigh waves in a weakly nonhomogeneous laminated medium, *Vestnik Leningrad. Univ. Ser. Mat.* **30**, 32–38 [Russian original: (1975) *Vestn. LGU* **8**, 13–21].
22. Babich, V.M. and Lukyanov, V.V. (1998) Wave propagation along a curved piezoelastic layer, *Wave Motion* **28**, 1–11.
23. Tromp, J. and Dahlen, F.A. (1992) The Berry phase of a slowly varying waveguide, *Proc. Roy. Soc. London* **A 437**, 329–342.
24. Kirpichnikova, N.Ya and Kiselev, A.P. (1991) Depolarization of elastic surface waves in a vertically inhomogeneous half-space, *Sov. Phys. Acoustics* **36**, 173–175.
25. Kirpichnikova, N.Ya and Kiselev, A.P. (1990) Polarization spectral anomalies of surface waves in an almost layered elastic half-space, *J. Sov. Math.* **50**, 2001–2011.
26. Kirpichnikova, N.Ya and Kiselev, A.P. (1995) On the anomalous polarization of Love and Rayleigh elastic waves in a layered medium, *J. Math. Sci.* **73**, 383–384.
27. Kiselev, A.P. and Perel, M.V. (1998) Asymptotic theory of a local degeneration of Love and Rayleigh modes, *Day on Diffraction'98, Proceedings*, St.Petersburg University Press, 56–59.
28. Babič, V.M. and Kirpičnikova, N.Ya. (1979) *Boundary Layer Method in Diffraction Problems*, Springer–Verlag, Berlin.
29. Kirpichnikova, N.Ya. (1969) Rayleigh waves concentrated near a ray on the surface of an inhomogeneous elastic body, *Seminars in Mathematics in Steklov Math. Institute* **15**, 49–62.
30. Yomogida, K. (1985) Gaussian beams for surface waves in laterally slowly-varying media, *Geophys. J. Roy. Astr. Soc.* **82**, 511–533.

COMPLEX RAYS AND INTERNAL DIFFRACTION AT THE CUSP EDGE

M. DESCHAMPS and O. PONCELET
Laboratoire de Mechanique Physique, University Bordeaux 1, UMR CNRS 5469
351, Cours de la Libration, 33405 Talence Cedex, FRANCE

Abstract. It is well known that wave surfaces for anisotropic solids exhibit cuspidal behaviours in certain directions and that for isotropic solids these surfaces reduce to spheres. The wave surface is a locus of real rays which are normal to the slowness surface. Each ray is associated with a single homogeneous plane wave (or three ones for the cusp area), which is the only real solution (or three) for which the energy velocity is well oriented along the ray direction. For inhomogeneous plane waves, there exists an infinite number of solutions with a complex wavevector for a given ray direction. The damping vector (the imaginary part of the wavevector) is always normal to the energy direction and its amplitude can take any value. Although these complex rays exhibit correct directions of energy propagation they do not satisfy the Fermat's principle. However, some of them are very close to satisfying this principle. Basing on this remark, an equation can be obtained to calculate the associated energy velocity. This intrinsic equation only refers to complex Christoffel equation. Limiting the study to principal planes and plotting the associated complex wave surface, it can be shown that four energetic rays always exist in any direction for both isotropic and anisotropic media (either beyond or inside the cusp). In other words, it is always possible to define four closed wave surfaces. Depending on the angles of energy propagation, either four surfaces are associated to real rays (cusp area), or two surfaces are associated with two real rays and two others with two complex rays. These angular areas are continuous and the distinctions appear at the cuspidal edges of wave surfaces. These calculations can easily explain the physical phenomena classically observed in these particular areas.

Key words: nonlinear surface waves, elasticity, Rayleigh waves, scale invariance, nonlocality, spectral kernel, spatial kernel, Hilbert transform

1. Introduction

Extending beyond the cuspidal edges of the acoustic wave surfaces of anisotropic solids there are sharp but non-singular features in dynamic response functions. The arrival times of these features coincide with maxima in the

R.V. Goldstein and G.A. Maugin (eds.),
Surface Waves in Anisotropic and Laminated Bodies and Defects Detection, 131–142.

acoustic energy flux. These quasi-singular features show up prominently in point source/point receiver measurements, and have been commented on in a number of recent publications [1,2,3,6,9]. This is essentially a diffraction effect and it cannot be explained on the basis of true ray theory. In this paper, the solutions of Christoffel's equation are sought for inhomogeneous plane waves [4,5]. For a given direction of observation, there exists an infinite number of such waves that can propagate. Among these, some are very close to satisfying the Fermat's principle. Such rays contribute with a maximum of energy, as can be shown by calculating dynamic Green's functions, for example by the Cagniard-de Hoop method [8,10]. These waves can be associated with a complex ray, for which the physical sense will be explained below.

2. Evanescent plane wave: background

In this section, the basic equations for propagation of inhomogeneous plane waves in anisotropic solids are introduced. Let us consider the plane waves defined at any point in space located at $\mathbf{M}$ and time t, by the following acoustic field velocity [4]:

$$\mathbf{U} = a_0 \mathbf{P} \exp i\omega(t - (\mathbf{S} \cdot \mathbf{M})) \tag{1}$$

where the vectors $\mathbf{P}, \mathbf{S}$ and $\mathbf{M}$ are expressed in the fixed reference (X_1, X_2). In the above expression, a_0 is the complex amplitude, $\mathbf{P}$ represents the polarisation vector. The variable is the angular frequency. The complex vector $\mathbf{S}$ denotes the slowness vector (the so-called slowness bivector). For a lossless anisotropic medium, the stress tensor σ_{ij} is given by:

$$\sigma_{ij} = C_{ijkl}\varepsilon_{kl}, \tag{2}$$

where C_{ijkl} stand for the elastic stiffness constants and ε_{kl} is the strain tensor. The parameters s_i and u_i (for $i = 1...3$) denote, respectively, the slowness and the particle velocity components on the $\mathbf{X}_i$-axis. The propagation equation for this plane wave is given by Christoffel's equation

$$\Gamma_{ij} u_j = \varrho \delta_{ij} u_i \qquad \text{for} = 1\ldots3, \tag{3}$$

where $\Gamma_{ij} = C_{ijkl}s_k s_l$ represent the Christoffel matrix components (for $k, l = 1\ldots3$), ρ denotes the mass density and δ_{ij} is the Kronecker symbol.

An interesting equation can be obtained by multiplying Eq.(3) by the complex conjugate of the particle velocity component, u_j, see reference [7]. Hence follows the relation

$$\mathbf{S} \cdot \langle \mathbf{J} \rangle = 2\langle E_c \rangle \tag{4}$$

where $\langle \mathbf{J} \rangle = -\frac{1}{2}\boldsymbol{\sigma}\bar{\mathbf{U}}$ represents the time-averaged complex Poynting vector and $\langle \mathbf{E_c} \rangle = -\frac{1}{4}\rho\mathbf{U}\bar{\mathbf{U}}$ stands for the time-averaged kinetic energy density. Similarly, multiplying Eq. (2) by the complex conjugate of the strain tensor, $\bar{\varepsilon}_{kl}$, gives the expression,

$$\bar{\mathbf{S}} \cdot \langle \mathbf{J} \rangle = 2\langle E_p \rangle \tag{5}$$

where $\langle E_p \rangle = \frac{1}{4}\varepsilon_{ij}C_{ijkl}\bar{\varepsilon}_{kl}$ is the time-averaged strain energy density. It should be noted that both quantities $\langle E_p \rangle$ and $\langle E_c \rangle$ are real and that they are different for evanescent plane waves (similar relation has been obtained in reference [4]). Consequently, a sum of Eqs. (4) and (5) yields

$$\mathbf{S}' \cdot (\mathbf{c}'_e + i\mathbf{c}''_e) = 1 \tag{6}$$

and their difference leads to

$$\mathbf{S}'' \cdot (\tilde{\mathbf{c}}'_e + i\tilde{\mathbf{c}}''_e) = 1 \tag{7}$$

where the quantities with prime and double prime indicate, respectively, the real and the imaginary parts. In these expressions the different vectors represent velocities defined as:

$$\mathbf{c}'_e = \frac{\mathbf{J}'}{\langle E_c + E_p \rangle}, \quad \mathbf{c}''_e = \frac{\mathbf{J}'}{\langle E_c + E_p \rangle}, \quad \tilde{\mathbf{c}}'_e = \frac{\mathbf{J}'}{\langle E_c - E_p \rangle} \text{ and } \tilde{\mathbf{c}}''_e = \frac{\mathbf{J}''}{\langle E_c - E_p \rangle}$$

Among these, the vector $\mathbf{c}'_e$ is the energy velocity. By noting that, the vectors $\mathbf{c}'_e$ and $\mathbf{c}''_e$ are, respectively, collinear to the vectors $\tilde{\mathbf{c}}'_{\mathbf{e}}$ and $\tilde{\mathbf{c}}''_{\mathbf{e}}$, from Eqs. (6) and (7), it is immediate to obtain:

$$\mathbf{S}' \cdot \mathbf{c}'_e = 1, \tag{8}$$

$$\mathbf{S}' \cdot \mathbf{c}''_e = \mathbf{S}' \cdot \tilde{\mathbf{c}}''_e = 0, \tag{9}$$

$$\mathbf{S}'' \cdot \mathbf{c}'_e = \mathbf{S}'' \cdot \tilde{\mathbf{c}}'_e = 0, \tag{10}$$

and

$$\mathbf{S}'' \cdot \tilde{\mathbf{c}}'_e = 1, \tag{11}$$

Two important remarks may be drawn from Eqs. (8) and (9), which will be essentially used in the next section. First, the energy velocity $\mathbf{c}'_e$ is the phase velocity in the direction of the real Poynting vector $\langle \mathbf{J}' \rangle$. If this direction is given by the unit vector $\mathbf{n}$, Eq. (8) takes the form:

$$\mathbf{S}' \cdot \mathbf{n} = s'_n, \tag{12}$$

where $s'_n = |\mathbf{c}'_e|^{-1}$ is the real part of the slowness component on the $\mathbf{n}$-direction. Second, the damping vector given by $\mathbf{S}''$ is always normal to the energy direction given by $\mathbf{n}$, (see Eq. (10)). A more detailed discussion on energy equations for inhomogeneous plane waves may be found in reference [4].

3. Fermat principle

Let us assume now that the observation point is located at a distance r from the origin and that the vector $\mathbf{M}$ is oriented in the $\mathbf{n}$-direction. The question under consideration in this section is: what are the evanescent (or not) plane waves for which the energy velocity is oriented in the $\mathbf{n}$-direction? For an evanescent wave, the arrival time of the phase front at the point $\mathbf{M}$ is given by the relation $t = r\mathbf{S} \cdot \mathbf{n}$. This relation can rewritten as follows:

$$s'_n = \mathbf{S} \cdot \mathbf{n}, \tag{13}$$

where the slowness component on the $\mathbf{n}$-direction is $s'_n = t/r$, which is real since both t and r are real. Consequently, this component reduces to $s'_n = \mathbf{S}' \cdot \mathbf{n}$ and the damping slowness $\mathbf{S}''$ is normal to the $\mathbf{n}$-direction. In other words, by taking into account Eq. (10), this requires, if $\mathbf{S}'' \neq 0$, that the energy velocity is always oriented to the $\mathbf{n}$-direction. It should be noted that for homogeneous plane waves, i.e. $\mathbf{S}'' = 0$, the energy velocity and the observation direction are the same only for those slowness vectors, for which this unique direction is normal to the slowness curves, as is well known. This can be obtained, for example, by using the Fermat principle, which can be written in the general form as follows:

$$\frac{d\tau(\eta)}{d\eta} = 0, \tag{14}$$

where $\tau(\eta)$ represents the wave time arrival and η is some vector parameter that affects the propagation. In our problem, the function $\tau(\eta)$ can be replaced by s'_n and the parameter η can be chosen as the real wave vector $\mathbf{S}'$. Then, the Fermat principle gives

$$\frac{d\mathbf{S}'}{ds'_n} \to \infty. \tag{15}$$

For simplicity, let us now restrict the analysis to the propagation in the principal plane $(\mathbf{X}_1, \mathbf{X}_2)$ of a cubic crystal and to the two in-plane polarised waves. In this case, expressing s_2 from the relation

$$s'_n = s_1 \cos\theta + s_2 \sin\theta \tag{16}$$

deduced from Eq. (12) for the angle of observation θ, and inserting this expression into the Christoffel equation

$$(c_{11}s_1^2 + c_{66}s_2^2 - 1)(c_{66}s_1^2 + c_{22}s_2^2 - 1) - (c_{66} + c_{12})^2 s_1^2 s_2^2 = 0, \tag{17}$$

gives the polynomial form in terms of s_1, where s'_n plays the role of a time parameter. The coefficients of this polynomial depend on the parameter s'_n and on the elastic constants. Since both the elastic constants and s'_n are real, the solutions of this polynomial function are either real or complex conjugate values. As an illustration, Fig. 1 shows the variation of the solutions as the component s'_n varies for a given angle of observation θ.

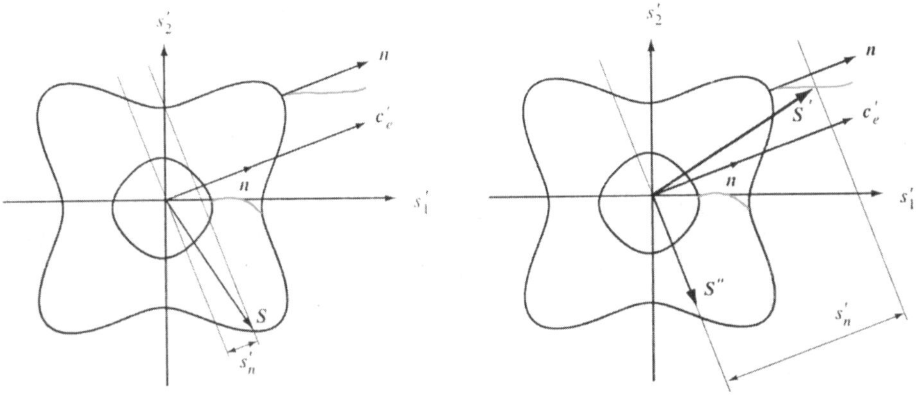

Figure 1. Nature of the solutions of the Christoffel equation for a fixed value s'_n. Real solutions (solid black line). Complex solutions (solid gray line) Left: four real solutions; right: four complex solutions

The real part of the slowness component s_2 is plotted as a function of the real part of the slowness component s_1. The real solutions are plotted in black while the complex solutions are displayed in gray. The stiffness constants are: $c_{11} = 110.6, c_{66} = 27.66$ and $c_{12} = 90$, which correspond to the anisotropy degree $\varepsilon = c_{11}(2c_{66} + c_{12})^{-1}$ of 0.761. The mass density is $\rho = 2.33$ kg/dm^3.

Let us first discuss Fig. 1. As the component s'_n changes, i.e. as the time changes, the line normal to the **n**-direction, the so-called Cagniard-de Hoop line, intercepts four, two or zero times the slowness curves. Obviously, this corresponds to the fact that four, two or zero solutions are real. Let us come back now to Eq. (15). These conditions are satisfied only for the homogenous plane waves for which the normal to the slowness curve is parallel to the **n**-direction. Singularities, i.e. infinite values at the wave arrival time, are then observed in the Green's function. This classically defines the wave surfaces of the ray theory and the true arrival waves (or true rays). Before or after these tangent points, the corresponding solution becomes complex and has an energy velocity vector always oriented along the **n**-direction, since the damping vector has been chosen normal to this direction. The associated rays are the so-called complex rays. Assuming that

Eq. (15) holds only for true arrival waves, for certain evanescent plane waves the derivative function given by this equation can still exhibits maxima instead of infinity for a specific value of s'_n. Wave arrival times can be then obtained by solving the following equation:

$$\frac{d^2\mathbf{S}'}{ds'^2_n} \tag{18}$$

This equation defines pseudo arrival time, in the sense that the Fermat principle is not satisfied. Keeping in mind that the energy directions of evanescent plane waves are necessarily following the **n**-direction, see Eq. (12), this is not inconvenient. As a matter of fact, on comparing the result with those obtained with the stationary phase method, Eq. (18) can be interpreted as constructive interferences. The essence of the stationary phase technique, for instance when approximating the acoustic field in the vicinity of a cusp, is to consider the interference of an infinite summation of homogeneous plane waves that propagate in different directions. The energy velocity of each wave is, of course, not on the right direction as well as the wave resulting from the summation. On the contrary, in our case, each evanescent plane wave is correctly oriented and the interferences take place in the time domain. From this point of view, in references [11], it can be found a comparison between the present approach, the quasi-stationary phase assumption, the solution for complex time of flight and experiments.

To emphasis our purposes, on the Figs. 2 to 4, the absolute value of the derivative function of the real slowness component on the $\mathbf{X}_2$-axis with respect to the slowness component on the **n**-direction are plotted versus the phase velocity along the **n**-direction. These functions are plotted for the two modes L and T. Only this component is plotted since it can be easily shown that both components exhibit the same maxima. Three cases are analysed which correspond to three typical examples. Note that the corresponding polar energy velocity curves will be presented in detail later on in Figs. 5-3 and 6-3.

First, both modes satisfy Eq. (18) for specific inhomogeneous plane waves, see Fig. 2. The anisotropy degree has been chosen such that $\varepsilon = 0,935$ and the angle of observation is $\theta = 1^o$. Clearly, in addition to the velocity of the true wave, which are about 7 and 3.6 $m/\mu s$, respectively, for the modes L and T, at an energy velocity of about 3.3 $m/\mu s$, a very large peak appears on the derivative function for both modes. Second, for the same anisotropy degree but for the angle of observation $\theta = 15^o$, only the mode L satisfy this equation and the cusp is not present due to the low anisotropy, see Figs. 3 and 5-3. In such a situation, the single peak is smoothed. Finally, for the anisotropic degree of $\varepsilon = 0.761$, again only the mode L satisfies Eq. (18) but for the angle of observation chosen the cusp exists, see Figs. 4 and 6-3.

It is of a great interest to plot the polar curves as the angle of observation changes. This is the purpose of the next section.

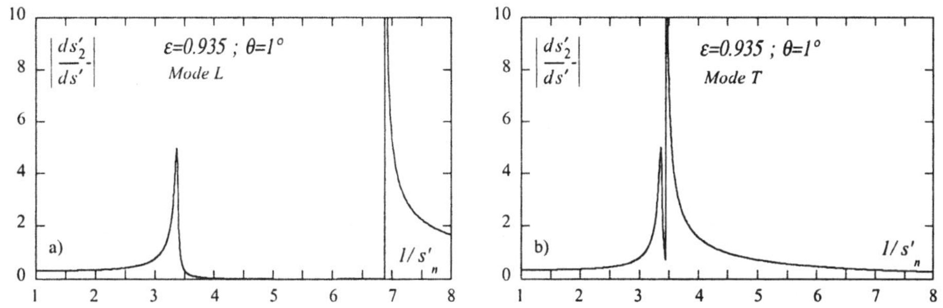

Figure 2. Absolute value of the derivative of the real slowness component on the $\mathbf{X}_2$-axis with respect the slowness component on the $\mathbf{n}$-direction versus the phase velocity along the $\mathbf{n}$-direction. anisotropic degree $\varepsilon = 0.935$ and angle of observation $\theta = 1^o$. a) Longitudinal mode; b) Transverse mode

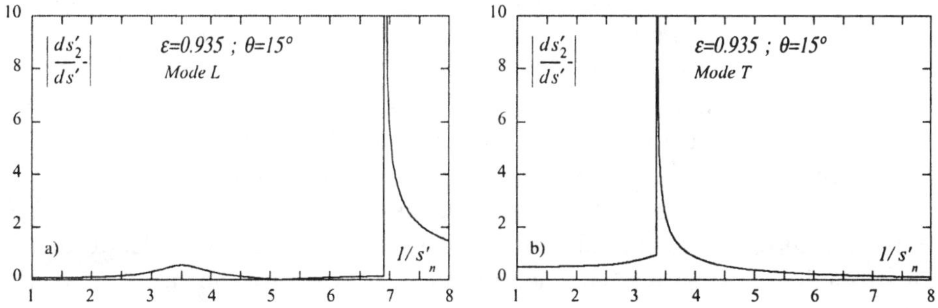

Figure 3. Absolute value of the derivative of the real slowness component on the $\mathbf{X}_2$-axis with respect the slowness component on the $\mathbf{n}$-direction versus the phase velocity along the $\mathbf{n}$-direction. anisotropic degree $\varepsilon = 0.935$ and angle of observation $\theta = 15^o$. a) Longitudinal mode; b) Transverse mode

4. Polar energy velocity curves

Numerical results concerning cubic materials with different degrees of anisotropy, are under considerations. This is to highlight physical phenomena arising as the anisotropy increases from isotopy up to high anisotropy, for which large cusps exist on the polar energy velocity curves. Let us calculate, for a fixed vector $\mathbf{n}$, the value of s'_n that satisfies Eq. (18).

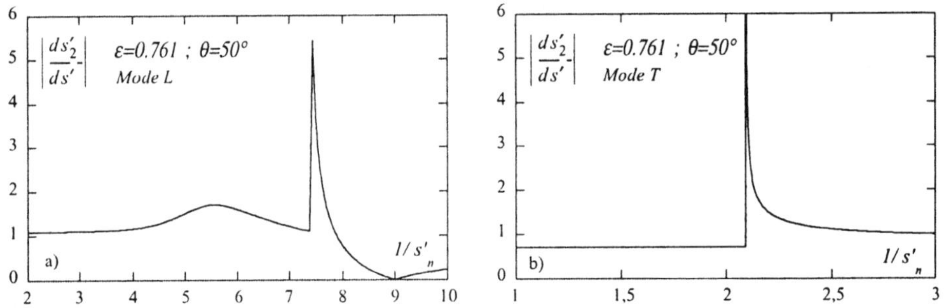

Figure 4. Absolute value of the derivative of the real slowness component on the $\mathbf{X}_2$-axis with respect the slowness component on the **n**-direction versus the phase velocity along the **n**-direction. anisotropic degree $\varepsilon = 0.761$ and angle of observation $\theta = 50^o$. a) Longitudinal mode; b) Transverse mode

It is easily obtained by using the numerical algorithm based on the Newton-Rapson method. Referring to Eq. (12), it is then immediate to calculate the energy velocity by the relation: $|\mathbf{c}_e'| = 1/s_n'$. As the direction of the vector **n** varies, the results are presented for various anisotropy degrees on Figs. 5 and 6. As the anisotropy degree varies, interesting phenomena show up in these figures, where the complex solutions are plotted (labelled L_e and T_e) in addition to the usual real solutions (labelled L and T). For simplicity, when, for a fixed value of ε, the two modes exhibit complex solutions (on Fig. 5 only), they are plotted on separate graphs, labelled a) and b), whereas the real solution remaining the same. Magnitude of the corresponding imaginary parts is graphically shown by the variation of the thickness of the gray line. The thicker it is, the greater the value of the damping vector is. For the examples given on Fig. 6, only the longitudinal mode has a complex solution.

Note first that, even for a weak anisotropy (see Fig. 5-1), there exist two complex solutions, which almost satisfy the Fermat's principle. When the anisotropy increases (see Figs. 5-2 and 5-3), the associated velocity of these two modes increases and the damping factor decreases. It should be noted, for the mode T_e and for specific angles of observation, the complex solution disappears, since the velocity cannot be greater than the velocity of the pure wave. In turn the velocity of the L_e mode becomes larger. Finally, when the cusp appears due to higher anisotropy, the mode T_e completely vanishes and the mode L_e takes place progressively in the continuity of the cusp, as clearly follows from comparing Figs. 6-1 and 6-2. Large anisotropy degrees have been chosen for the two last examples, see Figs. 6-3 and 6-4.

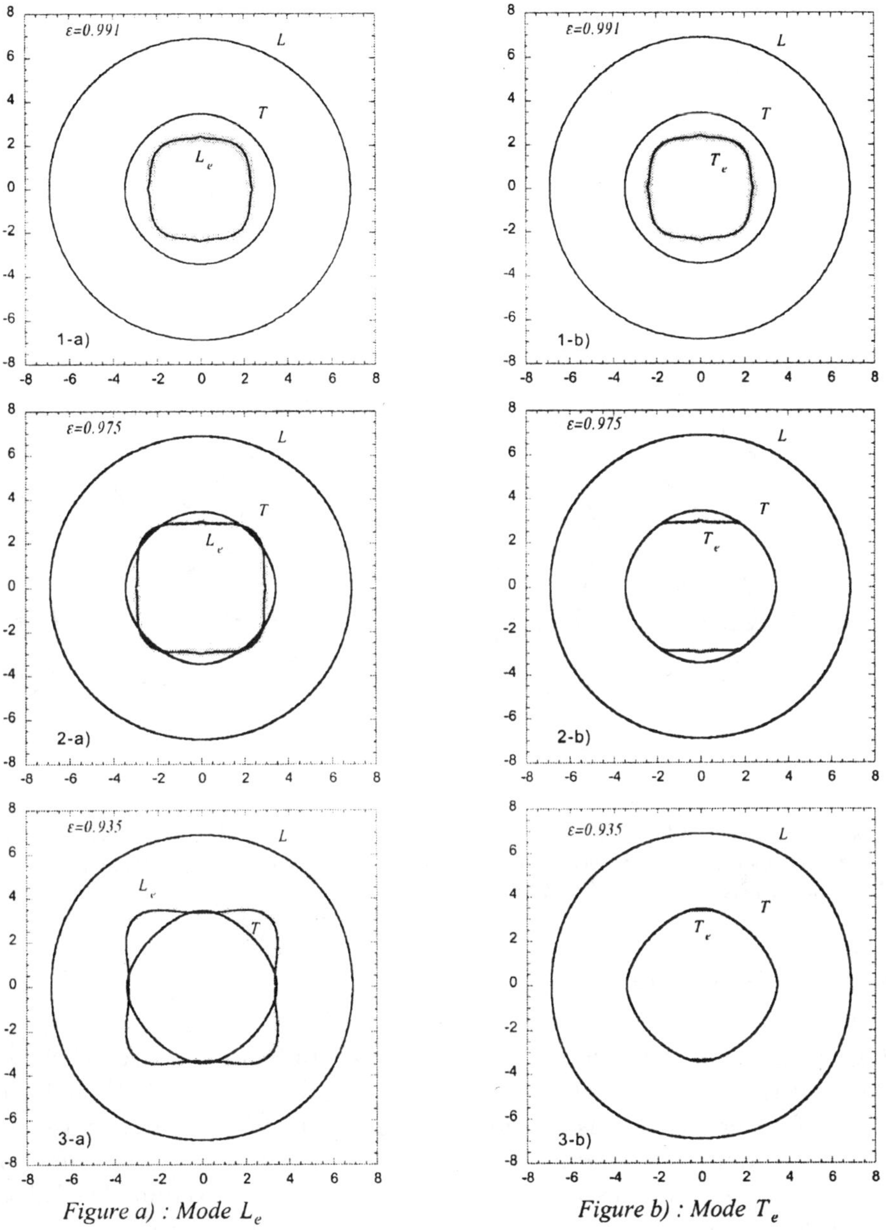

Figure 5. Group velocity $\mathbf{c}'_e$ in polar coordinates for various directions n and various anisotropy degrees. Real waves L and T; complex waves L_e and T_e; Both complex modes exist, the cusp is not present

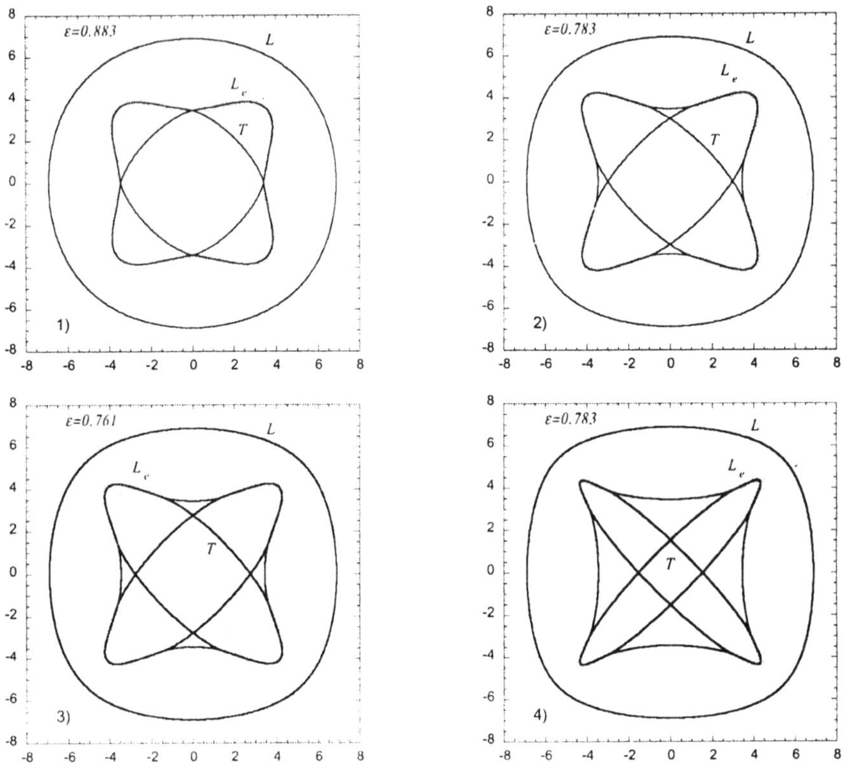

Figure 6. Group velocity $\mathbf{c}'_e$ in polar coordinates for various directions $\mathbf{n}$ and various anisotropy degrees. Real waves L and T; complex waves L_e; Only complex modes L_e exist, the cusp is present

5. Green's function.

Regarding the above results, the question naturally arises: is the new complex solution always observable? For answering this question, although with a loss of generality, let us consider the time response of a free space submitted to a delta force on the $\mathbf{X}_1$-direction. The associated Green function, in terms of the radial displacement, is calculated for a fixed angle of observation θ and at a distance $r = 1$ from the source. Using the Cagniard-de Hoop method, the waveform of the displacement along the direction θ is obtained and shown on Fig. 7, where the longitudinal and the transverse contributions are plotted, as well as the total field. The two anisotropy degrees, corresponding to Figs. 5-3 and 6-3, are analysed as they reveal to the most representative cases.

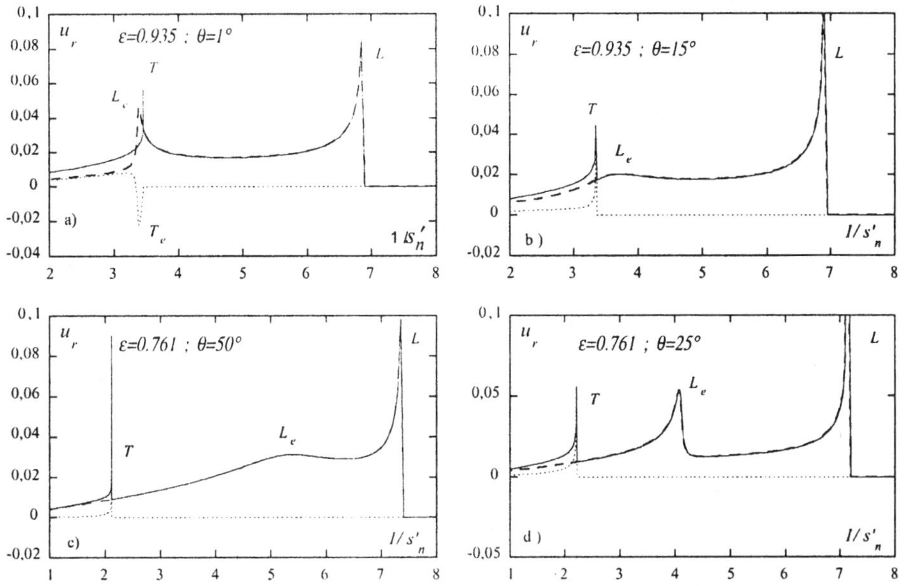

Figure 7. Response of a free space submitted to a delta force: radial displacement versus $1/s'_n$. Total contribution (—); Longitudinal contribution (- - -); Transverse contribution (···).

Two situations can be isolated. First, both the extra transverse T_e and longitudinal L_e modes exist. For $\theta = 1^o$ and $\varepsilon = 0.935$ this is the case, see Fig. 5-3. In such a case, as clearly seen in Fig. 7-a, although the waveform of the two partial contributions (dashed lines) presents a large peak associated with the extra solutions, they disappear on the total waveform (solid line). No wave arrival can be then observed. Second, only the extra longitudinal wave L_e is present. On keeping the same anisotropy degree and taking an angle of observation of $\theta = 15^o$, this condition is satisfied, see Fig. 5-3. The transversal partial mode does not exits before the true arrival time of the wave T, i.e., for a energy velocity approximately greater than 3.4 $m/\mu s$. Consequently, the extra longitudinal wave L_e cannot be cancelled since this is the only contribution at that time. A broad peak is then visible. Note that the broadening of the peak is directly connected to the imaginary part of the complex solutions plotted in Figs. 5 and 6. To emphasize this point, let us consider a larger anisotropy, i.e., $\varepsilon = 0.761$ and two angles of observation: $\theta = 50^o$ and $\theta = 25^o$, as shown in Figs. 7-c and 7-d.

6. Conclusion

This paper proposes a method to describe extension of the cuspidal edges of wave surfaces. This wave surface is deduced, without acoustic field calculation, by searching the solution of the Christoffel's equation, in terms of inhomogeneous plane waves, that almost satisfy the Fermat's principle. For homogeneous plane waves, it is necessary to satisfy the Fermat's principle exactly in order to have the Poynting vector oriented in the direction of observation. But for inhomogeneous plane wave, this is not necessary since this condition of energy orientation is taken into account by the fact that the damping vector is chosen normal to observation point.

References

1. Kim K. Y., Bretz K. C., Every A. G. and Sachse W. (1996), Ultrasonic imaging of the group velocity surface about the cubic axis in silicon, *J. Appl. Phys.* **79** (4), 1857-1863.
2. Hauser M. R., Weaver R. L. and Wolfe J. P. (1992), Internal diffraction of ultrasound in crystal: phonon focusing at long wavelength, *Phys. Rev. Lett.* **68**, 2604-2607.
3. Every A. G. (1981), Ballistic phonons and the shape of the ray surface in cubic crystals, *Phys. Rev.* **B 24**, 3456-3467.
4. Hayes M. (1980), Energy flux for trains of inhomogeneous pale waves, *Proc. R. Soc. Lond.* **A 370**, 417-429 .
5. Deschamps M. (1996), *Reflection and refraction of the inhomogeneous plane wave, Acoustic Interaction with Submerged Elastic Structures*, Part I, Edited by A. Guran, J. Ripoche and F. Ziegler, World Scientific Publishing Company, 164-206.
6. Corbel C., Guillois F., Royer D., Fink M. A. and De Mol R. (1993), Laser-generated elastic waves in carbon-epoxy composite, *IEEE Trans. Ultr. Fer. Freq. Contr.* **40**, 710-716.
7. Auld B.A. (1973), *Acoustic fields and waves in solids*, New York.
8. Mourad A., Deschamps M. and Castagnede B. (1996), Acoustic waves generated by an line impact in an anisotropic medium, *Acustica - Acta Acustica* **82**, 839-851.
9. Audoin B., Bescond C. and Deschamps M. (1996), Recovering of stiffness coefficients of anisotropic materials from point-like generation and detection of acoustic waves, *J. Appl. Phys.* **80**, 3760-3771.
10. Van Der Hijden J. H. M. T. (1987), *Propagation of transient elastic waves in stratified anisotropic media,* North-Holland series in Applied Mathematics and Mechanics, Elsevier Science Publishers, Amsterdam, vol. **32**.
11. Poncelet O., Deschamps M., Every A.G. and Audoin B. (2000), Extension to cuspidal edges of wave surfaces of anisotropic solids: treatment of near cusp behavior, *Review in Quantitative Nondestructive Evaluation* Vol. **20A**, ed. by D.O. Thompson and D.E. Chimenti, 51-58.

EDGE WAVES IN THE FLUID BENEATH AN ELASTIC SHEET WITH LINEAR NONHOMOGENEITY

R. V. GOLDSTEIN AND A. V. MARCHENKO
Institute for Problems in Mechanics
Prospect Vernadskogo, 101 Moscow 119526 Russia

Key words: nonlinear surface waves, elasticity, Rayleigh waves, scale invariance, nonlocality, spectral kernel, spatial kernel, Hilbert transform

Introduction

Edge waves are inherent to nonhomogeneous media with open wave-guides. The term open wave-guide means that perturbations from a surrounding medium can penetrate in to wave-guide region. At the same time there are perturbations such that their energy is localized in the wave-guide region and amplitude is rapidly decaying away the wave-guide. There are many examples of open wave-guides in different fields of physics. Faraday (1832) in laboratory experiments observed capillary-gravity waves propagating along submerged edge of a vertical plate, which amplitudes are decaying away the plate. The waves were forced by the influence of vibrations of the plate. Stokes considered gravity waves on the surface of the layer of ideal fluid with inclined bottom and found the solution of linearized equations of hydrodynamics describing periodic waves propagating along the coast and exponentially decaying far off the coast (Stokes, 1846). Later infinite number of edge wave modes has been found in this problem (Ursell, 1952). Solutions with the same properties describing electro-magnetic waves propagating along the edge of a plasma wedge have been studied in (Ignatov, 1982). Sezawa and Kanai (1939) investigated edge waves propagating on ocean shelf-sill and decaying on a large depth. They noted similarity of some properties of these edge waves with Love's waves propagating along linear nonhomogeneity of an elastic medium (1935). On the other side, the velocity field induced by a shelf edge wave depends on three space

R.V. Goldstein and G.A. Maugin (eds.),
Surface Waves in Anisotropic and Laminated Bodies and Defects Detection, 143–157.

variables, while the velocity field of an elastic medium perturbed by Love's wave depends on two space variables only.

Two-dimensional solutions of elastic plate theory describing the waves, which energy is localized near plate edge, are found in (Konenkov, 1960). Three dimensional edge waves propagating along the edge of elastic plate covering the surface of half-infinite compressible fluid are found in (Kouzov and Luk'ianov, 1972). Three dimensional edge waves propagating along the edge of elastic plate submerged in unbounded compressible fluid are investigated in (Abrahams and Norris, 2000). Three-dimensional solutions of linearized equations of hydrodynamics describing flexural-gravity edge wave propagation along a crack in a floating elastic sheet are found in (Goldstein et. al., 1994, Marchenko and Semenov, 1994). Parametric excitation of such counter-propagating waves by incoming plane wave is investigated in (Marchenko, 1999).

Coriolis's effect inputs anisotropy in the statement of above considered problems about gravity waves near the shore and causes the existence of additional solutions describing wave perturbations which energy is localized in the shelf region (Efimov et al., 1985). Typical examples are related to Kelvin waves (Thompson, 1897) and sub-inertial double Kelvin waves (Longuet-Higgins, 1968). Main goal of the present work is related to the investigation of the properties of flexural-gravity waves in shelf zone taking into account Coriolis's effect. The paper is organized as follows. In the first part flexural-gravity edge waves propagating along a crack in an elastic sheet and along a rod joining with sheet edges are investigated for the case of deep water. In the second part edge waves propagating in rotating shallow water beneath an elastic sheet near the shore are investigated. In the third part edge and shelf waves propagating in rotating shallow water with shelf-sill are considered. It is assumed that water is covered by ice partially in the shelf region. In conclusions an influence of ice cover on the properties of waves in shelf regions is discussed.

1. Edge waves in deep water beneath an elastic sheet

Equations describing the potential motion of deep water beneath an elastic plate are written in dimensionless variables as

$$\begin{aligned} &(\Delta + \partial_{zz})\varphi = 0, \ \ z \in (0, -\infty), \qquad (1.1)\\ &\partial_t \eta = \partial_z \varphi, \ \ \partial_t \varphi + (1 + \Delta^4)\eta = 0, \ \ z = 0,\\ &\varphi \to 0, \ \ z \to \infty; \Delta = \frac{\partial^2}{\partial x^2} + \frac{\partial^2}{\partial y^2}, \end{aligned}$$

where φ is the velocity potential, η is the elevation of water surface, t is the time, x, y are horizontal coordinates, axis z is upwards directed. Symbol

∂ with subscripts denotes partial derivations with respect to the variables indicated by the subscripts. Typical space and time scales are equal to l and $\sqrt{l/g}$ respectively, where

$$l^4 = \frac{Eh^3}{12\rho_w g(1-\nu^2)}, \tag{1.2}$$

E, ν and h are the Young modulus, Poisson ratio and plate thickness, ρ_w is the water density, and g is the gravity acceleration. The characteristic value of the velocity potential is chosen as $a\sqrt{gl}$, where a is the characteristic wave amplitude.

It is assumed that there is the linear discontinuity in the plate coinciding with line $x = 0$ and influencing the stress field inside the plate. In this case equations (1.1) should be completed by contact conditions defining cutting forces and bending moments at the plate edges at $x \to \pm 0$. Let us consider the plate edge which normal is directed along axis x. Generalized cutting force R_x, acting on the edge in the vertical direction,is defined as (Timoshenko and Woinowsky-Kriger, 1959)

$$R_x = -\partial_x(\partial_{xx} + \nu'\partial_{yy})\eta, \ \nu' = 2 - \nu. \tag{1.3}$$

Bending moment $M_x x$ is equal to

$$M_{xx} = -(\partial_{xx} + \nu\partial_{yy})\eta. \tag{1.4}$$

Contact conditions at the edges of the crack coinciding with line $x = 0$ have the following form

$$\lim_{x\to\pm 0} R_x = 0, \ \lim_{x\to\pm 0} M_{xx} = 0. \tag{1.5}$$

Let us construct contact conditions for the case when an elastic rod of infinite length is joined with edges of two floating half-infinite plates. In this case the edges of the plates have no relative displacements and bending moments are equal to zero along them

$$\lim_{x\to\pm 0} \eta = \eta_0, \ \lim_{x\to\pm 0} M_{xx} = 0. \tag{1.6}$$

The equation $z = \eta_0(t, y)$ defines displacement of the rod axis. A torsion of the rod doesn't cause deformations of the plates because of their hinge-type joining.

The equation of bending oscillations of the rod has the following form (Landau and Lifshiz, 1965)

$$(m_r\partial_{tt} + i_x\partial_{yyyy} + \mu)\eta_0 + \lim_{x\to -0} R_x - \lim_{x\to +0} R_x = 0, \tag{1.7}$$

where m_r and i_x are dimensionless density and rigidity of the rod, parameter μ characterizes an influence of the buoyancy force. In the case of the rod with the circular cross-section parameters m_r, i_x and μ are defined by formulas

$$m_r = 12\pi(1-\nu^2)\frac{\rho_r g(rl)^2}{Eh^3}, \; i_x = 3\pi(1-\nu^2)\frac{r^4 E_r}{lh^3 E}, \; \mu = \frac{2r}{l}, \qquad (1.8)$$

where r is the rod radius, E_r and ρ_r are the Young modulus and density of the rod. Equations (1.6) and (1.7) are contact conditions at the plate edges joining with the rod.

Let us consider solutions of equations (1.1) periodically depending on the time t and coordinate y

$$\varphi = \psi(x,z)e^{i(\omega t + k_y y)} + \text{c.c.}, \; \eta = \zeta(x)e^{i(\omega t + k_y y)} + \text{c.c.}. \qquad (1.9)$$

For periodic waves it follows

$$\psi = \psi_0 e^{ikx+\lambda z}, \; \zeta = \zeta_0 e^{ikx}. \qquad (1.10)$$

Substituting formulas (1.9) and (1.10) into equations (1.1), we obtain

$$\kappa(\omega,\lambda) \equiv \omega^2 - \lambda(1+\lambda^4) = 0, \qquad (1.11)$$

where $\lambda^2 = k^2 + k_y^2$. Dispersion equation (1.11) has one real root $\lambda = \lambda_0 \geq 0$ and two couple conjugated roots $\lambda = \lambda_e$, $\lambda = \lambda_e*$ and $\lambda = \lambda_b$, $\lambda = \lambda_b^*$ (star denotes complex conjugation), satisfying to conditions $\text{Re}\lambda_e > 0$ and $\text{Re}\lambda_b < 0$. Note that solution (1.9) and (1.10) doesn't satisfy to the decaying condition for the velocity potential at $z \to -\infty$, when $\lambda = \lambda_b$ or $\lambda = \lambda_b^*$.

The condition of the existence of real values defined by the formula $k = \sqrt{\lambda^2 - k_y^2}$ has the form

$$\omega > \omega_*(k_y), \; \omega_*^2 = |k_y|(1+k_y^4). \qquad (1.12)$$

The line $\omega = \omega_*(k_y)$ at the plane separates two regions I and II (fig. 1). In region I inequality (1.12) is satisfied, while in region II the opposite inequality holds, and wave number is pure imaginary. Thus plane wave solutions exist in region I only. In region II formulas (1.9) and (1.10) describe the solution exponentially increasing at $|x| \to \infty$.

Edge wave solutions describing wave perturbations, propagating in y -direction and exponentially decreasing at $|x| \to \infty$, can exist in region II. Such solutions are written as (Marchenko, 1999)

$$\psi = i\omega \int_{-\infty}^{\infty} \frac{P_3(k)}{\kappa(\omega,\lambda)} e^{ikx+\lambda z} dk, \; \zeta = \int_{-\infty}^{\infty} \frac{\lambda P_3(k)}{\kappa(\omega,\lambda)} \qquad (1.13)$$

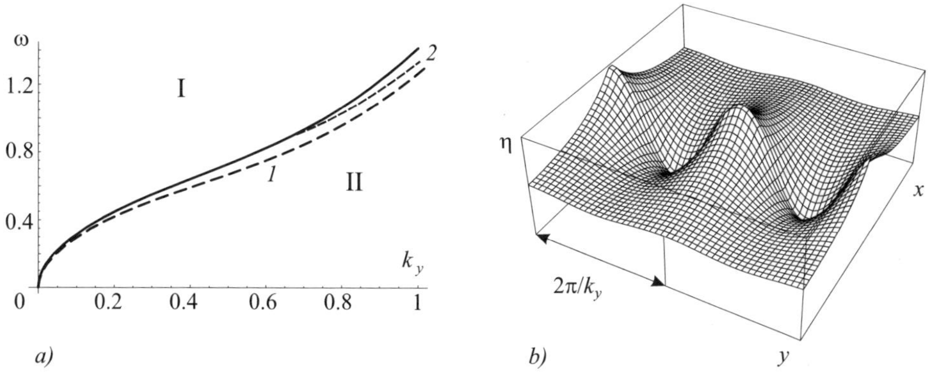

Figure 1. Dispersion curves of edge waves propagating along the crack (line 1 at figure a) and along the rod (line 2 at figure b). The shape of water surface deformed by edge wave in the vicinity of the linear discontinuity.

where $P_3(k) = C_0 + C_1 k + C_2 k^2 + C_3 k^3$ is an arbitrary polynomial of third order. Function $\lambda(k) = \sqrt{k^2 + k_y^2}$ is defined on a two-sheeted Riemann surface in the plane of complex variable k. It follows from the decay of the velocity potential at $z \to -\infty$ that we should consider the sheet of Riemann surface where $\mathrm{Re}\lambda \geq 0$. The cuts of the sheet coincide with half-axes $\mathrm{Re} = 0$, $|\mathrm{Im}k| \geq k_y$.

Substituting formulas (1.13) into contact conditions we obtain a system of linear algebraic equations for searching for constants C_j. A nonzero solution exists in region II in the case when determinant of the system is equal to zero

$$\Delta(\omega, k_y) = 0. \tag{1.14}$$

Condition (1.14) is the dispersion equation of edge waves propagating along the crack or the rod and exponentially decaying far off them. The coefficients of the system are symmetric with respect to changes $\omega \to -\omega$ and $k \to -k$. Hence, the same symmetry is inherent to dispersion curves of edge waves.

Numerical calculations, using the calculation of integrals (1.13) by residues, show that there are one branch of the dispersion curve of edge waves near the crack (contact conditions (1.5)) and one branch of the dispersion curve near the rod (contact conditions (1.6) and (1.7)). In both cases function $\zeta(x)$ is symmetric with respect to line $x = 0$, and coefficients C_1, C_3 are equal to zero. Typical shapes of dispersion curves of edge waves are given in figure 1 by curves 1 (edge waves near the crack) and 2 (edge waves near the rod), respectively. Dispersion curve 1 is following through origin

$k_y = 0$, $\omega = 0$. Dispersion curve 2 starts from a point located at the curve $\omega = \omega_*(k_y)$ at $k_y > 0$.

A typical shape of the function $\eta(t, x, y)$, describing the shape of the fluid surface deformed by the edge wave is given in figure 1b. Features of the fluid surface shape deformed by the edge wave near the crack are given in figure 2 for the frequencies $\omega = 0.6$ and $\omega = 1$. Dependencies of bending moments M_{xx} and M_{yy} in the plate with crack from x-coordinate are given in figure 3 for these values of the frequency. One can see, that there are two characteristic space scales λ_1 and λ_2, characterizing the behavior of the function $\zeta(x)$ in x-direction. The existence of scale $\lambda_1 \gg l$, characterizing the damping of wave amplitudes far off the crack, is related to the closeness of curve 1 in figure 1 and curve $\omega = \omega_*(k_y)$. Scale λ_2, given in figure 3, characterizes the location of maxima of bending moments M_{xx} and M_{yy}. One can see that $\lambda_2 \approx 4l$ at $\omega = 0.6$, and $\lambda_2 \approx 5l$ at $\omega = 1$.

From the symmetry of above considered edge waves relative to axis $X = 0$ it follows the existence of edge waves, propagating along vertical wall when the edge of the plate is not fixed at the wall. The influence of fluid depth becomes important for small wave numbers k_y (curve 1 in figure 1a). Flexural-gravity edge waves propagating in the layer of shallow water bounded by coastline are investigated in what follows.

2. Edge waves in shallow water of constant depth

Dimensionless equations of shallow water beneath an elastic plate are written as

$$\begin{aligned} &\partial_t u - v = -(1 + D\Delta^2)\partial_x \eta, \;\; \partial_t + v = -(1 + D\Delta^2)\partial_y \eta, \\ &\partial_t \eta + \partial_x(\mu u) + \partial_y(\mu v) = 0, \end{aligned} \tag{2.1}$$

where η is the elevation of the water surface and $\mathbf{u} = (u, v)$ is the two-dimensional vector of the water velocity. Dimensionless parameters D (rigidity of the elastic plate) and μ are defined as

$$D = \frac{Eh^3}{12\rho_w g(1 - \nu^2)l^4}, \;\; \mu = \frac{gH}{f^2 l^2}, \tag{2.2}$$

where l is the characteristic horizontal scale, H is the water depth and l is Coriolis parameter. It is assumed that the characteristic value of the water velocity U is related to the characteristic value of the wave amplitude a by the formula $U = ag(lf)^{-1}$. Parameter $D = 1$ when the length l is defined by formula (1.2). Equations (2.1) are used for the description of wave motions in shallow water regions of ice covered seas. Typical values of the Young modulus, Poisson ratio and Coriolis parameter in the Arctic seas are equal

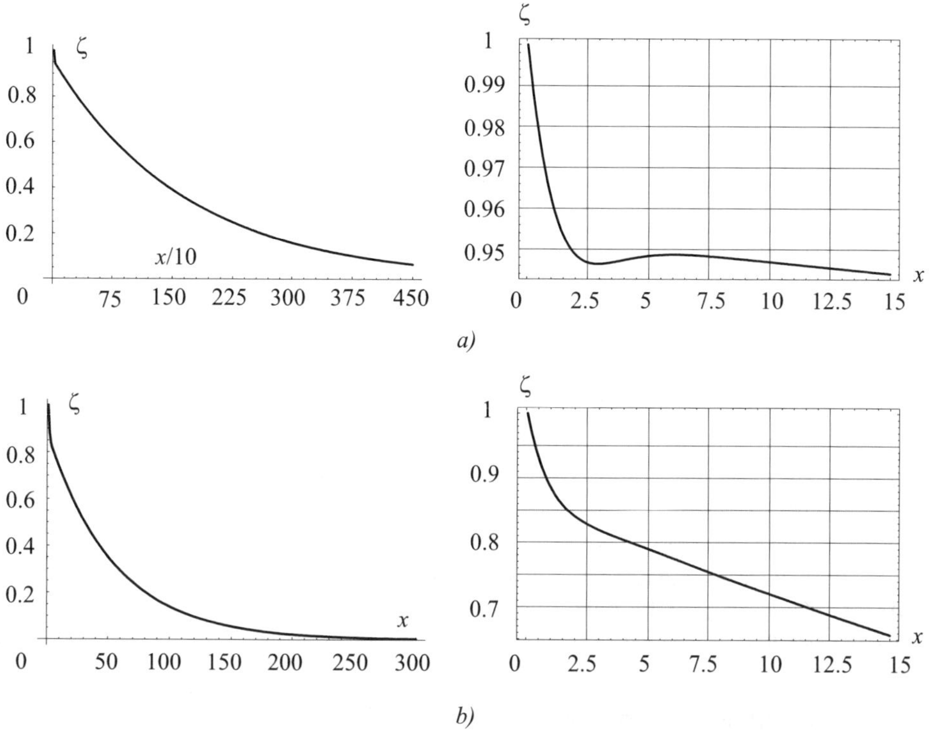

Figure 2. Function $\zeta(z)$ defining the shape of water surface deformed by edge wave of unity amplitude in the vicinity of the crack. Dimensionless wave frequency is equal to $\omega = 0.6$ (a) and $\omega = 1$ (b).

to (Bogorodskii and Gavrilo, 1980)

$$E = 3 \cdot 10^9 \, Nm^{-2}, \ \nu = 0.34, \ f = 1.4 \cdot 10^{-4} \, s^{-1}. \tag{2.3}$$

Assuming ice thickness $h = 1m$, one finds that length l defined by formula (1.2) is close to $12.9m$. Therefore, it follows that ice cover influence becomes important for sufficiently short waves of length of order several tens of meters. Assuming water depth $H = 30m$, one finds that $\mu = 10^8$. The high order of the parameter μ shows a small influence of the Coriolis force on such waves. On the other side, substituting $l = 100km$ and $H = 30m$ in the second formula (2.2) one finds that $\mu = 1$. Hence, the Coriolis force is important for waves of length close to several hundreds kilometers. Considered estimates explain difficulties of the analysis of dispersion properties of gravity and flexural-gravity waves in the wide spectral range of wave numbers.

Let us consider solutions of equations (2.1) periodically depending on

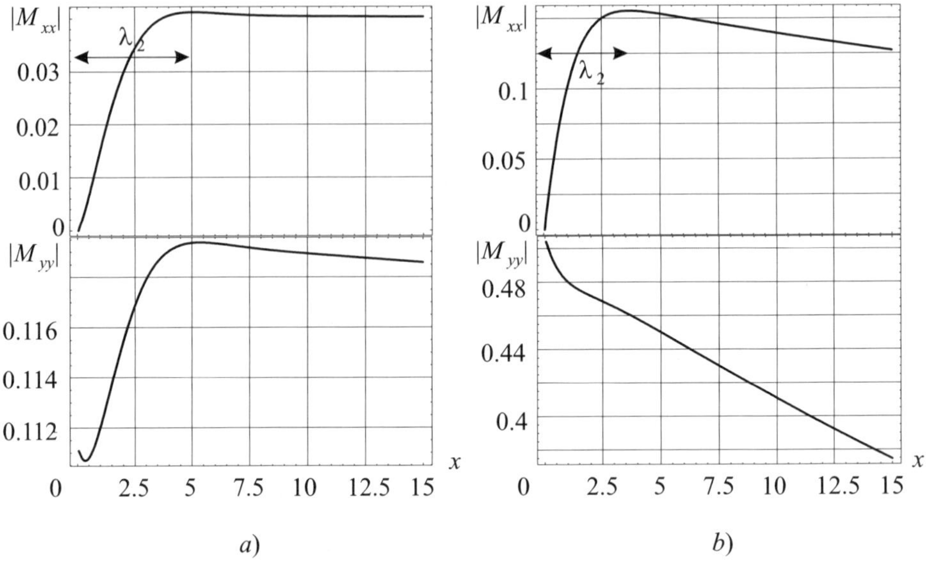

Figure 3. Absolute values of bending moments M_{xx} and M_{yy} formed in the elastic plate by edge wave of unity amplitude in the vicinity of the crack. Dimensionless wave frequency is equal to $\omega = 0.6$ (a) and $\omega = 1$ (b).

the time t and space variable y

$$u = w_x(x)e^{i(\omega t + k_y y)} + \text{c.c.}, \ v = w_y(x)e^{i(\omega t + k_y y)} + \text{c.c.}, \qquad (2.4)$$
$$\eta = \zeta(x)e^{i(\omega t + k_y y)} + \text{c.c.},$$

Substituting formulas (2.4) into equation (2.1) at $\mu = \text{const}$, we obtain the ordinary differential equation

$$(1 - \omega^2)\zeta - \mu(1 + D\Delta^2)\Delta\zeta = 0, \ \Delta = \frac{d^2}{dx^2} - k_y^2, \qquad (2.5)$$

and formulas for the definition of w_x and w_y

$$(1 - \omega^2)w_x = -i(\omega\partial_x + k_x)(1 + D\Delta^2)\zeta, \qquad (2.6)$$
$$(1 - \omega^2)w_y = (\omega k_y + \partial_x)(1 + D\Delta^2)\zeta.$$

Consider the wave motions in a shallow water layer of constant depth ($\mu = \text{const}$) bounded by a rigid wall (coastline) located at $x = 0$. Assuming the existence of tidal cracks oriented along coastline (Wadhams, 1980) we set that the plate edge is free near the coastline. Corresponding boundary conditions are written as follows

$$w_x = 0, \ \lim_{x \to -0} R_x = 0, \ \lim_{x \to -0} M_{xx} = 0, \qquad (2.7)$$

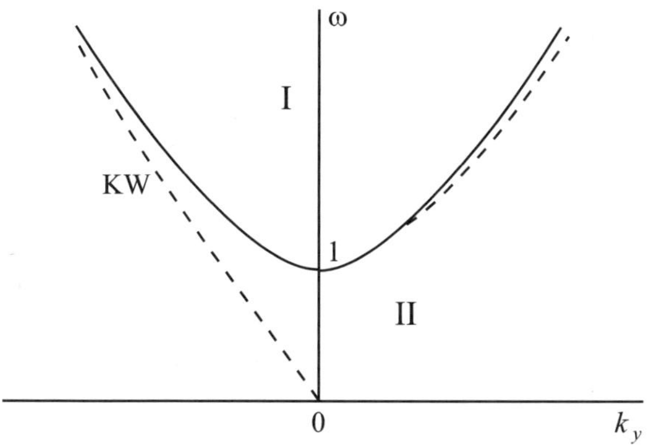

Figure 4. Dispersion curves of edge waves propagating in shallow sea covered by elastic ice sheet propagating along the coastline

where cutting force R_x and bending moment M_{xx} are defined by formulas (1.3) and (1.4).

The general solution of equation (2.5) has the form

$$\zeta = C_1 e^{ik_0 x} + C_2 e^{-ik_0 x} + C_3 e^{ik_e x} + C_4 e^{-ik_e x} + C_5 e^{ik_e^* x} + C_6 e^{-ik_e^* x}, \quad (2.8)$$

where C_j are arbitrary constants, and quantities $\pm k_0(k_y, \mu)$, $\pm k_e(k_y, \mu)$ and $\pm k_e^*(k_y, \mu)$ are the roots of the dispersion equation (star denotes complex conjugation)

$$\omega^2 - 1 = \mu\lambda^2(1 + D\lambda^4), \ \lambda^2 = k^2 + k_y^2. \quad (2.9)$$

The curve

$$\omega = \omega_*(k_y, \mu), \ \omega_*^2 = 1 + \mu k_y^2(1 + Dk_y^4), \quad (2.10)$$

separates plane (k_y, ω) on two regions I and II similar to curve (1.12) (fig. 4). Roots $\pm k_0$ are real when $\omega^2 > \omega_*^2$ (region I), and roots $\pm k_0$ are pure imaginary quantities when $\omega^2 \in (0, \omega_*^2)$(region II). Roots $\pm k_e$ and $\pm k_e^*$ are complex quantities for any values of ω. Later on for the sake of definiteness it is assumed that $\mathrm{Im}k_0 \geq 0$, $\mathrm{Re}k_e \geq 0$ and $\mathrm{Im}k_e \geq 0$.

Let us consider the case $\omega^2 \in (0, \omega_*^2)$. From (2.8) it follows that the solution is bounded in the region $x < 0$ when $C_1 = 0$, $C_3 = 0$ and $C_6 = 0$. In this case the water surface function $= zeta(x)$ equals

$$\zeta = C_2 e^{-ik_0 x} + C_4 e^{-ik_e x} + C_5 e^{ik_e^/ astx}. \quad (2.11)$$

Substituting (2.11) into boundary conditions (2.7) one finds the homogeneous system of three linear algebraic equations for searching for constants

C_2, C_4 and C_5. The nonzero solution exists in the case when the determinant of the system is equal to zero

$$\Delta(\omega, k_y, \mu) = 0. \tag{2.12}$$

Condition (2.12) is the dispersion equation of edge waves propagating along the wall and exponentially decreasing far off the wall. One can see that formulas (2.6) are not symmetric relative to the substitution $k_y/to{-}k_y$. Therefore, the dispersion curves are also not symmetric with respect to this substitution.

The numerical analysis of the function $\Delta(\omega, k_y, \mu)$ shows the existence of two branches of dispersion curves of edge waves in the region $\omega > 0$ of the plane (k_y, ω) (fig. 4). The dispersion curve KW located in the region $k_y < 0$ starts from the origin. This dispersion curve is transformed into the dispersion curve of Kelvin waves propagating along the wall bounded half-infinite shallow water layer when $D \to 0$. The beginning of dispersion curve located in the region $k_y > 0$ is located on the line $\omega = \omega_*(k_y, \mu)$ and tends to the infinity at $D \to 0$. Both branches of the dispersion curves become symmetric with respect to axis ω when $f \to 0$. In this case they coincide with the dispersion curves of the edge waves propagating along the crack in an elastic plate floating on the surface of shallow water (Marchenko and Semenov, 1994).

3. Edge waves in shallow water with depth discontinuity

It is assumed that water depth is equal to H_1 near the coastline and is increased abruptly up to H_0 on some distance l from the coast. Let us choose this distance as the characteristic horizontal scale. In dimensionless variables this shelf break is related to the discontinuity of the parameter μ. Thus we set $\mu = \mu_0$ when $x < x_0$, $\mu = \mu_1$ when $x \in (x_0, x_1)$, and $x_1 - x_0 = 1$. It is assumed that water is covered by an elastic ice sheet at $x \in (0, x_1)$. The bottom configuration and location of ice cover are shown in fig. 5.

Water motions are described by equations (2.5) and (2.6), where $D = 0$ at $x < 0$, $\mu = \mu_0$ at $x < x_0$, and $\mu = \mu_0$ at $x \in (x_0, x_1)$. Contact conditions along lines related to abrupt changing of parameters D and μ should be formulated. Contact conditions on the line $x = x_0$ follow from the continuity of the water mass flux and water pressure

$$\mu_0 \lim_{x \to x_0 - 0} w_x = \mu_1 \lim_{x \to x_0 + 0} w_x, \quad \lim_{x \to x_0 - 0} \zeta = \lim_{x \to x_0 + 0} \zeta. \tag{3.1}$$

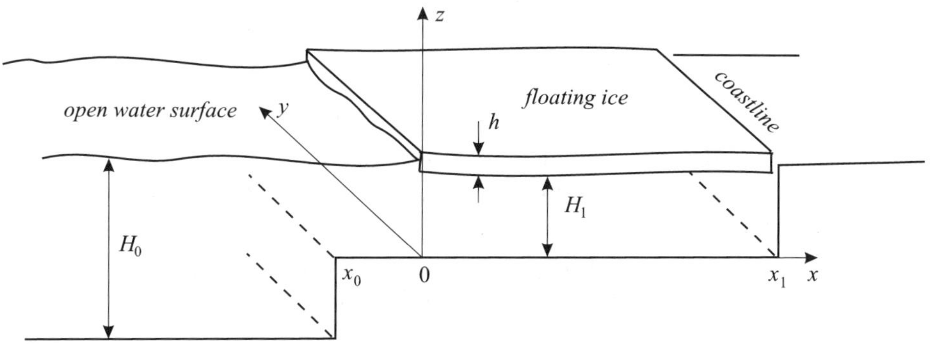

Figure 5. Scheme of sea bottom relief in nearcoastal region. Sea surface is covered by elastic ice sheet when $x \in (0, x_1)$

Contact conditions on the line $x = 0$ follow from the continuity of the water mass flux, water pressure and conditions at the free edge of a plate

$$\lim_{x\to -0} \zeta = \lim_{x\to +0} (1 + D\Delta^2)\zeta, \ \lim_{x\to -0} w_x = \lim_{x\to +0} w_x, \tag{3.2}$$
$$\lim_{x\to +0} R_x = 0, \ \lim_{x\to +0} M_{xx} = 0.$$

We set zero mass flux in the normal direction to the coastline and zero cutting force and bending moment at the ice edge near the coast

$$\lim_{x\to x_1} w_x = 0, \ \lim_{x\to x_1} R_x = 0, \ \lim_{x\to x_1} M_{xx} = 0. \tag{3.3}$$

Last two conditions are related to the existence of tidal cracks parallel to the coast.

Water surface elevation is described by formula (2.8) when $x \in (0, x_1)$. In this region quantities $\pm k_0$, $\pm k_e$ and $\pm k_e^*$ are the roots of dispersion equation (2.9) with $\mu = \mu_1$. The water surface elevation is described by formulas when $x < 0$

$$\zeta = A_1 e^{ik_0^{(1)}x} + A_2 e^{-ik_0^{(1)}x}, \ x \in (x_0, 0);$$
$$\zeta = B_1 e^{ik_0^{(0)}x} + B_2 e^{-ik_0^{(0)}x}, \ x < x_0, \tag{3.4}$$
$$k_0^{(0)} = \sqrt{\frac{\omega^2 - 1}{\mu_0} - k_y^2}, \ k_0^{(1)} = \sqrt{\frac{\omega^2 - 1}{\mu_1} - k_y^2}.$$

Let us define three curves at the plane (k_y, ω) as follows

$$\mathrm{A}: \ \omega = \omega_*^{(1)}(k_y, \mu_1), \ \omega_*^{(1)} = \sqrt{1 + \mu_1 k_y^2 (1 + D k_y^4)}, \tag{3.5}$$

$$\mathrm{B}: \ \omega = \omega_*^{(2)}(k_y, \mu_1), \ \omega_*^{(2)} = \sqrt{1 + \mu_1 k_y^2}, \tag{3.6}$$

$$\mathrm{C}: \ \omega = \omega_*^{(3)}(k_y, \mu_1), \ \omega_*^{(2)} = \sqrt{1 + \mu_0 k_y^2}. \tag{3.7}$$

These curves separate five regions I, II, III, IV and V, given in fig. 6. In region IV roots k_0, $k_0^{(0)}$ and $k_0^{(1)}$ are pure imaginary. In region III roots k_0 and $k_0^{(0)}$ are pure imaginary, while the root $k_0^{(1)}$ is real. In region II roots k_0 and $k_0^{(1)}$ are real, while the root $k_0^{(0)}$ is pure imaginary. In region I all roots k_0, $k_0^{(0)}$ and $k_0^{(1)}$ are real quantities. In region V roots $k_0^{(0)}$ and $k_0^{(1)}$ are real, while the root k_0 is pure imaginary. It is assumed that $\text{Im}k_0 \geq 0$, $\text{Im}k_0^{(0)} \geq 0$ and $\text{Im}k_0^{(1)} \geq 0$. Coordinates of the points O_1 and O_2 separating regions II and IV are equal to (k_O, ω_O) and $(-k_O, \omega_O)$, respectively, where $k_O^4 = (\mu_0/\mu_1 - 1)D^{-1}$ and $\omega_O = 1 + \mu_0 k_O^2$. In dimensional variables the wave number k_O equals

$$k_O = \frac{1}{l}\left(\frac{H_0 - H_1}{H_1}\right)^{1/4}, \tag{3.8}$$

where l is defined by formula (1.2).

The existence of edge waves propagating along the coastline and exponentially decreasing far off the coast is possible in regions II, III and IV. In this case it follows from the boundedness of the solution that $B_1 = 0$. The solution depends on 9 constants A_1, A_2, B_2 and C_1,..., C_6, which should be found from 9 contact conditions (3.1) - (3.3), which are reduced to homogeneous system of 9 linear algebraic equations.

Nonzero solution of the system exists when the determinant is equal to zero

$$\Delta(\omega, k_y, \mu_0, \mu_1) = 0. \tag{3.9}$$

Condition (3.8) is the dispersion equation of edge waves propagating along the coastline and exponentially decreasing far off the coast. As before dispersion curves of edge waves are not symmetric with respect to the substitution $k_y \to -k_y$, since formulas (2.6) are not symmetric relative to this substitution.

The numerical analysis of the function $\Delta(\omega, k_y, \mu_0, \mu_1)$ shows the existence of the Kelvin wave analog (curve KW in fig. 6) and sub-inertial shelf mode (curve DKW in fig. 6). Dispersion curve KW tends asymptotically to curve C at large negative k_y, dispersion curve DKW has the horizontal asymptote at $k_y \to -\infty$ similarly to the dispersion curve of the double Kelvin wave propagating along a rectilinear depth discontinuity in unbounded rotating shallow water (Longuet-Higgins, 1968). There is denumerable number of edge waves, which dispersion curves start on the curve B and asymptotically tend to curve at $|k_y| \to \infty$, when $x_0 < 0$. Dispersion curves of the first three edge wave modes are denoted as 1, 2, 3 in fig. 6a.

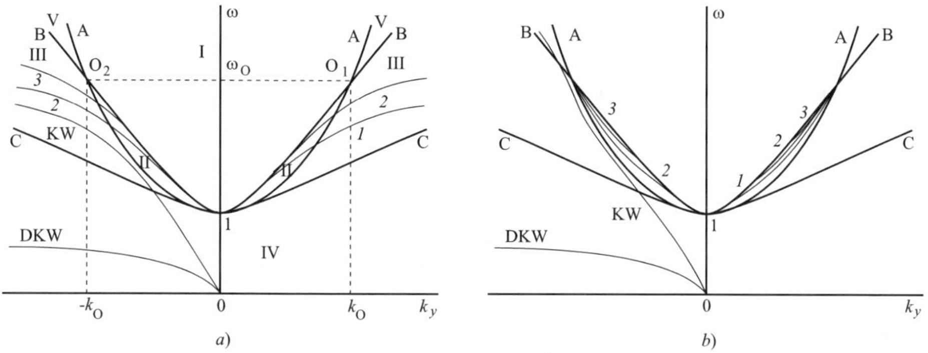

Figure 6. Dispersion curves of subinertial shelf waves(DKW) and shelf edge waves. The first mode of shelf edge waves turned in Kelvin waves when ice thickness tends to zero is denoted as KW, other edge wave modes are denoted as 1,2,3...; $x_0 < 0$ (a) and $x_0 = 0$ (b).

Figure 6b shows dispersion curves in the limiting case $x_0 \to 0$. With respect to the case $x_0 < 0$ there are following changes. Dispersion curve KW tends asymptotically to curve B at $k_y \to -\infty$. There is finite number of edge wave modes which dispersion curves are located inside region II. The beginnings of these dispersion curves are at curve B and their ends are in the point O_1 or O_2.

Conclusions

Edge waves propagating along rectilinear discontinuities in elastic sheet floating on the surface of deep water are investigated. Dispersion curves were found for symmetric edges waves propagating along a crack in an elastic sheet and along an elastic rod joined with the edges of an elastic plate. It is shown that the frequency of edge waves near the crack can be arbitrary small, while the frequency of edge waves near the rod is greater some limiting value depending on the stiffness and density of the rod. It is shown that the shape of the plate deformed by the edge wave near the crack is characterized by two space scales, one of them is related to the damping of wave amplitude far off the crack, and the other characterizes the behavior of bending moments in the vicinity of the crack. The second scale is much smaller the first.

Edge waves propagating in rotating half-infinite layer of shallow water beneath an elastic plate are investigated. It is shown that dispersion curve of the Kelvin wave is transformed into the dispersion curve of the edge wave, propagating along a crack in an elastic plate floating on the surface of shallow water, when the wave frequency tends to the infinity.

The influence of a depth discontinuity and floating ice cover on dispersion properties of shelf waves and edge waves propagating along the coastline is investigated. It is shown that the shapes of dispersion curves of the sub-inertial shelf waves, Kelvin waves and low frequency edge waves are similar to the dispersion curves of such waves propagating in shallow water with free surface. When the length of ice cover is smaller the distance between the coastline and shelf break there is infinite number of shelf modes. There is only finite number of edge wave modes when ice cover totally occupies region between the coast and the depth discontinuity or when the length of ice cover is greater the distance between the coast and shelf break. Wave numbers of edge waves are smaller limiting value k_O, which is defined by formula (3.6). One finds that when relative changing of water depth is $H_0/H_1 - 1 \approx 0.1$ and ice thickness is 1m. Period of this wave is equal to 8.7 sec when $H_1 = 30m$. Therefore, it is possible to conclude that the influence of floating ice on the waves propagating in coastline region of the depth several tens meters when wave frequency is smaller 0.1 Hz is small.

Acknowledgement

The work was supported by the RFBR 020100729 grant.

References

1. Abrahams, I.D., Norris, A.N., 2000. On the existence of flexural edge waves on submerged elastic plates. Proc. Soc. Lond. A, **456**, 1559-1582.
2. Faraday's Diary, **I**, G.Bell and Sons, LTD, London 1832.
3. Stokes, G.G., 1846. Report on recent researches in hydrodynamics. Rep. 16th meet. Brit. Assoc. Adv. Sci., L.: Murray, 1-20.
4. Thompson, W. (Lord Kelvin), 1879. On gravitational oscillations of rotating water. *Proc. Roy. Soc. Edinburg*, **10**, 92-100.
5. Ursell, F., 1952. Edge waves on a sloping beach. *Proc. Roy. Soc. London*, **A214**, 79-97.
6. Efimov, V.V., Kulikov, E.A., Rabinovich, A.B., Fain, I.V., 1985. Waves in boundary regions of ocean. L.: Gidrometeoizdat, 280 p.
7. Konenkov, Yu. K., 1960. A Rayleigh-type flexural wave. *Sov. Phys. Acoust*, **6**, 122-123.
8. Kouzov, D.P., Luk'ianov, V.D., 1972. On the waves propagating along plates edges. *Sov. Phys. Acoust.*, **4**, 549-555.
9. Longuet-Higgins, M.S., 1968. Double Kelvin waves with continuous depth profiles. *J. Fluid Mech.*, **34**, Part. 3, 417-434.
10. Marchenko, A.V., Semenov, A.Yu., 1994. Edge waves in a shallow water beneath an ice cover with a crack, *Izv. AN SSSR, Mekh. Zhidk. i Gaza*, **1**, 185-189. (in Russian)
11. Goldstein, R.V., Marchenko, A.V., Semenov, A.Yu., 1994. Edge waves in a fluid beneath an elastic sheet with a crack. *Dokl. Akad. Nauk*, **339**, N 3: 331-334. (in Russian)

12. Marchenko A.V., Parametric excitation of flexural-gravity edge waves in the fluid beneath an elastic ice sheet with a crack. *European Journal of Mechanics. B/Fluids.* 1999. **18**, N 3, 511-525.
13. Sezawa, K., Kanai, K., 1939. On shallow water waves transmitted in direction parallel to a sea coast, with special reference to Love-waves in heterogeneous media. *Bull. Earthquake Res. Inst., Univ. Tokyo*, **17**, 685-694.
14. Bogorodskii, V.V., Gavrilo, V.P., 1980. Ice, physical properties. Modern methods of glaciology. L.: Gidgrometeoizdat, 384 p. (in Russian)
15. Timoshenko, S., Woinowsky-Kriger, S., 1959. Theory of elastic plates and shells. 2nd ed., McGraw-Hill, New-York, U.S.A.
16. Landau, L.D., Lifshic, E.M., 1965. Theory of elasticity, M.: Nauka, 204 p.
17. Ignatov, A.I.,1982. Potential waves on arbitrary surface of isotropic dispersion medium. *Bull. of the Lebedev Phys. Inst.*, **5**, 40-45. (in Russian)
18. Wadhams, P., 1980. Ice characteristics in the seasonal sea ice zone. *Cold. Reg. Sci. and Technol.*, **2**, 37-87.

ON CONTINUUM MODELLING OF WAVE PROPAGATION IN LAYERED MEDIUM; BENDING WAVES

K. B. USTINOV
Institute for Problems in Mechanics of Russian Academy of Sciences.
Prospect Vernadskogo, 101. b.1., Moscow, 119526, Russia

Abstract. A continuum model of layered media with frictional interfaces is suggested. The model has been build as a variant of the non-symmetric theory of elasticity. The new media parameters related to the non-symmetric and moment stresses are expressed in terms of elastic and geometric parameters of the original media. The wave propagation in the model medium is considered. The considered examples show the applicability of the model. The comparison with the existing models has been made. It has been shown that the suggested model may be reduced to model [5] if the friction between the adjusted layers is small.

Key words: nonlinear surface waves, elasticity, Rayleigh waves, scale invariance, nonlocality, spectral kernel, spatial kernel, Hilbert transform

1. Introduction

For material consisting of a large number of layers the continuum approximation approach seems appropriate to describe the mechanical processes, such as deformation and wave propagation. In the frame of this approach the real structured media are replaced with new homogeneous media possessing some effective properties. In case of the perfect cohesion of the layers such a homogeneous medium is just an anisotropic elastic one. The general solution of the problem of the effective elastic parameters determination for such a medium was obtained by Lifshitz and Rozentsveig [1, 2]. However, if the medium allows significant relative sliding of the layers, the bending may occur at the places of significant stress gradients. The bending is accompanied by such effects as the violation of the shear stress parity rule, and, as a result, the moment stresses appear. A similar situation may take place even in the case of the perfect cohesion between the layers: when one group of the layers is sufficiently thin and compliant to be considered as

R.V. Goldstein and G.A. Maugin (eds.),
Surface Waves in Anisotropic and Laminated Bodies and Defects Detection, 159–171.

slippery interfaces. For example, such a situation can take place in layered composite materials with comparatively weak interfaces between the layers. The peculiarity distinguishing these cases from the anisotropic elasticity is the presence of *the additional degree of freedom*, associated with the relative movement of the adjusted layers. The role of bending for the body composed by multitude of elastic layers with slippery interfaces between them was emphasized and treated by considering such a body, as a continuum in the works of Sonntag [3] and Salganik [4, 5], where the layers were considered as beams and plates respectively in the frame of the traditional approach neglecting the influence of the shear stresses acting within the layers to the bending.

The idea of the additional degree of freedom has been explicitly, or implicitly used to describe the behavior of layered media. Sun et al. [6] considered the media composed by the relatively rigid and soft layers. Expanding the displacement field of each set of layers into Legandre series and keeping the constant and linearly changing terms only they obtained a system of differential equations describing such a media. Bolotin [7] suggested to estimate the contribution of bending, shear and elongation into the potential energy for each group of layers separately and to retain the essential terms only. Within this assumption he created the theory of multilayered plates. The model was developed further in the book by Bolotin and Novichkov [8]. Molotkov [9] considered the media composed by the relatively rigid and soft layers also and employed the method of matrix averaging to develop the continuum theory. Zvolinskii and Shkhinek [10] created a variant of the theory in 2-D by considering the layered structure as a particular case of non-symmetric elastic medium. They applied the formalism of the non-symmetric elastic (micropolar) theory and found the necessary constant related to non-symmetric and moment stresses from the better agreement of the model with a set of exact analytical results. The model was developed extensively in works [11-14] and generalized to the case of the media composed by the relatively rigid and soft layers. A model, which may be considered as one of possible generalizations of Sonntag approach, was proposed by Ustinov [15]. The model after correction [16] may be derived as a particular case of non-symmetric elastic theory, namely micromorphic theory contrary to micropolar theory used in [10]. The additional parameters related to the non-symmetric and moment stresses are expressed in terms of elastic and geometric parameters of the original layered media. Hereafter, the dynamic variant of the theory is considered.

The additional degree of freedom leads to appearance of the new type of waves, which nature is very close to the bending waves in beams and plates. Such waves were considered by Zvolinskii and Shkhinek [9] and Salganik

[4]. In the frame of the considered model, the particular parameters of such waves for important particular cases have been derived and compared to the ones for other models available. It has been shown that in case of asymptotically vanishing cohesion between the layers the solution for the wave propagating along the layers approaches the solution for bending wave in plates of Timoshenko's [17] type.

2. Volume elements, measures of deformation, compatibility

Let us consider a region of the media possessing a cubic lattice structure. Let us introduce the volume element ΔV, which is small to compare of the area dimensions, but not less then the size of the lattice structure (microelement). In the Cartesian co-ordinates the volume element may be denoted by $\Delta V = \Delta x_1 \Delta x_2 \Delta x_3$. The undeformed element is shown in Fig. 1.

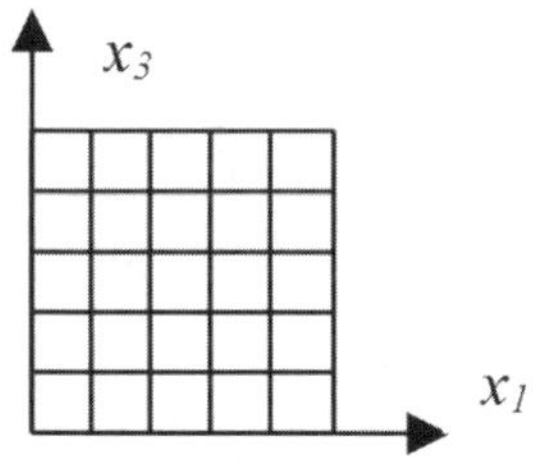

Figure 1. Undeformened element

Generally, two types of deformation may be distinguished: the deformation of the lattice (Fig. 2) where the particles neighbouring before deformation remains neighbouring after deformation; and the relative movements of the neighbouring particles (Fig. 3). The first type of deformation will be called the elastic deformation, while the second one will be referred to as the plastic deformation. The term "plastic" will be used throughout the text in that particular meaning, rather then implying an energy dissipation.

Geometry of the deformed body may be described by the tensor of elastic distortion β, with Cartesian components β_{ij}, and the components β_{ij}^P of plastic distortion. Geometries, corresponding to various components of elastic and plastic distortion, are shown in the figures 2 and 3. The sum of the elastic and plastic distortion is called the total distortion:

$$\beta_{ij}^T = \beta_{ij} + \beta_{ij}^P \tag{1}$$

Everywhere throughout the text the upper index "P" corresponds to the plastic fields, the upper index "T" corresponds to the total fields, while the variables without the upper index corresponds to the elastic fields.

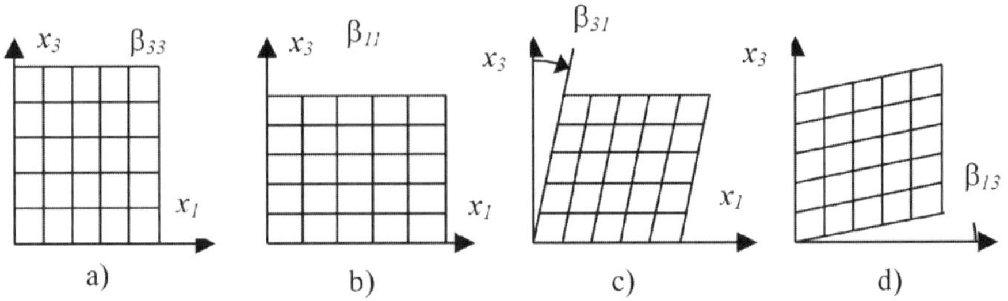

Figure 2. Homogeneous elastic distortion. a) β_{33}, b) β_{11}, c) β_{31}, d) β_{13}

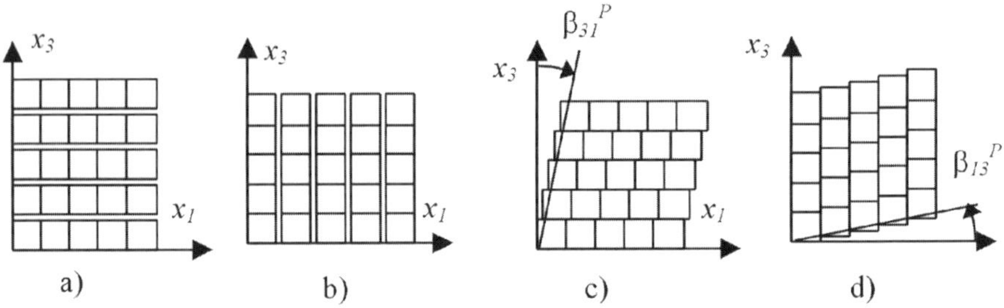

Figure 3. Homogeneous plastic distortion. a) β^P_{33}, b) β^P_{11}, c) β^P_{31}, d) β^P_{13}

It is supposed herein that the total distortion is compatible (the more general case is considered within the theory of disclination and is beyond the frame of the current study). The compatibility of the total distortion implies the existence of the total displacement vector U^T, with Cartesian components U_i^T. Therefore, the total distortion is the gradient of the total displacement:

$$\beta^T_{ij} = \partial_i U^T_j \tag{2}$$

In general case of non-homogeneous distortion, the elastic and plastic distortions are non-compatible, hence, the elastic and plastic displacements do not necessarily exist. The existence of such vectors is the peculiarity rather then the rule. The elastic distortion β_{ij} is usually called micro-distortion in the theory of non-symmetric elasticity in order to stress that it associates with the intact microelement. The total distortion β^T_{ij} is usually called macro-distortion in the theory of non-symmetric elasticity in order to stress that it associates with the whole element.

The integrability condition for the total distortion may be written in the form:

$$e_{ikl}\partial_k\beta^T_{lj} = 0 \ \ (\text{curl}\beta^T = 0) \tag{3}$$

Here e_{ijk} is the alternating tensor. For the elastic and plastic distortions we may write

$$e_{ikl}\partial_k\beta^P_{lj} = -e_{ikl}\partial_k\beta_{lj} = \alpha_{ij} \ \ (\text{curl}\beta^P = -\text{curl}\beta = \alpha) \tag{4}$$

Here the new tensor α with Cartesian components α_{ij} has been introduced. It is usually called the tensor of dislocation density in the theory of dislocation. It is seen from the above, that it may be interpreted as a quantitative measure of the plastic (and elastic) incompatibility. An obvious sequence of the definition of α is the equation

$$\partial_i\alpha_{ij} = 0 \ \ (\text{div}\alpha = 0) \tag{5}$$

Let us introduce the new variables, namely the gradient of distortion (total, elastic and plastic, respectively):

$$\begin{aligned} \kappa^T_{ijk} &= \partial_i\beta^T_{jk} \\ \kappa_{ijk} &= \partial_i\beta_{jk} \\ \kappa^P_{ijk} &= \partial_i\beta^P_{jk} \end{aligned} \tag{6}$$

It directly follows from (1) and (6) that

$$\kappa^T_{ijk} = \kappa_{ijk} + \kappa^P_{ijk} \tag{7}$$

It is obvious, that any parameter having the order of the second derivative of the displacement such as dislocation density α (for details see eg Kroner [18]) may be expressed in terms of any pair of three functions κ^T_{ijk}, κ_{ijk}, κ^P_{ijk} (or in a more lucky case in terms of only one function).

Let us now consider the layered media with x_3-axis being the normal to the layers. Media may deform, and relative sliding of the adjusted layers is allowed, but no separation of the layers. The situation corresponds to the case shown at Fig. 3c, where relative sliding and separation between the adjusted along x_1 and x_2 axii elements is prohibited. Hence, the only remained components of plastic distortions are β^P_{31}, β^P_{32}.

3. The energy variation

A common and rather general approach of deriving the governing equations consists in considering the variation of the potential energy density of the deformed media.

Within the frame of the classical theory of micromorphic materials (Mindlin, [19]) the potential energy is supposed to be the function of total and plastic distortion and gradient of elastic distortion. However, without loosing the generality, we may use any pair of distortion tensors as arguments of the potential energy. Let us chose the elastic and plastic distortion:

$$W = W(\beta_{ij}, \beta_{ij}^P, \kappa_{ijk}) \tag{8}$$

Including the gradients into the set of parameters may be considered as done just for the sake of generality; however, from the physical point of view, these gradients may cause some bending of the microelements, which leads to additional accumulation of energy. It should be noted that in the frame of this theory gradients of neither total nor plastic distortion are included in the list of arguments.

If we use α instead of κ (which may be composed from k), we came to the form used by Krener [18] in the theory of dislocation. If we drop the last argument in (8) we came to the symmetrical theory of dislocation plasticity used by Mokhel and Salganik [20].

For the variation of potential energy density we may write

$$\begin{aligned} \delta W = &\tau_{ij}\delta\beta_{ij} + \sigma_{31}\delta\beta_{31}^P + \sigma_{32}\delta\beta_{32}^P \\ &+\mu_{131}\delta\kappa_{131} + \mu_{132}\delta\kappa_{132} + \mu_{231}\delta\kappa_{231} + \mu_{232}\delta\kappa_{232} \end{aligned} \tag{9}$$

where it was supposed that the energy depends only on some components of the strain gradient, and it was set:

$$\tau_{ij} \equiv \frac{\partial W}{\partial \beta_{ij}}; \quad \sigma_{ij} \equiv \frac{\partial W}{\partial \beta_{ij}^P}; \quad \mu_{ijk} \equiv \frac{\partial W}{\partial \kappa_{ijk}}; \tag{10}$$

Here τ_{ij} have the meaning of components of tensor of stress acting within microelements; σ_{ij} may be called interface stresses; μ_{ijk} are usually called the moment stresses.

If the material of the media is pure elastic, and the sliding of the adjusted layers obeys to the shear spring-type law, then it is natural to suppose that the constitutive law possesses the following form:

$$\begin{aligned} &\tau_{ij} = a_{ijkl}\beta_{kl} \\ &\sigma_{31} = k\beta_{31}^P \quad \sigma_{32} = k\beta_{32}^P \end{aligned} \tag{11}$$

where a_{ijkl} are the elastic tensor or the material of the layers; k is the constant having the dimension of stress describing the relative sliding of the adjusted layers. For the isotropic layers the elastic tensor a_{ijkl} has the form:

$$a_{ijkl} = \lambda\delta_{ij}\delta_{kl} + \mu(\delta_{ik}\delta_{il} + \delta_{il}\delta_{jk}) \tag{12}$$

where λ and μ are the Lame constants of the material.

The simplest way to define the constitutive law for the moment stress is just to borrow it from the theory of bending of plates. For isotropic case it reads as follows in our notation:

$$
\begin{aligned}
\mu_{131} &= \frac{D}{2h}(\kappa_{131} + \nu\kappa_{232}) \\
\mu_{232} &= \frac{D}{2h}(\kappa_{232} + \nu\kappa_{131}) \\
\mu_{132} = \mu_{231} &= \frac{D}{2h}\tfrac{1-\nu}{2}(\kappa_{132} + \kappa_{231})
\end{aligned}
\tag{13}
$$

Here D stands for the flexural rigidity:

$$
D = \frac{8(\lambda+\mu)}{3(\lambda+2\mu)}\mu h^3 = \frac{2Eh^3}{3(1-\nu^2)} \tag{14}
$$

where E and ν stand for the Young's modulus and Poisson's ratio, respectively.

The standard procedure of variation of the energy (9) leads to five equations of equilibrium and five boundary conditions:

$$
\begin{aligned}
&\partial_1\tau_{11} + \partial_2\tau_{21} + \partial_3\sigma_{31} + f_1 = 0 \\
&\partial_1\tau_{12} + \partial_2\tau_{22} + \partial_3\sigma_{32} + f_2 = 0 \\
&\partial_1\tau_{13} + \partial_2\tau_{23} + \partial_3\tau_{33} + f_3 = 0 \\
&\sigma_{31} - \tau_{31} + \partial_1\mu_{131} + \partial_2\mu_{231} + \mu^v_{31} = 0 \\
&\sigma_{32} - \tau_{32} + \partial_1\mu_{132} + \partial_2\mu_{232} + \mu^v_{32} = 0
\end{aligned}
\tag{15}
$$

$$
\begin{aligned}
&n_1\tau_{11} + n_2\tau_{21} + n_3\sigma_{31} = T_1 \\
&n_1\tau_{12} + n_2\tau_{22} + n_3\sigma_{32} = T_2 \\
&n_1\tau_{13} + n_2\tau_{23} + n_3\tau_{33} = T_3 \\
&n_1\mu_{131} + n_2\mu_{231} = M_{31} \\
&n_1\mu_{132} + n_2\mu_{232} = M_{32}
\end{aligned}
\tag{16}
$$

Here f_i are the components of volume force; T_i are the components of surface force (stress vector); μ^v_{ij} are the components of doubled forces per volumetric unit; M_{ij} are the components of doubled forces per surface unit. Diagonal elements of μ^v_{ij} and M_{ij} are moment free doubled forces, while non-diagonal elements of μ^v_{ij} and M_{ij} are doubled forces possessing the moments. Skew-symmetric part of $\mu^v_{[ij]}$ is the volumetric total moment. Skew-symmetric part of $M_{[ij]}$ is the surface total moment (moment stress of Cosserat). The first indexes of μ^v_{ij} and M_{ij} show the lever direction; the second indexes show the force direction. It is supposed that at the face with positive normal the positive force acts at the positive end of the lever. The components of μ_{ijk} may be interpreted as double forces per surface unit;

the first index shows the normal direction, while the last two have the same meaning as the indexes of μ_{ijk} and M_{ij} described above.

The stress tensor τ is symmetric, since only the symmetric part of the elastic distortion contributes into the potential energy.

If we chose another set of four independent arguments in (8), we will come to another set of equations. However, all of them should allow transforming one set to another by the appropriate transformation laws. Among three tensors β_{ij}, β^P_{ij}, β^T_{ij} only two are linearly independent due to (1). So, we may choose any pair of the first triple; therefore, three combinations may be chosen. In his theory of micro-morphic media Mindlin [19] used β^T_{ij}, β_{ij}, κ_{ijk} as the set of arguments for W.

Therefore we got the following system of equations: Six symmetrized kinematics equations (2); four kinematics equations for the strain gradiens in the form of (6); five equations of equilibrium (15); twelve constitutive equations (11), (13).

Together with the boundary conditions (16) the listed above 27 equations form the closed system with respect to the following 27 unknowns: three components of the displacement vector, U^T_i; six components of the elastic deformation (the symmetric part of the elastic distortion), $\beta_{(ij)}$; two components of the plastic distortion, β^P_{31}, β^P_{32}; four components of rotation, κ_{131}, κ_{132}, κ_{231}, κ_{232}; six components of the volume stress, $\tau_{(ij)}$; two components of the interface stress, σ_{31}, σ_{32}; four components of moments μ_{131}, μ_{132}, μ_{231}, μ_{232}.

4. Relation to Salganik' equations

In case of the absence of volumetric moments, by differentiating the 4-th equation of (15) with respect to x_1 and the 5-th equation of (15) with respect to x_2 and substituting the resuts into the 3-d eauation of (15) with the help of (13) and (6), we may find:

$$\partial_1\sigma_{13} + \partial_2\sigma_{23} + \partial_3\tau_{33} + f_3 = -\frac{D}{2h}(\partial_{11} + \partial_{22})(\beta_{31,1} + \beta_{32,1}) \tag{17}$$

If we neglect the influence of the shear strain within the layers to compare the shear strain due to relative sliding (which is the case of small k/μ) we may write

$$\frac{1}{2}(\beta_{13} + \beta_{31}) = 0 \qquad \frac{1}{2}(\beta_{23} + \beta_{32}) = 0 \tag{18}$$

Substituting (18) into (17) and taking into account the fact that since $\beta^P_{13} = \beta^P_{23} = 0$ then $\beta_{13} = \partial_1 u_3$, $\beta_{23} = \partial_2 u_3$, and we may write

$$\partial_1\sigma_{13} + \partial_2\sigma_{23} + \partial_3\tau_{33} + f_3 = \frac{D}{2h}(\partial_{11} + \partial_{22})^2 u_3 \tag{19}$$

which coinsides with the equation obtained by Salganik [5].

5. Equation of motion

The equation of motion may be obtained by introducing inertia terms. For the sake of simplicity we restricts ourselves with the 2-D:

$$(\lambda + 2\mu)\frac{\partial^2 U}{\partial x^2} + k\frac{\partial^2 U}{\partial z^2} + \lambda\frac{\partial^2 W}{\partial x \partial z} + k\frac{\partial \omega_x}{\partial z} - \rho\frac{\partial^2 U}{\partial t^2} = 0$$

$$\lambda\frac{\partial^2 U}{\partial x \partial z} + \mu\frac{\partial^2 W}{\partial x^2} + (\lambda + 2\mu)\frac{\partial^2 W}{\partial z^2} - \mu\frac{\partial \omega}{\partial x} - \rho\frac{\partial^2 W}{\partial t^2} = 0 \tag{20}$$

$$\frac{D}{2h}\frac{\partial^2 \omega}{\partial x^2} + \mu\frac{\partial W}{\partial x} - k\frac{\partial U}{\partial z} - (\mu + k)\omega - \rho\frac{J}{2h}\frac{\partial^2 \omega}{\partial t^2} = 0$$

Here U and W denote the displacements u_1 and u_3, ω denotes κ_{131}; $J = 2h^3/3$ is the moment of inertia for the plate of thickness $2h$. It is supposed that there are no volume forces but inertia.

The solution of the plane wave type for equation (20) will be looked for in the form of

$$\begin{aligned} U &= U_0 e^{iq(\alpha x + \beta z - ct)} \\ W &= W_0 e^{iq(\alpha x + \beta z - ct)} \\ \omega &= \omega_0 e^{iq(\alpha x + \beta z - ct)} \end{aligned} \tag{21}$$

where $\pi = 2p/L$ is the wave number; L is the wave length; $\alpha = \cos\varphi$; $\beta = \sin\varphi$; φ - is an angle between the layering and the surface of equal phase. By substituting (21) into (20) the system of ordinary linear equation with respect to three unknowns is obtained; its nontrivial solution exist if the determinant vanishes.

$$\begin{vmatrix} (\lambda + 2\mu)q^2\alpha^2 + kq^2\beta^2 - \rho q^2 c^2 & \lambda q^2\alpha\beta & kq\beta \\ \lambda q^2\alpha\beta & \mu q^2\alpha^2 + (\lambda + 2\mu)q^2\beta^2 - \rho q^2 c^2 & -\mu q\alpha \\ -kq\beta & \mu q\alpha & \rho\frac{J}{2h}q^2c^2 - \frac{D}{2h}q^2\alpha^2 - (\mu + k) \end{vmatrix} = 0 \tag{22}$$

Equation (22) with respect to c^2 has three nonnegative real roots. On finding phase velocity, c, the displacements may be found by substituting the obtained value into (22) and then into (22). Let us consider the particular cases.

Wave propagating along layering: $\varphi = \pi/2$. Equation (22) reduces to:

$$\begin{vmatrix} (\lambda+2\mu)q^2-\rho q^2c^2 & 0 & 0 \\ 0 & \mu q^2-\rho q^2c^2 & -\mu q \\ 0 & \mu q & \rho\frac{J}{2h}q^2c^2-\frac{D}{2h}q^2-(\mu+k) \end{vmatrix}=0 \quad (23)$$

it has the following solutions:

$$c_1^2=\frac{\lambda+2\mu}{\rho}$$

$$c_{2,3}^2=\frac{\mu J+D+\frac{2h}{q^2}(\mu+k)\mp\sqrt{\left(\mu J+D+\frac{2h}{q^2}(\mu+k)\right)^2-4\mu J\left(D+\frac{2h}{q^2}\right)}}{2\rho J} \quad (24)$$

The first solution corresponds to longitudinal wave, the corresponding displacement has the traditional form:

$$U=U_0e^{iq(x-c_1t)}; \qquad W=\omega=0 \quad (25)$$

Velocities c_2 and c_3 correspond to the wave with flexural component:

$$U=0$$

$$W=i\omega_0\frac{\mu}{q(\rho c^2-\mu)}e^{iq(x-c_{2,3}t)} \quad (26)$$

$$\omega=\omega_0e^{iq(x-c_{2,3}t)}$$

These waves posses dispersion. For the long waves, $L/h>>1$, (24) has the following asymptotics:

$$\begin{aligned} \rho c_2^2&=\frac{\mu k}{\mu+k}-\frac{\mu^3k}{(\mu+k)^3}\frac{J}{2h}q^2+\frac{\mu^2}{(\mu+k)^2}\frac{D}{2h}q^2+... \\ \rho c_3^2&=\frac{\mu+k}{\frac{J}{2h}q^2}+\frac{\mu^2}{\mu+k}+\frac{D}{J}+... \end{aligned} \quad (27)$$

It is seen from (27) that the wave with velocity c_2 asymptotically approaches to the shear wave in a media with the effective shear modulus determined by

$$\tilde{\mu}=\frac{\mu k}{\mu+k} \quad (28)$$

Velocity c_3 is unbounded with increasing the wave length.

For the case of zero friction ($k = 0$) the expressions (24), (27) reduce to the solution for bending wave in plates obtained by Timoshenko [17].

Fig. 4. gives the relation between the phase velocities (c_2^2) and the dimensionless wave number qh calculated according to various models ([6], [10] and the present theory)and the exact solution [10]. It is seen that for the present set of parameters ($\rho = 1$, $\lambda = \mu = 1$, $k/\mu = 0.01$) the relative divergence among the results is small (it becomes significant for bigger k/μ), and all theories yield good results.

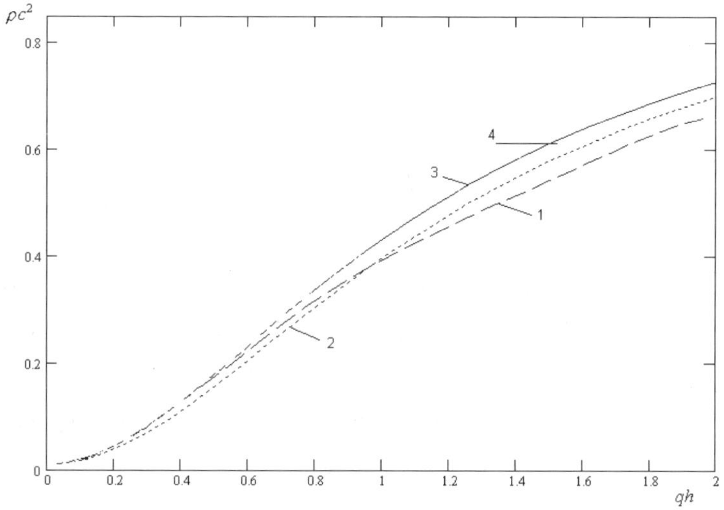

Figure 4. Phase velocities (c_2^2) versus the dimensionless wave number calculated according to various theories: (1)-exact solution [10]; (2)-[6]; (3)-[10]; (4)-the present theory.

Wave propagating normal to layering: $\varphi = 0$. Equation (19) reduces to:

$$\begin{vmatrix} kq^2 - \rho q^2 c^2 & 0 & kq \\ 0 & (\lambda + 2\mu)q^2 - \rho q^2 c^2 & 0 \\ -kq & 0 & \rho \frac{J}{2h} q^2 c^2 - (\mu + k) \end{vmatrix} = 0 \tag{29}$$

it has the following solutions:

$$c_1^2 = \frac{\lambda + 2\mu}{\rho}; \quad c_{2,3}^2 = \frac{\mu + k + k\frac{J}{2h}q^2 \mp \sqrt{\left(\mu + k + k\frac{J}{2h}q^2\right)^2 - 4\mu k \frac{J}{2h} q^2}}{2\rho \frac{J}{2h} q^2} \tag{30}$$

The first solution corresponds to longitudinal wave, the corresponding displacement has the traditional form:

$$W = W_0 e^{iq(z-c_1 t)}; \qquad U = \omega = 0 \tag{31}$$

Velocities c_2 and c_3 correspond to the wave with flexural component:

$$U = \frac{i\omega_0}{q(1-\rho c^2/k)} e^{iq(x-c_{2,3}t)}$$

$$W = 0 \tag{32}$$

$$\omega = \omega_0 e^{iq(x-c_{2,3}t)}$$

These waves posses dispersion. For the long waves, $L/h >> 1$, (30) has the following asymptotics:

$$\rho c_2^2 = \frac{\mu k}{\mu + k} + \frac{\mu k^3}{(\mu + k)^3} \frac{J}{2h} q^2 ...; \quad \rho c_3^2 = \frac{\mu + k}{\frac{J}{2h} q^2} + \frac{k^2}{\mu + k} + ... \tag{33}$$

It is seen from (32) that the wave with velocity c_2 asymptotically approaches to the shear wave in a media with the effective shear modulus determined by (28). Velocity c_3 is unbounded with increasing the wave length.

For the case of zero friction ($k = 0$) the mode corresponding to c_2 vanishes.

6. Conclusions

A new continuum model of the layered media is obtained on the base of a micromorphic elastic theory. The new media parameters are expressed in terms of elastic and geometric parameters of the original media.

The additional degree of freedom leads to the new type of waves, which nature is very close to the bending waves in beams and plates.

In case of asymptotically vanishing cohesion between the layers the solution for such a wave propagating along the layers approaches the solution for bending wave in plates of Timoshenko's type.

Author is grateful to Mr. A. Levitin for his kind assistance in preparing the manuscript. The investigation was supported by the Russian Fund of Basic Researches (Project 00-01-00376).

References

1. Lifshitz, I.M. and Rosenzweig, L.N. (1946) On the theory of elastic properties of polycrystals, *Jornal of Experimntal and Theoretical Physcs*, **16**, N 11.

2. Lifshitz, I.M. and Rosenzweig, L.N. (1951) Letter to the editor. On the theory of elastic properties of polycrystals, *Jornal of Experimntal and Theoretical Physcs*, **21**, No. 10.
3. Sonntag, G. (1957) Die in Schichten gleicher Dicke Reibungsfrei geschichtete Halbebene mit periodisch verteilter Randbeelastung. *Forsch. Geb. Ingenieur wesens*, B. **23**, N 1/2 3-8.
4. Salganik, R.L. (1987) Continuum approximation for describing the deformation of a laminar mass", *Mechanics of Solid*, **22**, N 3, 45-53.
5. Salganik, R.L. (1988) Deformation of a Straight Line Lamellar Massif, *Mechanics of Solid*, **23**, N 6, 20-28.
6. Sun, C.-T., Achenbach J.D., Herrmann, G. (1968) Continuum theory for a laminated medium *Trans. of the ASME J. of Applied Mechanics* **35** N 3 467-475.
7. Bolotin, V.V. (1964) Ob izgibe plit, sostoyzshih iz bolshogo chisla sloev. *Izv Ac Sci USSR Mechanica i mashinostroenie*, N 1, 61-66 (In Russian).
8. Bolotin, V.V., Novichkov Yu.N. (1980) *Mechanics of layered structures*, Moscow, Mashinostroenie (In Russian).
9. Molotkov, L.A. (1998) O metodah vyvoda, opisyvayshih effektivnye modeli sloistyh sred, *Zapiski nauchnyh seminarov POMI*, **250**, (In Russian).
10. Zvolinskii, N.V., Shkhinek, K.N. (1984) Continual model of laminar elastic medium, *Mechanics of Solids*, **19** (1) 1-9.
11. Muhlhaus, H.-B. (1993) Continuum models for layered and blocky rock. In *Comprehensive Rock Eng.*, Invited Chapter for Vol. II: Analysis and Design Methods, Pergamon Press, 209-230
12. Muhlhaus, H.-B. (1995) A relative gradient model for laminated materials. In Muhlhaus, H.-B. (ed) *Continuum Models for Materials with Micro-Structure*, Chap. 13
13. Adhikary, D.P., Dyskin A.V., Jewell, R.J. (1996) Numerical modelling of flexural deformation of foliated rock slopes. *Int. J. Rock Mech. Min. Sci. and Geomech. Abstr.*, **33**, N 6, 595-606.
14. Adhikary, D.P., Dyskin A.V. (1997) A Cosserat continuum model for layered media. *Computers and Geotechnics*, **20**, N 1, 15-46.
15. Ustinov K.B. (1999) On continuum modeling of layered medium, *Preprint No. 644 of Institute for Problems in Mechanics Russian Academy of Sciences* (In Russian).
16. Ustinov K.B. (2001) K voprosu postroenia kontinualnoi modeli sloistoi sredy, In *Thesises of VIII Russian National Congress on Theoretical and Applied Mechanics* (In Russian).
17. Timoshenko, S.P. (1921) On the correction for shear of the differential equation for transverse vibrations of prismatic bars, *Philosophical magazine*, Ser. 6, **41**, 744-746.
18. Kroner, E. (1980) Continuum Theory of Defects, In Balian, R, et al. eds., *Physics of Defects, Amsterdam, North-Holland*, 215-315.
19. Mindlin, R.D. (1964) Micro-structure in linear elasticity, *Archive of Rational Mechanics and Analysis*, N 1, 51-78.
20. Mokhel, A.N., Salganik, R.L. (1993) Development of slip concept in the theory of plasticity of materials and rocks. *Journal of the Mechanical Behavior of Material*, **4**, 343-351.

EDGE LOCALISED BENDING WAVES IN ANISOTROPIC MEDIA: ENERGY AND DISPERSION

D. D. ZAKHAROV
Inst. for Problems in Mechanics,
101-1 Vernadsky avenue, 117526 Moscow, Russia

Abstract. Existence, dispersion properties, velocities and energy of waves, localized near the stress-free edge of thin anisotropic plates, are investigated. Some qualitatively new effects are shown: velocity of Rayleigh type waves can be not minimal between bending waves; wave decay takes place with oscillations; under some type of anisotropy power flow can equal zero, and can change the sign. The famous Leontovich-Lighthill theorem does not hold any longer; despite the same sign of the phase and group velocities the power flow can have the same and opposite sign.

Key words: nonlinear surface waves, elasticity, Rayleigh waves, scale invariance, nonlocality, spectral kernel, spatial kernel, Hilbert transform

1. Introduction

Since the first paper of Yu. A. Konenkov (1960) [1] the bending waves of Rayleigh type (BWRT) have been treated by many authors (see, for example, [2-6]). But in contrast to the usual in-plane Rayleigh waves [7] which enable some non-destructive testing, seismic monitoring, etc, the interest was not very high due to the relatively small decay factor (of the order ν^4, where ν is Poisson's ratio) for isotropic media. So, the investigation was focused to a mathematical formalism. Next progress in this field concerned two directions. First devoted to the waves at the "interface" (edge by edge contact between plates [8,9]), which represent an analogue of the Stonely waves at bending.Second dealt with the edge waves in plates, immersed in fluid, interesting for the applications in hydroelasticity and mechanics of ice [10-13].

Actually, with the wide spread of different composite materials, many of which are highly anisotropic, the following questions become actual both for the theoretical and practical viewpoints:

- Do BWRT exist in media with general anisotropy;

R.V. Goldstein and G.A. Maugin (eds.),
Surface Waves in Anisotropic and Laminated Bodies and Defects Detection, 173–186.

- If they exist, what are their properties (energy, factor of the exponential decay, etc);
- What is the influence of layup in laminates to the wave properties (i.e., symmetrical layup or asymmetrical layup with coupled bending and stretching [14]). In the present paper two of the mentioned questions are considered and answered. Despite some interest, shown in [5,6] for the orthotropic materials with simplest orientation, the detailed investigation was absent.

2. Statement of the problem

Let us consider a thin laminate with symmetrical layup, made of perfectly joint anisotropic plies whose geometry is shown in Figure 1. Notice the total thickness of laminate as $2h$ and introduce the Cartesian coordinates

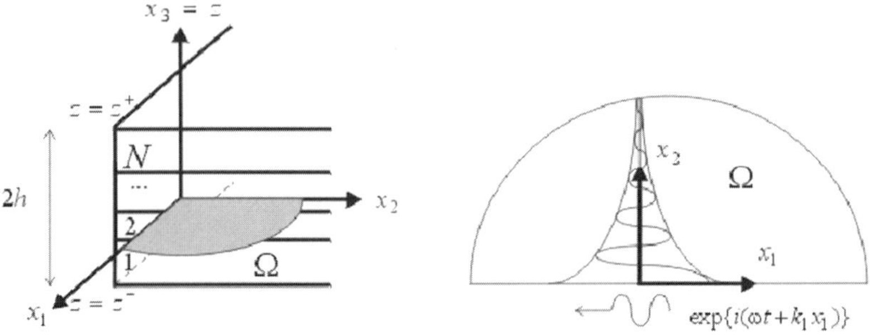

Figure 1. Laminate and layup (left) and BWRT in the coordinate half plane (right).

x_1, x_2, and $x_3 = z$ normalized over h. The internal stress-strain state (SSS) of the laminate is supposed to be long-wave, i.e. it satisfies the classical relations of the 2D anisotropic plate bending theory [15]. In what follows the dimensionless quantities are considered, namely the elastic moduli are normalized over the maximal Young's modulus of the plies, and the mass densities are normalized in a similar manner over the maximal mass density of the plies.The main relations and the equation of motion for the normal deflection w, slopes θ_α, longitudinal displacements u_α, torques $M_{\alpha\beta}$, and transversal forces $Q_{\alpha z}$ look as follows

$$\theta_\alpha = -\partial_\alpha w, \quad u_\alpha = z\theta_\alpha \quad (\alpha = 1,2),$$

$$M_{11} = -\left(d_{11}^3\partial_1^2 + d_{16}^3\partial_1\partial_2 + d_{12}^3\partial_2^2\right) w,$$

$$M_{12} = -\left(d_{16}^3\partial_1^2 + d_{66}^3\partial_1\partial_2 + d_{62}^3\partial_2^2\right) w,$$

$$Q_{\alpha z} = \partial_1 M_{1\alpha} + \partial_2 M_{2\alpha} \quad (1 \leftrightarrow 2), \quad \partial_\alpha Q_{\alpha z} = \rho \partial_t^2 w,$$

where $\mathbf{D} = \left\| d_{pq}^3 \right\|$ is a matrix of bending stiffness, t is time and ρ is dimensionless integral mass density. Investigate the harmonic oscillations of a semi-infinite laminate occupying a region $\Omega : x_2 \geq 0, -\infty < x_1 < \infty$ in its plane. The edge of laminate $x_2 = 0$ is supposed to be free of external loading.

Remark. The term *stress free edge* is used under known concept for thin plates, when the main SSS is subdivided into the internal SSS and the boundary layer (BL), not considered here. So, the internal SSS is described by the classical Kirchhoff's model of thin plate with respective integral boundary conditions, and all corrections are included in BL. Thus, Kirchhoff's theory is valid at the distance of a few h near the edge.

Then the normal bending $w = w_*(x_1, x_2)e^{i\omega t}$, normalized over h, satisfies the equation of oscillation with frequency ω

$$\left\{ L_3(\partial_1, \partial_2) - \rho\omega^2 \right\} w_* = 0, \tag{2.1}$$

$$L_3(\partial_1, \partial_2) \equiv d_{11}^3 \partial_1^4 + 4d_{16}^3 \partial_1^3 \partial_2 + 2(d_{12}^3 + 2d_{66}^3)\partial_1^2 \partial_2^2 + 4d_{26}^3 \partial_1 \partial_2^3 + d_{22}^3 \partial_2^4.$$

At the edge the following boundary conditions should be satisfied

$$M(\partial_1, \partial_2)w_* = 0, \quad F(\partial_1, \partial_2)w_* = 0 \tag{2.2}$$

$$\begin{bmatrix} M \\ F \end{bmatrix} = -\begin{bmatrix} d_{12}^3 \partial_1^2 + d_{26}^3 \partial_1 \partial_2 + d_{22}^3 \partial_2^2 \\ 2d_{16}^3 \partial_1^3 + (d_{12}^3 + 4d_{66}^3)\partial_1^2 \partial_2 + 4d_{26}^3 \partial_1 \partial_2^2 + d_{22}^3 \partial_2^3, \end{bmatrix}$$

where M and F are operators, responsible for the normal moment M_{22} and for the transversal Kirchhoff's shear force $P_{2z} = 2\partial_1 M_{12} + \partial_2 M_{22}$. Let us consider the existence of solutions, propagating along the edge with exponential decay when going far from the edge inside the laminate, i.e. desired BWRT

$$w_* = Ae^{i(k_1 x_1 + k_2 x_2)}, \quad A = Const, \quad \text{Im } k_2 < 0.$$

Since $x_2 \geq 0$ the latter inequality provides the exponential decay along the half-axis $x_2 \geq 0$ (see Figure 1). Introduce the notations

$$d_{pq} = \frac{d_{pq}^3}{d}, \quad s^4 = \frac{\rho\omega^2}{dk_1^4}, \quad \xi = \frac{k_2}{k_1},$$

where d is a certain bending stiffness, chosen for normalization (for example, a maximal one). For the definiteness, set $k_1 > 0$. Substitution of w_* into (2.1) yields the characteristic equation with constant coefficients for the variable ξ

$$L(1, \xi) - s^4 = 0, \tag{2.3}$$

$$L(1,\xi) \equiv d_{11} + 4d_{16}\xi + 2(d_{12} + 2d_{66})\xi^2 + 4d_{26}\xi^3 + d_{22}\xi^4$$

When using the normalized coefficients d_{pq} for the symbol of operator from (2.1) another notation $L(1,\xi)$ is introduced. The roots of equation (2.3) describe all kind of monochromatic waves and only the conditions (2.3) should be satisfied additionally. The following theorems hold.

Theorem 1. When the twist coupling stiffnesses are absent ($d_{16} = d_{26} = 0$) equation (2.3) can have pure imaginary roots ξ : Re$\xi = 0$; in opposite case roots are real or complex.

Under presence of the twist coupling stiffnesses the equation (2.3) immediately leads to a contradiction since for the pure imaginary ξ the real and imaginary part of the left hand side must be equal zero.

Theorem 2. All complex roots of equation (2.3) are conjugated; two pairs of complex conjugated roots (and respective BWRT) can exist only at

$$s < s_*, \quad s_*^4 = \underset{\xi\in R}{\infty} L(1,\xi).$$

In fact, the characteristic polynomial of forth order $L(1,\xi)$ is positively determined for real values ξ [14]. Since this polynomial is a smooth function, the equation (2.3) at $s = s_*$ has at least one real root of multiplicity 2. At $s > s_*$ the number of real roots is not less than 2. Thus, at $s \geq s_*$ not more than one pair of complex conjugated roots exists, and this is insufficient to satisfy two boundary conditions (2.2) with simultaneous exponential decay along the axis x_2.

Theorem 3. The BWRT phase velocity $V_R = -\omega/k_1$ has the upper bound

$$|V_R| < V_*, \quad V_* = s_*^2 k_1 \sqrt{d/\rho}.$$

This is a simple physical corollary of *Theorem 2.*

Remark. Notice that the natural direction of propagation for chosen wave is against the direction of axis x_1.

Denote the desired pair of complex roots as $\xi_{1,2}(\text{Im}\xi_{1,2} > 0\ )$. Boundary conditions (2.2) acquire the form

$$\det \mathbf{\Delta}(s) = 0, \tag{2.4}$$

$$\mathbf{\Delta}(s) \equiv \begin{bmatrix} d_{12} + 2d_{26}\xi_1 + d_{22}\xi_1^2 & d_{12} + 2d_{26}\xi_2 + d_{22}\xi_2^2 \\ f(\xi_1) & f(\xi_2) \end{bmatrix},$$

$$f(\xi) = 2d_{16} + (d_{12} + 4d_{66})\xi + 4d_{26}\xi^2 + d_{22}\xi^3,$$

$$\frac{A_2}{A_1} = -\frac{d_{12} + 2d_{26}\xi_1 + d_{22}\xi_1^2}{d_{12} + 2d_{26}\xi_2 + d_{22}\xi_2^2}, \tag{2.5}$$

$$w_*(x_1,x_2) = \{A_1 e^{i\xi_1 k_1 x_2} + A_2 e^{i\xi_2 k_1 x_2}\} e^{ik_1 x_1}.$$

Finally the question of BWRT existence is reduced to the investigation of the roots s of equation (2.4) at the branches $\xi_1(s), \xi_2(s)$.

3. Particular cases

In the particular case of an orthotropic medium, whose principal axes coincide with the axes x_1, x_2, the situation is essentially simplified. Then the twist coupling stiffnesses $d_{16} = d_{26} = 0$ and the characteristic equations (2.3), (2.4) have only pure imaginary roots $\xi_1(s), \xi_2(s)$. On choosing the coefficient $d = d_{22}$, from (2.3) and (2.4) we obtain the relations

$$\xi_{1,2} = i\left\{C \mp \sqrt{D+s^4}\right\}^{1/2}, \quad D = C^2 - \frac{d_{11}}{d_{22}}, \quad C = \frac{d_{12}+2d_{66}}{d_{22}}, \quad E = \frac{2d_{66}}{d_{22}},$$

$$f(s) \equiv \left\{\frac{E+\sqrt{D+s^4}}{E-\sqrt{D+s^4}}\right\}^2 \left\{\frac{C-\sqrt{D+s^4}}{C+\sqrt{D+s^4}}\right\}^{1/2} = 1 \; (f(s) = 1 \Leftrightarrow \det \mathbf{\Delta}(s) = 0).$$

At $s^4 \in \left[E^2 - D, C^2 - D\right]$ the function $f(s)$ varies from 0 to $+\infty$, i.e. the real root s of equation (3.2) exists and equals to

$$s = \left\{-D + CE\left(2 - 3a^2 + 2\sqrt{2(a^2 - \frac{1}{2})^2 + \frac{1}{2}}\right)\right\}^{1/4}, \quad a^2 = \frac{E}{C}, \qquad (3.1)$$

where the positive definiteness of the radicals follows from the positive definiteness of the tensor of elastic constants. The magnitude ratio (2.5) is evidently positive. The physical quantities are given as the real part of the solution $\mathrm{Re}\left\{w_*(x_1,x_2)e^{i\omega t}\right\}$, and for the slopes θ_1, θ_2 and longitudinal displacements u_1, u_2 the following expressions are obtained

$$\theta_\alpha = -\partial_\alpha \mathrm{Re}\left\{w_*(x_1,x_2)e^{i\omega t}\right\}, \quad (\alpha = 1,2)$$

$$u_1 = -z\mathrm{Re}\left\{ik_1 A_1\left[e^{-k_1|\xi_1|x_2} + \frac{A_2}{A_1}e^{-k_1|\xi_2|x_2}\right]e^{i(\omega t + k_1 x_1)}\right\},$$

$$u_2 = z\mathrm{Re}\left\{k_1 A_1\left[|\xi_1|\, e^{-k_1|\xi_1|x_2} + |\xi_2|\,\frac{A_2}{A_1}e^{-k_1|\xi_2|x_2}\right]e^{i(\omega t + k_1 x_1)}\right\}.$$

Thus, for a real magnitude A_1 the displacements u_1, u_2 (and the slopes θ_1, θ_2) are harmonic functions with the phase difference $-\pi/2$. During a full period the trajectory of an arbitrarily chosen point (x_1, x_2) is an ellipse, whose direction is counterclockwise and the semi axes decay exponentially when going far away from the edge.

In the particular case of an isotropic (and of a transversely isotropic) medium the equality (3.1) leads to the same relation, as found by Konenkov [1]

$$s = \left\{(1-\nu)\left(3\nu - 1 + 2\sqrt{1-2\nu+2\nu^2}\right)\right\}^{1/4},$$

where ν is Poisson's ratio.

It is also interesting what is the qualitative behavior of the phase velocity V_R and its ratio over the velocity V_B of the ordinary bending wave. Consider the wave vector $\mathbf{k}$ with the angle disclination φ with respect to the axis x_1. After replacing in equation (2.1) ∂_1 and ∂_2 by $|\mathbf{k}|\cos\varphi$ and $|\mathbf{k}|\sin\varphi$, respectively, the velocities V_B, V_R and their ratio $r(\varphi)$ are given by formulas

$$\begin{aligned} V_B &\equiv \frac{\omega}{|\mathbf{k}|} = \left[\frac{\omega^2 L_3(\cos\varphi,\ \sin\varphi)}{\rho}\right]^{1/4}, \\ V_R &= -\frac{\omega}{k_1} = -s\left[\frac{\omega^2 d_{22}^3}{\rho}\right]^{1/4}, \\ r(\varphi) &= \frac{|V_R|}{V_B} = s\left\{\frac{d_{22}^3}{L_3(\cos\varphi,\ \sin\varphi)}\right\}^{1/4}. \end{aligned} \tag{3.2}$$

In the case of isotropic materials the ratio $r(\varphi)$ is always less than unit [1]. Another important characteristic is the value of the averaged power flow $\Im$ across the section of a plate, normal to the direction of propagation x_1. As usual it is introduced using the integral of the product of stresses resultants and couples and respective speeds $u_\alpha^\bullet$, $\theta_\alpha^\bullet$ and $w^\bullet$, averaged over the period of oscillations [16]. Due to the definition, for the complex form of physical quantities

$$\Im = -\frac{\omega}{2\pi}\int_0^{\frac{2\pi}{\omega}} dt \int_0^{+\infty} \{\mathrm{Re}\theta_1^\bullet \mathrm{Re}M_{11} + \mathrm{Re}\theta_2^\bullet \mathrm{Re}M_{12} + \mathrm{Re}w^\bullet \mathrm{Re}Q_{1z}\}\, dx_2. \tag{3.3}$$

4. Some numerical results

For numerical illustration we choose two types of orthotropic materials: material T300/epoxy (T) with Young's moduli $E_1 = 130000, E_2 = 9750$, shear modulus $G_{12} = 6000$ $[N/mm^2]$, Poisson's ratio $\nu_{12} = 0.27$ and mass density $\rho = 1.58$ $[g/cm^3]$; and material E-glass (E) with constants $\rho = 2$, $E_1 = 45000, E_2 = 13000, G_{12} = 4400, \nu_{12} = 0.29$. The principal axes coincide with the coordinate axes (T/0) or are rotated about the angle $\pi/2$ (T/90); the thickness is $2h = 1[mm]$, coefficient $d = \max(d_{11}, d_{22})$.

The characteristic values s and the normalized power flow of BWRT are given in the Table1.

TABLE 1.

	T/0	T/90	E/0	E/90
s	0.9983	0.7323	0.9920	0.5234
$\min(\operatorname{Im}\xi_1,\ \operatorname{Im}\xi_2)$	0.01743	0.04483	0.04103	0.02254
$\dfrac{\Im}{\omega d \left\|k_1 A_1\right\|^2}$	-1.341	-0.3715	-1.734	-0.9601

The plots of the resultant velocity ratio $r(\varphi)$ are shown in Figure 2. As seen for different materials $r(\varphi)$ can be smaller or greater than unit.

Thus, for the standard orthotropic material orientation we may conclude, that

- The BWRT exists;

- The decay factor $\min(\operatorname{Im}\xi_1, \operatorname{Im}\xi_2)$ can be higher than for isotropic medium, where this value is no more than 10^{-2};

- In contrast to the case of isotropic medium [1] the phase velocity of BWRT is no longer minimal between all possible bending waves (see Figure 2).

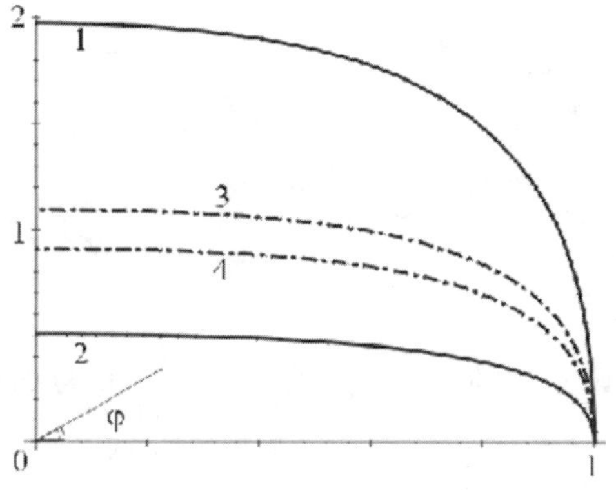

Figure 2. Plots $r(\varphi)$ in polar coordinates for materials T/0 (curve 1), T/90 (2), E/0 (3) and E/90 (4).

5. Case of arbitrary anisotropy

In the case of general anisotropy ($d_{16}, d_{26} \neq 0$) the analytical formulas for the roots $\xi_1(s), \xi_2(s)$ and characteristic value s are no longer available and found numerically. The procedure looks as follows: the parameter s is set and from (2.3) two branches $\xi_1(s), \xi_2(s)$ are calculated and substituted into (2.4) (real and imaginary part of $\det \mathbf{\Delta}(s)$). Let us show results for materials (T) and (E), whose principal axes x'_1, x'_2 are obtained by a rotation of the axes x_1, x_2 about the angle $0 < \psi < \pi/2$.

Plots of the functions $s(\psi)$ are shown in Figure 3. The respective branches $\xi_1(\psi), \xi_2(\psi)$ are presented in Figures 4-5.

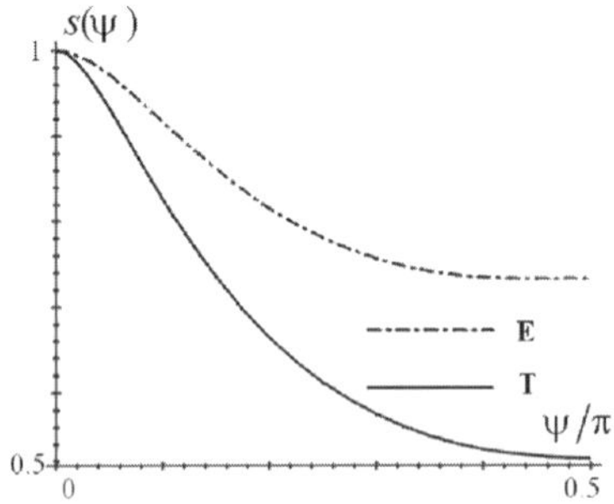

Figure 3. Curves of frequency parameter for T- and E-materials

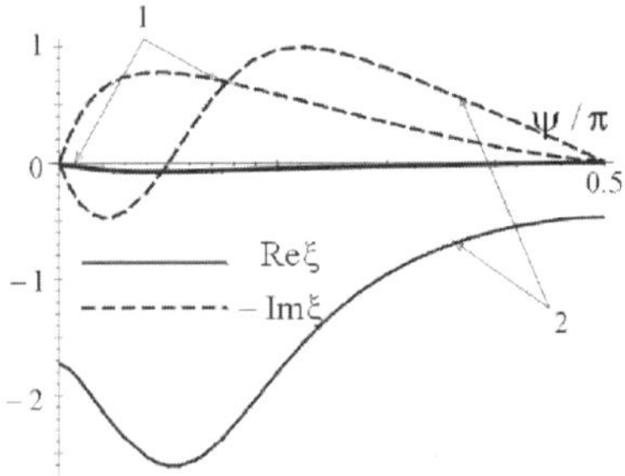

Figure 4. Roots $\xi_1(\psi), \xi_2(\psi)$ for T-materials

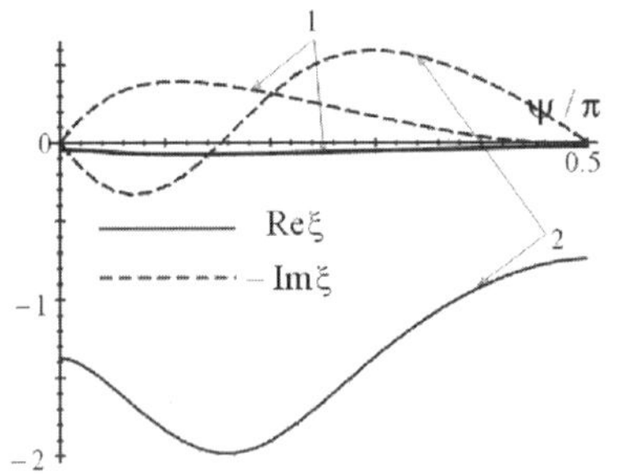

Figure 5. Roots $\xi_1(\psi), \xi_2(\psi)$ for material E

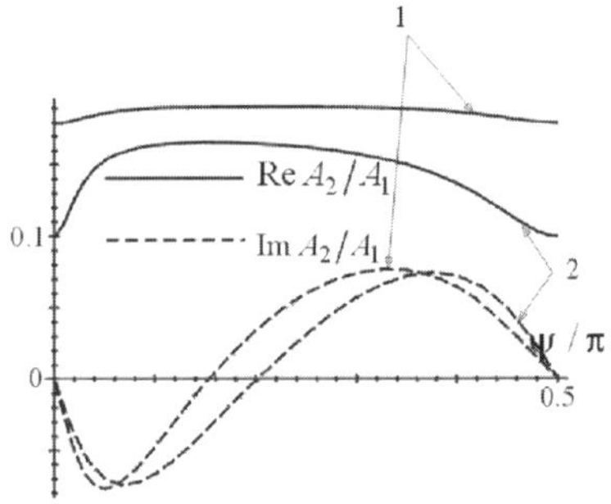

Figure 6. The magnitude ratio A_2/A_1 function of ψ for materials T (index 1) and E (index 2)

The magnitude ratio (2.5) is shown in Figure 6 and confirms that none of components can be neglected and the decay factor is defined by $\Im\xi_1(\psi)$. This decay factor is shown in Figure 4, 5 by solid curves 1. As seen for (T) and (E) materials the exponential decay factor can acquire the values of the order 10^{-1} (at $\psi \approx 0.11\pi$ (T) and $\psi \approx 0.162\pi$ (E), respectively), which are ten times more that for the isotropic case [1].

6. Average power flow behaviour

The behavior of the power flow $\Im(\psi)$ is of special interest and plots of this normalized function are shown in Figure 7. For $k_1 > 0$ the power flow seems to be negative. However, for both materials there exists a critical value $\psi_* = 0.1249\pi(\mathrm{T}), 0.1767\pi(\mathrm{E})$ of the orientation angle for which the desired wave becomes steady, and then ($\psi > \psi_*$) it changes the direction of energy transfer. At the next critical angle $\psi_{**} = 0.1848\pi(\mathrm{T}), 0.2268\pi(\mathrm{E})$ the wave becomes steady again and under $\psi > \psi_{**}$ the sign of the power flow is the same as the initial one. This fact is new and observed only for a medium with *general anisotropy* ($d_{16}, d_{26} \neq 0$). For isotropic media and for orthotropic media with standard orientation such a fact cannot be realized in principle.

To clarify the effect consider the generalized acoustical impedances I_m and I_p introduced as follows

$$I_m^* = \frac{M_{11}}{\theta_1^\bullet} = -\frac{I_m}{V_R}, \quad I_p^* = \frac{P_{1z}}{w^\bullet} = -\frac{k_1^2 I_p}{V_R}, \quad I = I_m + I_p,$$

$$\theta_1 = -ik_1 w, \quad \theta_1^\bullet = \omega k_1 w, \quad w^\bullet = i\omega w.$$

Integration by parts the right hand side of (3.5) leads to the formulas

$$\Im = \Im_0 + \mathrm{Re}\left\{\mathrm{Re}w^\bullet \mathrm{Re}M_{12}\right\}\big|_{x_2=0}, \tag{6.1}$$

$$\Im_0 \equiv -\frac{\omega}{2\pi}\int\limits_0^{\frac{2\pi}{\omega}} dt \int\limits_0^{+\infty} \left\{\mathrm{Re}\theta_1^\bullet \mathrm{Re}M_{11} + \mathrm{Re}w^\bullet \mathrm{Re}P_{1z}\right\} dx_2 =$$

$$= \frac{\omega^2 k_1^2}{2V_R}\int\limits_0^{+\infty} |w|^2\, \mathrm{Re}\left(I_m + I_p\right) dx_2.$$

Remark. The way to introduce the generalized impedances is not unique here. The chosen style deals with such quantities as bending moment and Kirchhof's force, which directly participate in the formulation of boundary conditions. It also allows us to proceed to the magnitudes of the normal deflection for J_0 .

The main contribution into power flow $\Im$ gives the variable J_0. Due to the simplest analogue for the harmonic oscillator with dissipation energy, subjected to the action of any external force F, consider a point of mass m on the spring with stiffness c under linear viscous friction with coefficient b. Its generalised acoustical impedance is

$$I^* \equiv \frac{F}{y^\bullet} = \frac{my^{\bullet\bullet} + by^\bullet + cy}{y^\bullet} = i\left(m\omega - \frac{c}{\omega}\right) + b, \tag{6.2}$$

and one can expect a positive value of $\operatorname{Re} I^*$ and the possible change of the sign of the imaginary part. So, for BWRT the most naturally would be the constant sign of $\operatorname{Re}(I_m + I_p)$ and possible sign changing of $\operatorname{Im}(I_m + I_p)$.

Plots of the real and of the imaginary parts of the generalized impedances $I_m, I_p, I_m + I_p$ are shown in Figures 8-10 for three values of the angle ψ :outside the interval $[\psi_*, \psi_{**}]$ (Figures 8, 10) and inside this interval (Figure 9). As seen, outside the interval $[\psi_*, \psi_{**}]$ the analogue with (6.2) holds, but for the intermediate value of ψ one can single out an active zone near the edge with opposite direction of energy flow. This is a physical reason of the sign changing of the integral $\Im$. It should be also noticed, that zeros $\Im$ and of $\operatorname{Re}(I_m + I_p)$ do not coincide and their shifting is explained by the presence of additional terms in (6.1).

Such situation occurs for any intermediate value of the angle ψ. In particular, on considering $\operatorname{Re}(I_m + I_p)$ in the point $x_2 = 0$ under different ψ (see Figure 7, this function is always positive at the origin) it is seen that this function changes its sign at other critical values (e.g.$\psi' \approx 0.062\pi, \psi' < \psi_*$ and $\psi'' \approx 0.241\pi, \psi'' > \psi_{**}$ for material (T)). Inside the interval $[\psi', \psi'']$ the value of $\operatorname{Re}(I_m + I_p)$ remains negative and determines the sign changing of the total power flow.

The overall situation looks as follows: at small ψ the density of power flow (sub integral function in (3.5)) at $x_2 = 0$ is positive, then at a certain value of angle it equals zero and remains negative for larger ψ. Obviously, for $x_2 >> 1$ the power flow density is always positive, so an intermediate zone, where the power flow density is negative, exists near the edge. Since the total power flow is obtained by integration for all x_2, its sign will change when this zone is large enough to give the leading contribution into $\Im$. Hence, such zone with a reverse power flow appears at $\psi < \psi_*$ and disappears at $\psi > \psi_{**}$, because for ψ near $\pi/2$ the plate behavior is qualitatively similar to one at small ψ.

7. Leontovich-Lighthill theorem

The obtained results confirm the importance of the adequate formulation of the energy radiation principle for the problems of dynamic bending and do not contradict to the classical energy relations. Remember the known formulation of the energy law in differential form

$$\partial_t e + div\, \mathbf{p} = 0, \tag{7.1}$$

where e is a density of total energy and $\mathbf{p} = (p_1, p_2)$, $p_\alpha = \operatorname{Re}\theta^\bullet_\beta \operatorname{Re} M_{\alpha\beta} + \operatorname{Re} w^\bullet \operatorname{Re} Q_{\alpha z}$ are components of the Umov-Pointing vector [16]. On varying the frequency and the wave number

$$\omega = \omega_0 + i\delta\omega, \quad k_1 = k_0 + i\delta k, \tag{7.2}$$

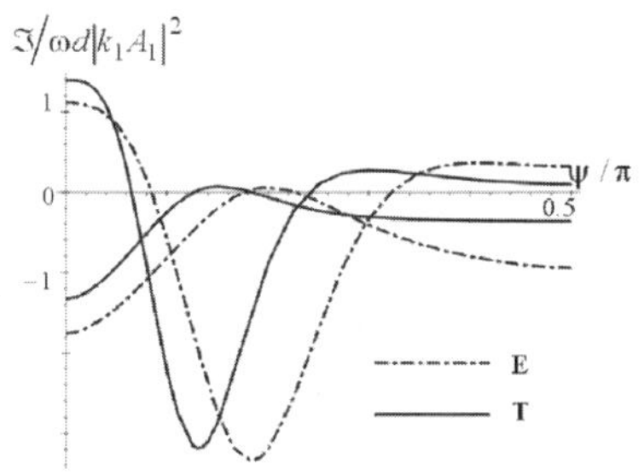

Figure 7. Normalized averaged power flow $\psi < \psi_*$ (T)

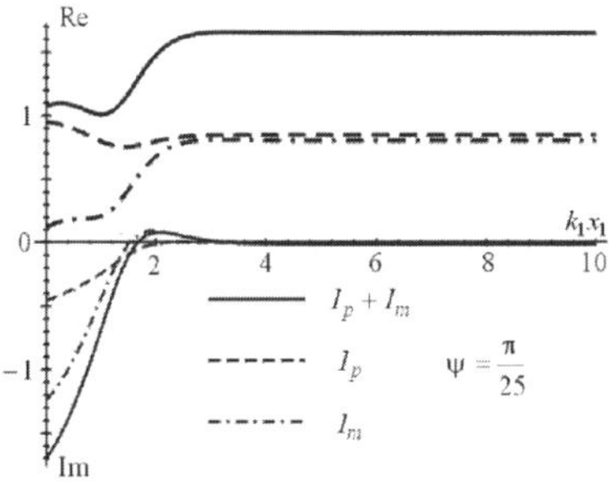

Figure 8. Normalized impedances for and acoustic impedances $\mathrm{Re}\,(I_m + I_p)$ for $x_2 = 0$ (the latter are positive in the origin)

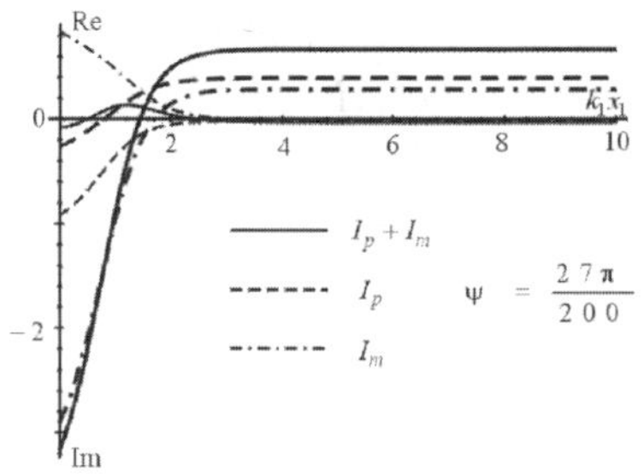

Figure 9. Normalized impedances for $\psi_* < \psi < \psi_{**}$

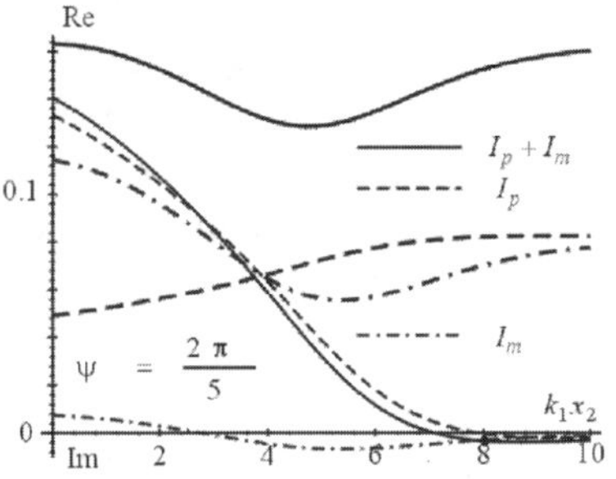

Figure 10. Normalized impedances for $\psi > \psi_*$

and on introducing the group velocity $V_G = -\delta\omega/\delta k$, after some transformations [17] and after integration and averaging (3.5), it is easy to obtain

$$\Im - V_G\tilde{e} = \frac{1}{2}\,\mathrm{Re}\left\{w^{\bullet}\bar{M}_{12}\right\}\big|_{x_2=0}, \text{where} \quad \tilde{e} = \frac{\omega}{2\pi}\int_0^{\frac{2\pi}{\omega}} dt \int_0^{+\infty} e dx_2. \qquad (7.3)$$

Thus, $V_G \neq \Im/\tilde{e}$ any longer. In fact, when averaging over period and using complex representation for variables like $A = A_* e^{i(\omega t + k_1 x_1 + k_2 x_2)}$ and $B = B_* e^{i(\omega t + k_1 x_1 + k_2 x_2)}$ the final contribution into the averaged value of product $\mathrm{Re}(A)\mathrm{Re}(B)$ is given only by $1/2\mathrm{Re}(A\bar{B})$ ($\delta\omega \to 0$). It leads to the appearance of the exponential term $\exp\left\{-2\left(t\delta\omega + x_1\delta k + x_2\mathrm{Im}k_2\right)\right\}$ in each product and to the replacement for the outer derivatives by $\partial_t \to -2\delta\omega$,

$\partial_1 \to -2\delta k$ for components of e and p_1. Since by virtue of the exponential decay at the infinity and the boundary conditions at the edge

$$p_2 = \text{Re}\theta_2^{\bullet}\text{Re}M_{22} + \text{Re}w^{\bullet}\text{Re}P_{2z} - \partial_1\left(\text{Re}w^{\bullet}\text{Re}M_{12}\right),$$

$$\int\limits_0^{+\infty} \partial_2 p_2 dx_2 = p_2|_0^{+\infty} = -\left.p_2\right|_{x_2=0} = \partial_1\left.\left(\text{Re}w^{\bullet}\text{Re}M_{12}\right)\right|_{x_2=0},$$

from equation (7.1) we obtain

$$-2\delta\omega e - 2\delta k p_1 + \partial_2 p_2 + (...) = 0.$$

After integration over x_2 and averaging by time the terms in brackets disappear and we finally arrive at (7.3).

As a result the group velocity does not coincide with the ratio of the averaged power flow over the averaged energy density (7.3), what is the essence of *Leontovich-Lighthill theorem* and of the group velocity criterion [16]. Thus, Leontovich-Lighthill theorem *does not hold* with classical formulation for plane waves. The definition of the variable s, which depends only on the stiffness, yields the expression for group velocity

$$V_G = -\frac{d\omega}{dk_1} = \frac{2k_1 s^4 d}{\rho V_R},$$

which confirms the sign coincidence of the phase and group velocities. In the meantime it justifies the correctness of the chosen variation (7.2) for the frequency and wave number.

In principle, the situation with a changeable sign of the average power flow is familiar for layered structures. Even some of the classical plane Lamb's waves in isotropic layer with stress free faces possess such property [16]. However, this fact remains in accordance with the changeable sign of the group velocity and with the Leontovich-Lighthill theorem. Cases are also known where this theorem is inapplicable. For example, for 3D Lamb's waves it does not hold [17], but for the waves with a small front curvature (or at infinity, where in each point the front curvature of the 3D wave is neglected) it holds for a leading part of the energy and of the average power flow [17]. Hence, the energy radiation principle can be formulated equally for the average power flow and for the group velocity of waves. *The case of plane waves, considered in presented paper is different.* The sign of group velocity is no longer equal to the sign of power flow due to the formulation of boundary conditions in the Kirchhoff's theory of thin plates. And the sign changing of the power flow itself becomes a property of the anisotropy orientation. Notice again that this effect cannot be realized for isotropic materials.

8. Conclusion

So, we can conclude that bending waves of Rayleigh type in anisotropic plates can be qualitatively different from similar waves in isotropic plates, as well as from the classical Rayleigh waves under plane strains or under plane stresses. These properties are caused by the bending stiffnesses (and especially by the twist coupling stiffnesses) and by the boundary conditions in the Kirchhoff's theory of thin plates. Consequently, in our consideration only integral values of the stiffnesses are important and all main effects hold for the plates, made of one or more plies (with symmetrical layup).

Acknowledgement

This work is performed under partial support of INTAS project 96-2306, which is gratefully acknowledged. References

References

1. Yu. K. Konenkov (1960). A Rayleigh-type flexural wave. *Soviet Acoust. Physics* **6**, 124-126.
2. R. N. Thurston and J. McKenna (1974). Flexural acoustic waves along the edge of the plate. *IEEE Transaction of Sonics and Ultrasonics* **21**, 296-297.
3. B. K. Sinha (1974). Some remarks of propagation characteristics of ridge guides for acoustic waves at low frequencies. *J. Acoust. Soc. of America* **56**, 16-18.
4. C. Kauffmann (1998). A new bending wave solution for the classical plate equation. *J. Acoust. Soc. of America* **104**, 2220-2222.
5. M. V. Belubekyan and I. A. Engibaryan (1996). Waves, localized along the stress-free edge of plate with cubic symmetry. *Mechanics of Solids [MTT]* **6**, 139-143.
6. A. N. Norris (1994). Flexural edge waves. *J. of Sound and Vibrations* **174**, 571-573.
7. P. Chadwick and G. D. Smith (1977). Foundations of the theory of surface waves in anisotropic elastic materials. In: *Advances in Applied Mechanics.* Academic Press, New York, 17, 303-376.
8. A. S. Zilbergleit and I. B. Suslova (1983). Contact flexural waves in thin plates. *Soviet Acoust. Physics* **29**, 186-191.
9. D. P. Kouzov, T. S. Kravtsova and V. G. Yakovleva (1989). On the scattering of the vibrational waves on a knot contact of plates. *Soviet Acoust. Physics* **35**, 392-394.
10. D. P. Kouzov and V. D. Louk'yanov (1972). On the waves propagating along the edges of plates. *Soviet Acoust. Physics* **18**, No. 4, 129-135.
11. R. V. Goldstein and A. V. Marchenko (1983). The diffraction of plane gravitational waves by the edge of an ice corner. *J. Appl. Maths. Mechs. [PMM]* **53**, 731-736.
12. D. Abrahams and A. N. Norris (2000). On the existence of flexural edge waves on submerged elastic plates. *Proceed. of Royal Soc. London* A **456**, 1559-1582.
13. A. N. Norris, V. V. Krylov and I. D. Abrahams (2000). Flexural edge waves and Comments on "A new bending wave solution for the classical plate equation". *J. Acoust. Soc. of America* **107**, 1781-1784.
14. D. D. Zakharov and W. Becker (2000). 2D problems of thin asymmetric laminates. *ZAMP* **51**, 555-572.

15. S. G. Lekhnitskii (1968). *Anisotropic plates.* Gordon & Breach Scientific Publishers, New York.
16. B. A. Auld (1990). *Acoustic fields and waves in solids.* 1,2. Krieger Publishers, Malabar.
17. D. D. Zakharov (1988). Generalized orthogonality relations for eigenfunctions in 3D dynamic problem for elastic ply. *Mechanics of Solids [MTT]* **6**, 62-68.

Part III

Experimental, numerical and semi-analytical methods for analysis of surface waves

SURFACE ELECTROMAGNETIC PERTURBATIONS INDUCED BY UNSTEADY-STATE SUBSURFACE FLOW

P. M. ADLER
Institut de Physique du Globe de Paris
Tour 24, 4, Place Jussieu, 75252 Paris, Cedex 05, France

V. M. ENTOV
Institute for Problems in Mechanics
of the Russian Academy of Sciences
101-1, prospekt Vernadskogo, 119526, Moscow, Russia

Key words: nonlinear surface waves, elasticity, Rayleigh waves, scale invariance, nonlocality, spectral kernel, spatial kernel, Hilbert transform

1. Introduction

The electrokinetic effect consists in generation of electric current by fluid flow through porous media and in the reverse effect of inducing flow by application of an electric field. Its primary cause is the difference in mobilities of ions, some of which are fixed at the surface of the solid matrix of the porous medium, while some dissolved counterions can move with the pore fluid, or force it to move, if an electric field is applied. Macroscopically, the flow and electric current are described by the equations

$$\mathbf{u} = -\frac{k}{\eta}\nabla p + \beta\nabla\psi, \quad \mathbf{I} = -S[\nabla\psi - C\nabla p], \quad C = \alpha/S. \tag{1.1}$$

Here, k is the medium permeability, η is the fluid viscosity, S is the medium electric conductivity customarily expressed in terms of the fluid conductivity σ_f and the medium formation factor F as $S = \sigma_f/F$ with F up to several tens. The Onsager relation holds between the coupling coefficients C and β: $\beta = \alpha \equiv CS$. Details are given in [26, 9, 8].

R.V. Goldstein and G.A. Maugin (eds.),
Surface Waves in Anisotropic and Laminated Bodies and Defects Detection, 189–204.

All these macroscopic quantities should be replaced by tensors when the porous medium is anisotropic at the local scale. Such a formal complication was deliberately avoided here and more details can be found in [2].

Both "streaming potentials", i.e., electric fields generated by fluid flow, and "electroosmotic flow", i.e., the flow driven by the electric potential gradient, have important applications. Here we will consider flows with significant pressure gradients and small variations of the electric potential, for which the flow field proves to be independent of the electric field, while the current and streaming potential are generated by flow. Then the *electrokinetic coupling coefficient* C turns out to be the most important macroscopic property of the system. This lump property depends on many fine details of geometry, surface properties of the matrix, *pH*, salinity and ionic composition of pore brine, etc. In the limit of thin double layers, Overbeek [26] expressed C in terms of the zeta-potential ζ of the matrix surface, of the fluid conductivity σ_f, of its viscosity η and of the dielectric constant ϵ as

$$C = \epsilon\zeta/(\eta\sigma_f). \tag{1.2}$$

Effects of nonplanar geometry and of finite thickness of the double layer are discussed in [8, 23, 2].

Values of the viscosity η and of the dielectric constant ϵ are relatively constant, and do not vary much: $\eta = 10^{-3}\ Pa \cdot s$, $\epsilon = 7 \times 10^{-10}\ F\ m^{-1}$. However, both the fluid conductivity and the zeta potential vary in wide ranges, so that estimates of the coupling coefficient C are rather ambiguous, and range for geologically significant situations from -4 mV/MPa to - 4905 mV/MPa [19-21]. For our purposes here, it is sufficient to have some general orders of magnitude and trends, namely:

i The order of magnitude of the ECC = Electrokinetic Coupling Coefficient is $C = (100 - 1000)\ mV/MPa$ for monovalent ions solutions, such as NaCl.

ii The ECC may be different for different porous media, and generally it decreases with the permeability or with the characteristic size of the pore space.

The main reason is that the internal resistance of the electrokinetic effect as a source of electric potential (or of an electromotive force) increases with decreasing permeability, and therefore bypassing currents due to surface conductivity of the rock matrix become relatively more significant. The electrokinetic signals should be sufficiently large (in fact, of order of several tens of mV) to be detected. This requirement provides an estimate of the pressure drop driving the flow; it should be of the order of 1 MPa or more, to be detectable; larger values are necessary for identification of the transient phenomena, such as earthquake precursors or co-seismic signals. Some characteristic examples of possible sources of electrokinetic signals are:

i Hydraulic fracturing, whether artificial (intentional fracturing of low permeable reservoirs to improve oil - and gas wells productivity or create pathways for extracting geothermal energy; unintentional reservoir fracturing during well cementing or water injection) or natural (magma intrusion into pre-existing fractures).

ii Earthquakes and volcanic activity : fast deformation, faulting and fracture of rocks to release accumulated elastic strains accompanied by large-scale ground water flows [1, 3, 5, 6, 11-13, 18, 24, 25, 27, 28-32].

iii Compaction of sediments, especially within accretion zones adjacent to subduction areas [16,17,19].

Here, we consider a specific situation that may prove to be applicable to dynamical tectonically generated flows, such as pre- and co-seismic ones.

The paper is organized as follows. First, it is shown in Section 2 that the current is necessarily zero in any porous domain bounded by an insulating boundary, provided that the electrokinetic coupling coefficient is uniform throughout the domain. Therefore, non-homogeneity and/or current leakage is a prerequisite of existence of current and magnetic field.

Then an unsteady-state flow which is caused by instantaneous slip along a vertical plane, the deformation pattern characteristic for earthquakes,is considered in Section 3. We assume that the slip is accompanied by normal (opening) displacement due to geometric non-conformity of two shores of the fracture or slip surface. Therefore, it results in void development along the fault surface that eventually should be filled with pore water. Therefore, the slip generates a flow within the saturated porous medium which corresponds to a distributed instantaneous sink. The transient flow, in its turn, generates transient electric and magnetic fields due to electrokinetic phenomena. We are going to estimate characteristic values of these fields for a simplest case. Namely, we assume that the flow is confined to a porous layer of finite thickness over a homogeneous conducting layer. A one-dimensional pressure wave propagates in the porous layer normally to the slip plane. This wave generates two-dimensional electric and magnetic fields (Sections 3 and 4). Their prediction is the main objective of this paper.

2. A general property of the electrokinetic equations

Consider the electrokinetic potential generated in saturated porous media neglecting by neglect the effect of the induced electric field on the flow. The following equations hold for the current and the electric potential:

$$\boldsymbol{I} = -S(\nabla\psi - C\nabla p); \tag{2.1}$$

$$\nabla \cdot \boldsymbol{I} = -\nabla \cdot S(\nabla\psi - C\nabla p) = 0. \tag{2.2}$$

The last equation follows from the basic assumption of absence of free charges; it is assumed to be fulfilled throughout the medium. Essentially, Eq.(2.2) is just an elliptic equation for the potential distribution, provided that the pressure distribution is known. When the electrokinetic coupling coefficient is constant, $C = const$, (2.2) has an obvious solution

$$\psi = Cp + const. \tag{2.3}$$

Obviously, there is no current in this case.

Now, we will discuss a generalization of this property. Let us consider an arbitrary (possibly time dependent) pressure field $p(\mathbf{x}, t)$ and the corresponding electrokinetic potential $\psi(\mathbf{x}, t)$ within a domain D of a porous medium with, generally, non-uniform properties $\boldsymbol{k}(\boldsymbol{x})$, $\boldsymbol{S}(\boldsymbol{x})$ where the permeability $\boldsymbol{k}$ and the conductivity $\boldsymbol{S}$ are symmetric second-order tensors with positive eigenvalues.

When C is constant, Eq.(2.2) can be written as

$$\nabla \cdot (\boldsymbol{S} \cdot \nabla\chi) = 0 \ ; \ \ \boldsymbol{I} = -\boldsymbol{S} \cdot \nabla\chi \tag{2.4}$$

where χ is referred to as "the current potential" in a medium with constant C, and it is defined by

$$\chi = \psi - Cp. \tag{2.5}$$

Equation (2.5) is an elliptic second order partial differential equation, and therefore its solutions share common properties of solutions to elliptic equations, in particular:

i χ assumes its maximum and minimum values at the boundary Γ of the flow domain D ("the maximum principle");

ii If the current potential is not constant throughout the flow domain, the normal derivative $\partial\chi/\partial n$ along the outward normal to the boundary is positive at the boundary points where the maximum value is reached, and negative, at the points, where the minimum value is reached; this is the Hopf maximum principle see, for example, [7, 14].

As a result,

$$\chi \equiv const; \quad S \cdot \nabla\chi = \mathbf{I} = 0 \tag{2.6}$$

for any flow within a domain D with an insulating boundary, i.e., a boundary along which

$$I_n(\mathbf{x}) = -\mathbf{n}\cdot\boldsymbol{S}\cdot\nabla\chi = 0, \quad \mathbf{x} \in \Gamma = \partial D. \tag{2.7}$$

This simple property implies that there is no current (and, hence, no magnetic field generated) within any porous domain bounded by an insulating boundary, provided that the electrokinetic coupling coefficient is uniform throughout the domain.

Inconsistency of boundary conditions for pressure and electric potential, as well as non-uniformity in distribution of the coupling coefficient C and conductivity S, are the most important sources of the electric field perturbations induced by the pressure perturbations to be studied further. This property was explicitly stated by Fitterman [11], who pioneered theoretical studies of electric and magnetic anomalies caused by pore fluid flow (see also [18,19]).

The problem, considered here, is chosen taking into account this property.

3. Electrokinetic potential induced by an unsteady flow

In an unsteady flow, the pressure distribution is described by a diffusion-type equation while the potential distribution is still described by an elliptic equation with an infinite propagation rate. This opens additional possibilities for generation of transient electric and magnetic signals. However, as it follows from the general conclusion of Section 2, if the flow domain is bounded by an insulating boundary, and if the distribution of the coupling coefficient in this domain is uniform, the electric current vanishes identically, while the electrokinetic potential just follows the pressure distribution. Non-trivial effects may be observed only if at least one of these limitations is removed. Some of these cases are analyzed below.

Essentially, we are considering two horizontal layers. The upper layer of thickness H is both conducting and permeable; the lower one of thickness H_2 is conducting, but impermeable (or separated from the upper layer by an impervious plane. As a limiting case $H_2 \to \infty$, we consider the lower layer of infinite thickness (or just halfspace). The flow to be analyzed corresponds to instantaneous suction of a volume of fluid into plane $x = 0$ across the entire thickness of the permeable layer. It is a feasible flow pattern induced by fast slip along the plane $x = 0$ (fault plane), or an earthquake. Physically, the suction is caused by localized dilatancy (or void development) during fast slip due to geometrical incompatibilty of opposite faces of fault after fast slip. A rough estimate of the magnitude of the effective 'fracture opening' can be derived from observed relation between magnitude of the slip L and the thickness of the gouge zone v (see.[29]) $v \approx 0.01L$. So it is reasonable to assume, that v can be of order of 10^{-3} – 10^{-1} m. As it is shown below, such induced suction may cause substantial electric and magnetic field perturbations. Direct large-scale hydrological consequences of earthquakes are discussed in detail in [24].

3.1. PRESSURE WAVE

Consider an uniform saturated horizontal porous layer of a thickness H: $-\infty < x < \infty; \quad -\infty < y < \infty; \quad 0 \leq z \leq H$, underlayed by an impermeable layer of thickness H_2: $-\infty < x < \infty; \quad -\infty < y < \infty; \quad -H_2 \leq z \leq 0$. We assume that the flow can be described by the pressure diffusion equation corresponding to unsteady-state flow in a slightly deforming porous medium (cf.[4]).

Under assumptions stated, the pressure distribution is described by the z-independent solution of the equation

$$\frac{\partial p}{\partial t} = \kappa \left(\frac{\partial^2 p}{\partial x^2} + \frac{\partial^2 p}{\partial z^2} - q\delta(x)\delta(t) \right) \tag{3.1}$$

with initial and boundary conditions:

$$p(x,z,0) = 0, \quad \frac{\partial p}{\partial z} = 0|_{z=0,H} - \infty < x < \infty. \tag{3.2}$$

Here, $\kappa = kK/m\eta$, is the pressure diffusivity, expressed in terms of the medium permeability k and porosity m, the fluid viscosity η and effective bulk modulus K, $q = v\mu/k$; v is the fluid volume sucked instantaneously into the slip surface $x = 0$ per unit area, or effective fracture opening.

Besides providing a fundamental solution of the problem, this source function also mimics an instantaneous suction produced along a fault (fracture) during a tectonic slip in saturated porous media.

The Fourier-Laplace transform of the solution satisfies

$$\frac{\lambda}{\kappa}\bar{\bar{P}} = \frac{d^2\bar{\bar{P}}}{dz^2} - \omega^2\bar{\bar{P}} - q, \quad \bar{\bar{P}}'(z) = 0, \quad z = 0, H. \tag{3.3}$$

Here, λ and ω are parameters of the Laplace and Fourier transforms, respectively. Therefore,

$$\bar{\bar{P}} = -\frac{q}{\omega^2 + \frac{\lambda}{\kappa}}. \tag{3.4}$$

It corresponds in original variables to the usual Gaussian exponent,

$$p(x,t) = -\frac{q\sqrt{\kappa}}{2\sqrt{\pi t}} \exp\left(-\frac{x^2}{4\kappa t}\right). \tag{3.5}$$

3.2. ELECTROKINETIC POTENTIAL

The electrokinetic potential distribution at any time $t > 0$ satisfies the following equations in the permeable layer

$$\frac{\partial^2 \psi}{\partial x^2} + \frac{\partial^2 \psi}{\partial z^2} = C\Delta p = \frac{C}{\kappa}\frac{\partial p}{\partial t} + Cq\delta(x)\delta(t), \quad 0 < z < H, \tag{3.6}$$

and in the adjacent conducting, but impermeable layer

$$\frac{\partial^2 \psi}{\partial x^2} + \frac{\partial^2 \psi}{\partial z^2} = 0, \quad -H_2 < z < 0. \tag{3.7}$$

While the potential equation is time-independent, and the delta-functions in the r.h.s. of Eq.(3.6) vanish identically for $t > 0$, this term is crucial. Essentially, it accounts for the instantaneous redistribution of the potential immediately after fluid injection (or suction) into the porous medium, and it is equivalent to the introduction of a localized initial charge. We assume also symmetry conditions at the z-axis corresponding to vanishing electrokinetic current and zero-current condition on the ground surface:

$$\frac{\partial \psi}{\partial x} = C\frac{\partial p}{\partial x}|_{x=0} = 0, \quad 0 < z < H; \quad \frac{\partial \psi}{\partial x}|_{x=0} = 0, \quad -H_2 < z < 0. \tag{3.8}$$

The upper boundary located at the ground surface is assumed to be both impervious and insulating, while at the interface between the permeable layer and the bottom rock, the potential and the current are continuous so that

$$\frac{\partial \psi}{\partial z} = 0, \ z = 0; \ \psi(x, +0) = \psi(x, -0); \ S_1\frac{\partial \psi}{\partial z}|_{z=+0} = S_2\frac{\partial \psi}{\partial z}|_{z=-0}. \tag{3.9}$$

For the Fourier-Laplace transforms of the potential, we obtain

$$\begin{aligned}
\frac{d^2\bar{\bar{\Psi}}}{dz^2} - \omega^2\bar{\bar{\Psi}} &= \frac{\lambda C}{\kappa}\bar{\bar{P}} = -\frac{(\lambda/\kappa)Cq}{\omega^2 + \lambda/\kappa} + \frac{Cq}{\omega^2} = \frac{\omega^2 Cq}{\omega^2 + \lambda/\kappa}; \quad 0 < z < H; \\
\frac{d^2\bar{\bar{\Psi}}}{dz^2} - \omega^2\bar{\bar{\Psi}} &= 0; \quad -H_2 < z < 0.
\end{aligned} \tag{3.10}$$

Taking into account the boundary conditions at $z = H$, we find

$$\bar{\bar{\Psi}} = -\frac{Cq\kappa}{(\kappa\omega^2 + \lambda)} + A\cosh\omega(z - H), \quad 0 < z < H. \tag{3.11}$$

Now, consider two extreme cases of boundary conditions at the lower boundary, $z = -H_2$, namely those of zero potential (A) or of vanishing current (B)

$$\bar{\bar{\Psi}}(-H_2) = 0, \quad (A); \qquad \bar{\bar{\Psi}}'(-H_2) = 0, \quad (B). \tag{3.12}$$

$$\bar{\bar{\Psi}} = B \sinh \omega(z + H_2), \quad (A); \quad \bar{\bar{\Psi}} = B \cosh \omega(z + H_2), \quad (B). \tag{3.13}$$

Satisfying the conditions at the interface $z = 0$ we find for the cases A and B solutions in the lower layer $-H_2 < z < 0$ as:

Case A:

$$\bar{\bar{\Psi}}(z) = -\frac{S_1}{S_2} \frac{Cq\kappa}{(\omega^2\kappa + \lambda)} \frac{\sinh \omega H \sinh \omega(z + H_2)}{D_A \cosh \omega H_2}; \tag{3.14}$$

$$D_A = \cosh \omega H + \frac{S_1}{S_2} \tanh \omega H_2 \sinh \omega H, \tag{3.15}$$

Case B:

$$\bar{\bar{\Psi}}(z) = -\frac{S_1}{S_2} \frac{Cq\kappa}{(\omega^2\kappa + \lambda)} \frac{\sinh \omega H \cosh \omega(z + H_2)}{D_B \sinh \omega H_2}; \tag{3.16}$$

$$D_B = \cosh \omega H + \frac{S_1}{S_2} \coth \omega H_2 \sinh \omega H; \tag{3.17}$$

It is readily seen that in terms of the Laplace transform

$$\frac{\kappa C q}{(\omega^2\kappa + \lambda)} \longleftrightarrow \frac{\sqrt{\kappa}}{2\sqrt{\pi t}} C q e^{-\omega^2\kappa t}. \tag{3.18}$$

Therefore, from Eqs.(3.15), we have, say, for the case A:

$$\psi(x, 0, t) = \frac{S_1}{S_2} \frac{Cq\kappa^{1/2}}{2\pi^{3/2} t^{1/2}} \int_0^\infty e^{-\omega^2\kappa t} \cos(\omega x) \frac{\sinh \omega H \sinh \omega H_2}{\cosh \omega H_2 D_A} d\omega; \tag{3.19}$$

$$\psi(x, H, t) = \frac{Cq\kappa^{1/2}}{2\pi^{3/2} t^{1/2}} \int_0^\infty e^{-\omega^2\kappa t} \cos(\omega x) \left[1 - \frac{1}{D_A}\right] d\omega; \tag{3.20}$$

$$I_z(x, 0, t) = \frac{S_1 Cq\kappa^{1/2}}{2\pi^{3/2} t^{1/2}} \int_0^\infty e^{-\omega^2\kappa t} \cos(\omega x) \frac{\sinh \omega H \cosh \omega H_2}{D_A \cosh \omega H_2} \omega d\omega. \tag{3.21}$$

Similar expressions apply to the Case B.

These expressions become much simpler if the conductivities of the two layers are equal, $S_1 = S_2$ (for the case A):

$$\psi(x, 0, t) = \frac{Cq\kappa^{1/2}}{2\pi^{3/2} t^{1/2}} \int_0^\infty e^{-\omega^2\kappa t} \cos(\omega x) \frac{\sinh \omega H \sinh \omega H_2}{\cosh \omega(H + H_2)} d\omega; \tag{3.22}$$

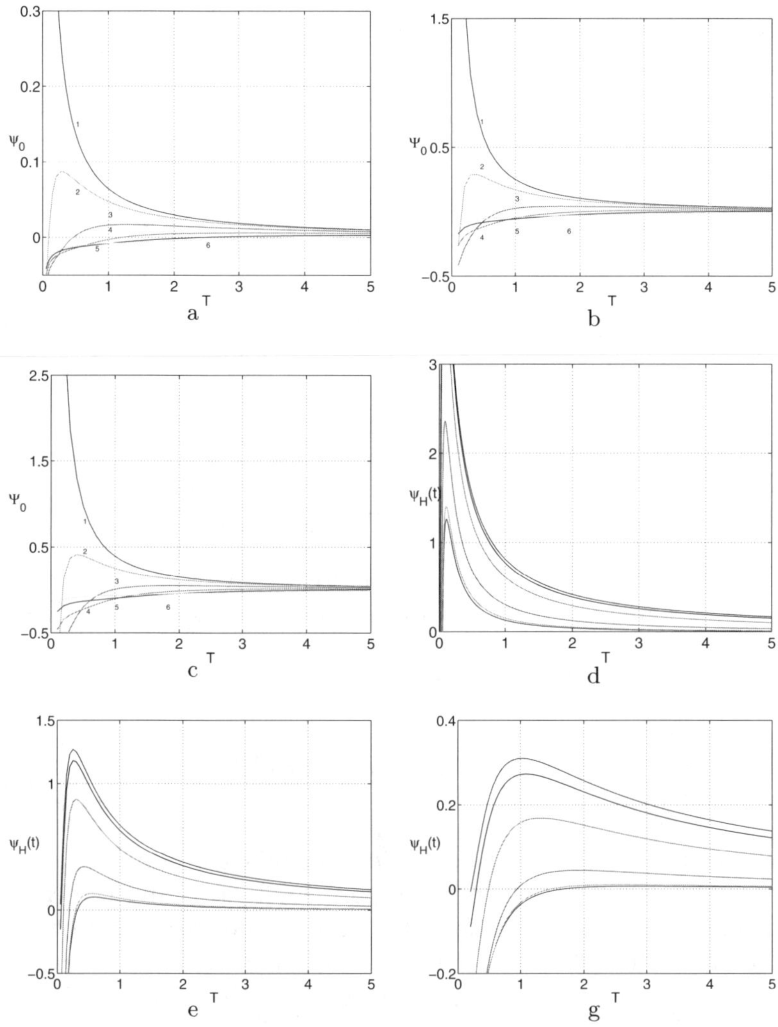

Figure 1. The transient electrokinetic potential Ψ_0 at the interface between the permeable layer and conductive bottom rock, H=1, H_2=10. Data are for: S_1/S_2=10 (a), 1 (b), 10^{-8} (c). Each curve corresponds to a given value of x/H which varies from 0 to 5 by steps equal to 1 for curves 1-6 respectively. Data for x/H= 0.5 (e), 1 (f), 2 (g). Each curve corresponds to a given value of $\lg(S_1/S_2)$ which varies from -2 to 3 by steps equal to 1 (from upper curve downward).

$$\psi(x,H,t)=\frac{Cq\kappa^{1/2}}{2\pi^{3/2}t^{1/2}}\int_0^\infty e^{-\omega^2\kappa t}\cos(\omega x)\left[1-\frac{\cosh\omega H_2}{\cosh\omega(H+H_2)}\right]d\omega; \quad (3.23)$$

$$I_z(x,0,t)=S_1\frac{Cq\kappa^{1/2}}{4\pi^{3/2}t^{1/2}}\int_0^\infty e^{-\omega^2\kappa t}\cos(\omega x)\frac{\sinh\omega H\cosh\omega H_2}{\cosh\omega(H+H_2)}\omega d\omega. \quad (3.24)$$

Simple particular case $H_2\to\infty$ results in the expressions, common for the both cases A and B:

$$\psi(x,0,t)=\frac{S_1}{S_2}\frac{Cq\kappa^{1/2}}{2\pi^{3/2}t^{1/2}}\int_0^\infty e^{-\omega^2\kappa t}\cos(\omega x)\frac{|\sinh\omega H|}{\cosh\omega H+\frac{S_1}{S_2}|\sinh\omega H)|}d\omega; (3.25)$$

$$\psi(x,H,t)=\frac{Cq\kappa^{1/2}}{2\pi^{3/2}t^{1/2}}\int_0^\infty e^{-\omega^2\kappa t}\cos(\omega x)\left[1-\frac{1}{\cosh\omega H+\frac{S_1}{S_2}|\sinh\omega H|}\right]d\omega; \quad (3.26)$$

$$I_z(x,0,t)=S_1\frac{Cq\kappa^{1/2}}{2\pi^{3/2}t^{1/2}}\int_0^\infty e^{-\omega^2\kappa t}\cos(\omega x)\frac{\sinh\omega H}{\cosh\omega H+\frac{S_1}{S_2}|\sinh\omega H)|}\omega d\omega (3.27)$$

Some typical results are presented in Figs.1. All these results are for $H_2/H=10$. Define dimensionless variables as follows:

$$X=\frac{x}{H};\ T=\frac{\kappa t}{H^2};$$

$$\Psi_0=\frac{2\pi^{3/2}HS_2\psi(x,0,t)}{\kappa^{1/2}CqS_1};\ \Psi_H=\frac{2\pi^{3/2}H\psi(x,H,t)}{\kappa^{1/2}Cq};\ I_z^0=\frac{2\pi^{3/2}I_z(x,H,t)}{\kappa^{1/2}S_1Cq} (3.28)$$

Figures 1 display the transient non-dimensional electrokinetic potential at the the surface interface between the permeable layer and the conductive bottom rock $z=0$ as a function of time for different values of the ratio S_1/S_2 equal to 10, 10^{-8} and 0.1 for Figs.1,a-c; plots 1-6 correspond to $X=0;1;2;3;4;5;6$ respectively. The greater the distance X, the less is the maximum value of the potential, and the later it is attained; the potential increase with decreasing S_1/S_2. Figures 1,d-f show the transient electrokinetic potential at the free surface ($z=H$) as function of time at three different locations ($X=5$, 1, and 2 for Figs.1,d,e,f respectively). Different plots correspond to different values of the conductivities ratio S_2/S_1, $\lg(S_2/S_1)=-2,-1,0,1,2$, and 3 for the plots shown (starting from the upper plot downward).

In both sets of results, the same trend is obvious: the greater the distance X, the less is the maximum value of the potential, and the later it is attained; the potential peak value increase with decreasing S_1/S_2.

Figures 1,e and f show transient potential at the same location ($z = H, x = H$) for different values of the conductivities ratio S_1/S_2. The magnitude of potential variation is increasing with the ratio S_1/S_2.

As it should be expected, the magnitude of the transient electrokinetic potential rapidly decreases with distance from the instantaneous source and strongly depends on the conductivity of the underlying bottom rock. Somewhat unexpectedly, at large distances from the source the induced potential changes its sign with time. As it follows from the presented results, the magnitude of the surface EK potential peak is of the order of

$$\psi^* = (.1 - 1) \times (S_1 C \kappa^{1/2} q / 2\pi^{3/2} S_2 H). \tag{3.29}$$

Using the estimate $q = \eta v/k$ for the case of a tectonic slip, and letting $d = 0.1\ m$, $C = 1V/MPa$; $\eta = 1\ mPa \cdot s$; $S_1/S_2 = 1$; $k = 10^{-17}\ m^2$; $\kappa = 0.1\ m^2/s$, $H = 10^3\ m$ we get $\psi^* = 30\ V$ (*sic!*) for a moderately-sized fault. It is readily seen that this unexpectedly large value of transient EK potential is primarily due to the small value of the permeability k used in the estimate.

4. Magnetic field

Generally, the magnetic field is given by the Biot-Savart law as the integral over the electric current field. However, in simple cases it is convenient to use directly the fundamental equation for the magnetic induction vector $\boldsymbol{B}$ expressed in terms of the vector potential $\boldsymbol{A}$ [22]

$$\nabla \times \boldsymbol{B} = \boldsymbol{I}; \quad \boldsymbol{B} = \nabla \times \boldsymbol{A}; \quad \nabla \cdot \boldsymbol{A} = 0; \quad \Delta \boldsymbol{A} = -\boldsymbol{I}. \tag{4.1}$$

We consider only the case of media of constant magnetic permeability, close to that of vacuum ($\mu_m = 1$). The current field $\boldsymbol{I}$ is a plane vector field of the (x, z)-plane and does not depend on y, $\boldsymbol{I} \equiv \{I_x(x, z), 0, I_z(x, z)\}$. It is readily seen that the corresponding magnetic field has only one non-vanishing component $\mathbf{B} = \{0, B_y = B(x, z), 0\}$ that satisfies the equations

$$\frac{\partial B}{\partial z} = -I_x; \quad \frac{\partial B}{\partial x} = I_z. \tag{4.2}$$

This allows us to find the magnetic field explicitly when the current field is known:

$$B_y(x, z) = \int_\Gamma (I_z dx - I_x dz) + B(x_0, z_0). \tag{4.3}$$

Here, Γ is an arbitrary curve connecting points (x_0, z_0) and (x, z); the integral is path-independent as $\nabla \times \boldsymbol{I} = 0$. In our case, as the pressure and

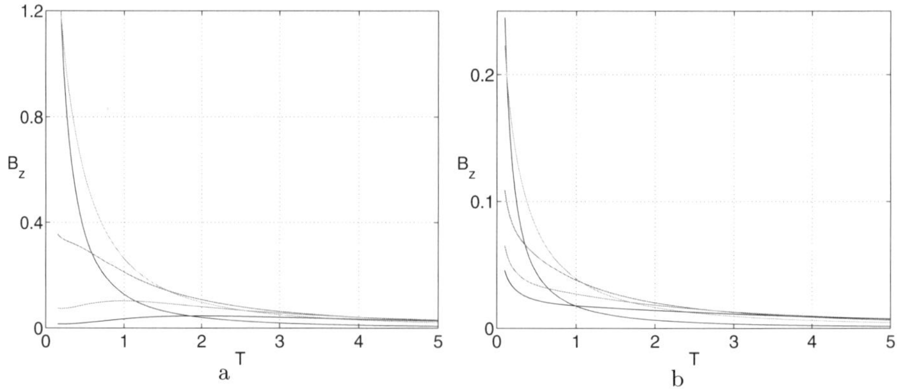

Figure 2. The gradient of the transient magnetic field at the surface of the permeable layer $z = H$; H_2/H_1=10; S_1/S_2=10. Each curve corresponds to a given value of x/H which varies from 0.5 (1) to 4.5 by steps equal to 1. Data are for: $S_1/S_2 = 10^{-8}$ (a), 10 (b).

potential fields depend on x and z only, the y -component of the induced caurrent vanishes and the magnetic field is aligned in along the y-direction (assuming uniform magnetic permeability). It is readily seen that in this case the magnetic field vanishes at the surface, $B(x, H) = 0$. Its gradient can be expressed as

$$\frac{\partial B}{\partial z}|_{z=H} = -S_1 \left(\frac{\partial \psi}{\partial x} - C \frac{\partial p}{\partial x} \right) |_{z=H}$$

$$= -S_1 C \frac{q\kappa^{1/2}}{2\pi^{3/2} t^{1/2}} \int_0^{\infty} \omega \sin \omega t \exp(-\omega^2 t) K_m(x, \omega) d\omega. \tag{4.4}$$

$$K_m(x, \omega) = (cosh(\omega H) + (S_1/S_2) \tanh(\omega H_2) \sinh(\omega H))^{-1}. \tag{4.5}$$

Corresponding results are presented in Fig.2. Here, B_z is dimensionless magnetic field gradient at the surface,

$$B_z = -\frac{2\pi^{3/2}}{S_1 C q \kappa^{1/2}} \frac{\partial B}{\partial z}|_{z=H}.$$

The figures correspond to $S_2/S_1 = 10^{-8}, 1$, and 10 respectively. In each figure five plots correspond to $X = 0.5; 1.5; 2.5; 3.5$, and 4.5; the greater X, the less is the maximum value of magnetic field perturbation, and the later it occurs. Essentially the same results are illustrated by Figs.3. Figures 3,a-c show non-dimensional values of magnetic field at the interface between

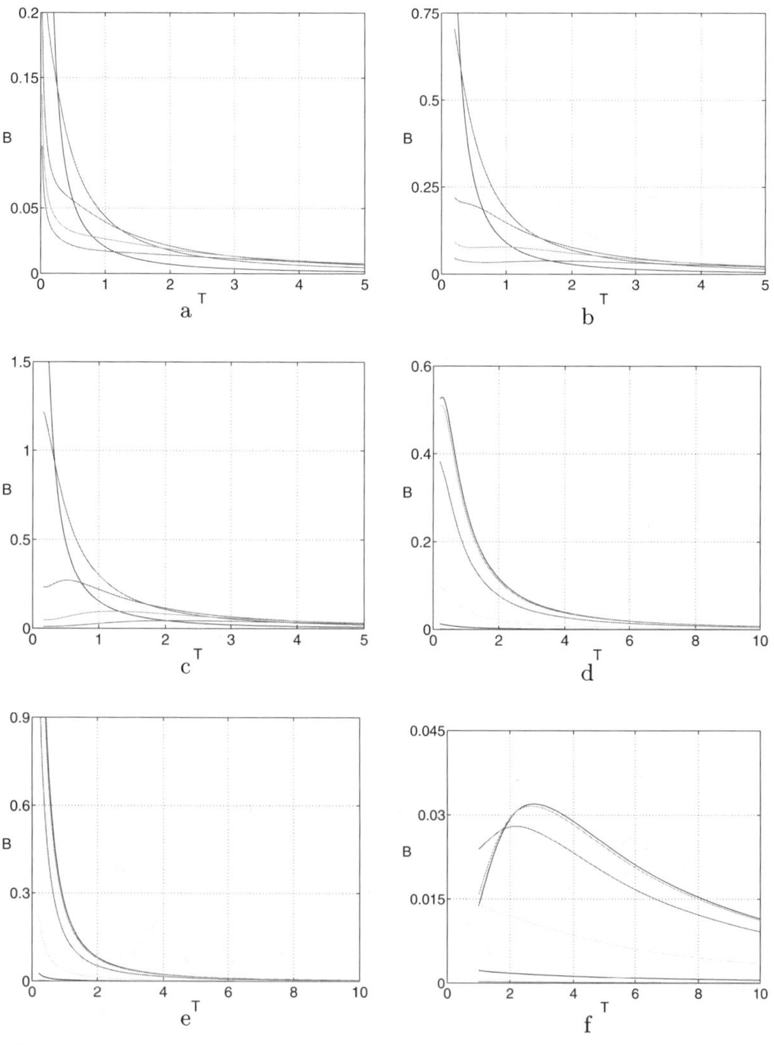

Figure 3. The transient magnetic field at the B at the interface between the permeable layer and conducting bottom rock $B(x, 0, t)$. Data are for: $S_1/S_2 = 10$ (a), 1 (b), 10^{-8} (c): each curve corresponds to a given value of x/H which varies from 0.5 (1) to 4.5 by steps equal to 1. Data for $x/H = 1$ (d), 2 (e), 5 (f). Each curve corresponds to a given value of $\lg(S_1/S_2)$ which varies from -2 to 3 by steps equal to 1 (from upper curve downward).

the permeable layer and conducting bottom rock as function of time for the same values of X and S_1/S_2 as Fig.2. Figures 3,d-e show the magnetic field at three different locations ($X = 1; 2; 5$) for different values of the conductivity ratio.

Roughly, an approximate relation holds true

$$B^0 = \frac{2\pi^{3/2} H}{S_1 C q \kappa^{1/2}} B(x, 0, t) \approx B_z. \tag{4.6}$$

Comparing expressions (3.19) and (4.4), it is immediately seen that an order of magnitude for the transient induced magnetic field is

$$B^*(x, 0, t) \sim S_1 \psi^*(x, H, t). \tag{4.7}$$

With $\psi^* \sim 30\ V$ and $S_1 = 0.01\ Ohm^{-1} \cdot m^{-1}$, this value is of the order of $\sim 0.3\ A/m$.

This provides some estimate of the transient magnetic field perturbations that may be associated with earthquake-induced porous water flows.

5. Acknowledgements

V.Entov gratefully acknowledge the support of a temporary professorship at IPGP, Paris. This work has been partly supported by an ACI Grant.

References

1. *A Critical Review of VAN Earthquake Prediction From Seismic Electrical Signals* (1996). Sir James Lighthill, ed; World Scientific Publ. Co., Singapore, New Jersey, London, Hong Kong.
2. Adler, P.M.,(2001) Macroscopic electroosmotic coupling coefficient in random porous media, *Mathematical Geology*, **33**, 63-93.
3. Adler, P.M., Le Mouel,J.-L. and Zlotnicki J. (1999) Electrokinetic and magnetic fields generated by flow through a fractured zone: a sensitivity study for La Fournaise volcano, *Geophys. Res. Lett.* **26**, 795-798.
4. Barenblatt, G.I.,Entov, V.M. and Ryzhik,V.M. (1990) *Fluids Flow through Natural Rocks*, Kluwer Publishers, Dordrecht.
5. Bernard, P. (1992) Plausibility of long distance electrotelluric precursors to earthquakes,*J. Geophys. Research*, **97**(B12), 17531-1754.
6. Bernard, P. and Le Mouel,J.L.(1996) On Electrotelluric Signals, in: *A Critical Review of VAN Earthquake Prediction From Seismic Electrical Signals*, Sir. James Lighthill, ed; World Scientific Publ. Co., Singapore, New Jersey, London, Hong Kong, 118-152.
7. Bers, L. (1958) *Mathematical Aspects of Subsonic and Transonic Gas Dynamics*, Wiley, N.Y.; Chapman and Hall, London.
8. Coelho, D., Shapiro,M. Thovert,J.-F. and Adler, P.M. (1996) Electroosmotic phenomena in porous media, *J. of Colloid and Interface Science*, **181**, 169-190.

9. Dukhin,S.S. and Deryagin,B.V. (1974) in: *Surface and Colloid Sci., Vol. 7: Electrokinetic phenomena*, E.Matijevich, ed.; J.Wiley, NY.
10. Fitterman, D.V. (1978) Electrokinetic and magnetic anomalies associated with dilatant regions in a layered Earth,*J. Geophys. Research*, **83**, 5923-5928.
11. Fitterman, D.V. (1979) Theory of electrokinetic-magnetic anomalies in a faulted half-space, *J. Geophys. Research*, **84**(B11), 6031-6040.
12. Fitterman, D.V. (1979) Calculations of self-potential anomalies near vertical contacts, *Geophysics*, **44**(2), 195-205.
13. Fitterman, D.V. (1981) Correction to 'Theory of electrokinetic-magnetic anomalies in a faulted half-space',*J. Geophys. Research*, **86**(B10), 9585-9588.
14. Goldstein, R.V. and Entov,V.M. (1994) *Qualitative Methods in Continuum Mechanics*, Pitman Monographs and Surveys in Applied Mathematics, **72**, Longman Scientific & Technical.
15. Gradshtein, I.S., and Ryzhik, I.M. (1994) *Tables of Integrals, Series and Products*, 5th Ed., Academic Press, L.
16. Ishido, T. and Mizutani, H. (1981) Experimental and theoretical basis of electrokinetic phenomena in rock-water system and its application to geophysics. *J. Geophys. Research*, **86**, 1763-1775.
17. Ishido, T. and Pritchett, J.W. (1999) Numerical simulation of electrokinetic potentials associated with subsurface fluid flow, *J. Geophys. Research*, **104**(B7), 15247-15259.
18. Johnston, M.J.S. (1997) Review of electric and magnetic phenomena accompanying seismic and volcanic activity, *Surveys in Geophys.*, **18**, 441-475.
19. Jouniaux, L., Pozzi,J.-P., Berthier,J. and Masse, Ph.(1999) Detection of fluid flow variations at the Nankai trough by electric and magnetic measurements in boreholes or at the seafloor, *J. Geophys. Research*, **104**(B12), 29293-29309.
20. Jouniaux, L. and Pozzi, J.-P. (1995) Permeability dependence of streaming potential in rocks of various fluid conductivity, *Geophys. Res. Lett.*, **22**, 485-488.
21. Jouniaux, L., Bernard,M.L., Zamora M. and Pozzi, J.-P.(2000) Streaming potential in volcanic rocks from Mount Pelée, *J. Geophys. Research*, **105**(B7), 8391-8401.
22. Landau, L.D. and Lifshitz, E.M. (1960) *Electrodynamics of continuous media*, Translated from the Russian by J.B. Sykes and J. S. Bell, Pergamon Press, Oxford, New York.
23. Marino, S., Coelho,D., Békri,S. and Adler, P.M. (2000) Electroosmotic phenomena in fractures, *J. of Colloid and Interface Science* **223**(2), 292 - 304.
24. Muir-Wood, R. and King, G.C.P. (1993) Hydrological signatures of earthquake strain, *J. Geophys. Research*, **98**, 22035-22068.
25. Nur, A. and Booker J.R., (1972) Aftershocks caused by pore fluid flow? *Science*, **175**, 885-887.
26. Overbeek, J.Th.G.(1952) Electrochemistry of the double layer, in: *Colloid Science*, edited by H.R.Kruyt, 115-193, Elsevier, New York.
27. Rice J.R. (1992) Fault Stress States, Pore Pressure Distributions, and the Weakness of the San Andreas Fault, in: *Fault Mechanics and Transport Properties in Rocks* (eds. B. Evans and T.-F. Wong), Academic Press, 475-503.
28. Roellofs, E. (1988) Hydrological precursors to earthquakes: A Review *Pageoph* **126**: 177-209.
29. Scholz, C.H. (1990) The Mechanics of Earthquakes and Faulting, Cambridge Univ. Press.
30. Sibson, R.H. (1981) Fluid flow accompanying faulting: Field evidence and models,

in: *Earthquake Prediction: An International Review*, Maurice Ewing Ser., vol. 4, edited by D.W. Simpson and P.G. Richards, pp. 593-603, AGU, Washington, D.C.

31. Sibson R.H. (1986) Earthquakes and rock deformation in crustal fault zones, *Ann. Rev. Earth. and Planet. Sci.*
32. Zlotnicki, J. and Le Mouel, J.-L. (1990) Possible electrokinetic origin of large magnetic variations at La Fournaise volcano, *Nature* **343**, 633-635,

RESONANT WAVES IN A STRUCTURED ELASTIC HALFSPACE

M. V. AYZENBERG-STEPANENKO
Department of Mathematics.
Ben-Gurion University of the
Negev. P.O.B. 653, Beer-Sheva 84105, Israel

Abstract. Development of resonant waves is examined when the surface load moves along a layered elastic system: thin plate-thick layer-halfspace. In steady-state formulation, the waveguide properties are analyzed. Multiple roots are revealed of the dispersion equation of the system, and then critical velocities of the moving load and the space frequency of excited resonant waves are established. It is shown that together with the well-known surface resonance developing in a free halfspace under the action of a load moving with the Rayleigh velocity, a set of additional resonances appear depending on parameters of structured halfspace members. In the unsteady-state formulation, asymptotical solutions are obtained determining the qualitative picture of the process established with time: the growth rate and the spectrum saturation of corresponding resonant perturbations. Computer simulations performed enable quantitative features of the process to be explored over the whole time interval. Numerical and asymptotic results are compared to reveal the range of the asymptotic solution acceptability. In distinction to the resonance in a free halfspace, the growth rate in the observed system turns out less than linear: due to dispersion, some part of the energy pumped in by the external load is diverted with time for non-growing waves.

Key words: nonlinear surface waves, elasticity, Rayleigh waves, scale invariance, nonlocality, spectral kernel, spatial kernel, Hilbert transform

1. Introduction

In the paper the waveguide properties of a structured halfspace are investigated and resonant regimes are described developing in the system under a moving surface load. The aim is to explore the specific character of wave dispersion in the system and to describe the growth rate and the spectrum structure of resonant perturbations. As was shown in the pioneer work by Goldshtein [1], when a load moves with the Rayleigh velocity along the surface of a free halfspace, first, a steady-state limit of the solution is absent; secondly, particle velocities and stresses near the surface rise

R.V. Goldstein and G.A. Maugin (eds.),
Surface Waves in Anisotropic and Laminated Bodies and Defects Detection, 205–215.

linearly. Then diverse aspects of resonant waves were studied for elongated structures, in general, for cylindrical shells interacting with acoustic media (see [2 6]). Special points in the case of a layered halfspace were examined in [7], asymptotic and computer solutions of flexural resonant waves are obtained in [8] for various forms of moving loads. The main contribution to the theory of resonant waves has come from Slepyan [2, 6], an approach to reversing the double Laplace-Forrier transforms in the vicinity of a moving wave ($x = Ct$, C is the wave speed, t is time), developed by him, enables asymptotic (for large values of time) solutions to be designed. Notably on this basis analytical solutions have been obtained in [3-5, 8, 9]. In [10] basic aspects of steady and unsteady waves relation have been explored and the principal problem of excitation of resonant waves in elastic waveguides by moving load has been solved. In the case of the step loading, a wave resonance is formed in the vicinity of the front, if the load velocity equals critical, V_{cr}, which coincides with phase, c, and group, c_g, velocities of a free wave propagating in the systems ($V = V_{cr} = c = c_g$): a load moving along the waveguide axis enables permanent pumping up of external energy into the energy of the oscillating waves arising with time *).The spectrum of such waves (or a set of waves) possessing the same phase and group velocities can be determined by special solutions of the dispersion equation representing dependence $c(q)$, where q is the wavenumber $q = 2\Pi/\lambda$ and λ is the wavelength. Taking into account that $c_g = c + q\partial c/\partial q$, it can be seen that special solutions (they are noted below as special points in dispersion curves) $q_*, c_* (c_* = c(q_*) = c_g(q_*))$ appear in the following spectrum parts:

(i) in the long-wave spectrum: $q = 0, c^* = c(0)$

(ii) in the middle-wave spectrum: extremum points of $c = c(q)$, and

(iii) the short-wave spectrum: $q \to \infty$.

The latter case (iii) is not discussed here, because the Bernoulli model for plate used below is not relevant for short waves. Special points in long-wave and middle-wave spectra are schematically depicted in Figure 1.

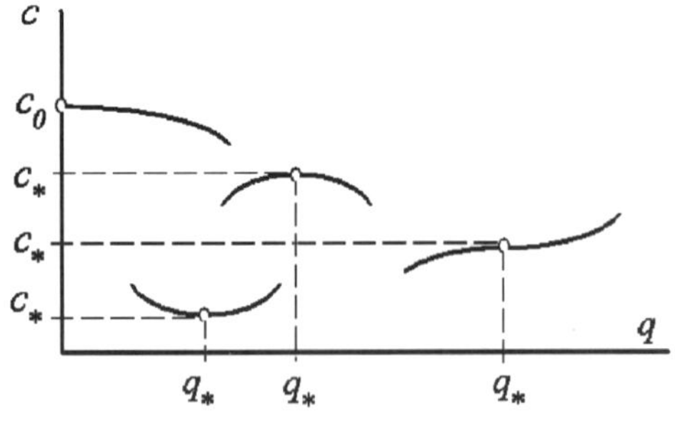

Figure 1. Special points of coordinates resulting in parameters of resonant waves

In the case of a free halfspace (i.e. a dispersionless surface waveguide in the terminology used here), the critical velocity is equal to the Rayleigh velocity c_R, which does not depend on the wavelength: $c(0) = c_g(0) = c_R$. If $V = c_R$, a surface resonant process is formed, in which the growth rate of perturbations is linear, proportional to t (t is time). Contrary to this, in dispersion waveguides resonances can arise at several separate wavelengths and their growth rate turns out less than linear: due to dispersion, some part of the energy pumped by the external load is diverted with time [1], for non- growing perturbations of wavelengths and velocities close to the resonant ones. Two qualitatively different processes can develop:

(i) long wave ($q \sim 0$) resonance propagates in the quasi-stationary way, and

(ii) medium-spectrum resonances propagate of quasi-stationary envelope and current frequency $\omega* = c_* q_*$.

Although the many questions of the problem under consideration have been solved, we note that surface resonances in a structured halfspace like the one discussed here have not yet been examined so far in spite of their evident theoretical and practical importance. Besides, an important question not revealed up to now relates to the applicability of asymptotic solutions in actual problems. It is shown below that the system examined here, thin plate-thick layer-halfspace, possesses, first, a special point in the longwave spectrum ($q = 0, c = c_g = c_R$): a long surface wave overlooks finite thickness structures covering the halfspace, and second, a set of special points (maximum, minimum, inflection) in a medium-wave spectrum, which appear due to the structured surface. In the present work, critical velocities of moving load and the spectrum of perturbations growing with time are analytically obtained on the basis of spectral analysis of a stationary problem. Unsteady-state wave propagation processes are analyzed analytically and numerically.

2. Problem Formulation

Let us consider a plane dynamic problem for the system: thin plate - thick layer - halfspace. On the external surface of the structure (that is on the plate) beginning at time $t = 0$ normal stresses of a given shape Q and constant velocity V move along the surface, that is along longitudinal direction x. In Figure 2, H is the Heaviside function, numbers $0, 1, 2$ refer to the plate, layer and halfspace respectively; c_0 is the velocity of longitudinal waves in

---

[1] As was shown in [10], resonant waves can be excited by an oscillating load of constant velocity $V = c$ and frequency $\omega = qc$ not depending on relation between phase and group velocities. Below we assume notably step loading, therefore the equality $c = c_g$ turns out essential.

the plate, ρ_j $(j = 0, 1, 2)$ are densities, $h_j (j = 0, 1)$ thicknesses, subindices l and s related to velocities of longitudinal and shear waves respectively:c_{jl}, and c_{js}.

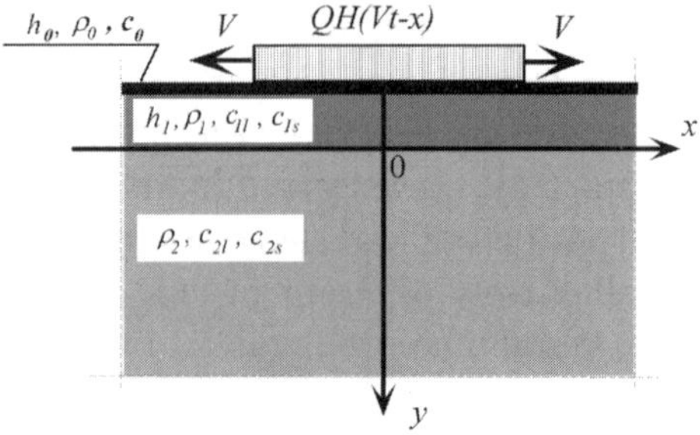

Figure 2. System: thin plate-layer-halfspace

The dynamic elasticity theory describes the motion of the layer $(0 < y < h_1 : j = 1)$ and halfspace $(y > h_1 : j = 2)$,

$$\begin{aligned} \ddot{u} &= c_{jl}^2 u''_{j,xx} + c_{js}^2 u''_{j,yy} + \left(c_{jl}^2 - c_{js}^2\right) w''_{j,xy}, \\ \ddot{w} &= c_{jl}^2 w''_{j,yy} + c_{js}^2 w''_{j,xx} + \left(c_{jl}^2 - c_{js}^2\right) u''_{j,xy} \end{aligned} \tag{1}$$

while the classic Bernoulli equation is used for plate dynamic bending:

$$(y = 0) : \ddot{w} + c_0^2 \left(h_0^2 / 12\right) w_x^{(IV)} = (P - R) / \rho_0 h_0. \tag{2}$$

Here $P = QH(Vt - x)$ is a moving step load, R is the normal reaction of the layer to the plate motion:

$$R(x, t) = \sigma_{yy}^{(1)}(x, 0, t) = \rho_1 \left[c_{1l}^2 w'_{1,y} + \left(c_{1l}^2 - 2c_{1s}^2\right) u'_{1,x}\right] \tag{3}$$

All the components of the composition are connected by a rigid contact excluding longitudinal connection between the plate and layer (or the halfspace in a reduced system: plate-halfspace), which are assumed to be absent. So the following relations are proved:

$$\begin{aligned} &y = 0 : w_1(x, 0, t) = \sigma_{xy}^{(1)} = 0, \\ &y = h_1 : w_1 = w_2, \mathrm{u}_1 = \mathrm{u}_2, \sigma_{yy}^{(1)} = \sigma_{yy}^{(2)}, \sigma_{xy}^{(1)} = \sigma_{xy}^{(2)}. \end{aligned} \tag{4}$$

Note that the load source is not fixed here; therefore the depicted load image can be seen somewhat conditionally. For example, in the case of an external pressure wave of an air blast in a far field or the action of a plane landing onto a strip, the parameters of loads can be considered as given. In the case of an internal source (for example, earthquake or underground explosion), the superposition wave theory method can be used to calculate the parameters of the moving surface loading. In the steady-state formulation, the solution of the problem is found as a superposition of Fourier harmonics, $\exp[iq(x \pm ct) - \zeta y]$, propagating along the x-axis and exponentially decaying at $y \to 0$. Factor ζ is calculated from Fourier-transforms of original equations for the halfspace. By using the boundary conditions, the dispersion equation connecting the phase velocity, c, and the wave number, q, is obtained. This equation is transcendental, it has a cumbersome structure, and its formal expression only is presented below:

$$L(q, c; h_0, c_0, c_{1l}, c_{2l}, \rho l, c_{2l}, c_{2s}, \rho_2) = 0 \tag{5}$$

3. Dispersion Analysis

3.1. THIN PLATE-HALFSPACE

The aim of analysis of Eqn. (5) is to find special points and examine the behavior of dispersion curves in their vicinities. First, we will consider a system: thin plate- halfspace, which is the simplest (single-mode) special case of the system. We introduce the following notation: $c_l = c_{1l} = c_{2l}, c_s = c_{1s} = c_{2s}, \rho = \rho_1 = \rho_2$, Then Eqn. (5) is written as follows:

$$\begin{gathered} \rho_0 q\sqrt{1-c^2}\,(c^2 - c_0^2 q^2/12)\,c^2 + c_s^4 L_R = 0, \\ L_R = (2 - c^2/c_s^2)^2 - 4\sqrt{1-c^2}\sqrt{1-c^2/c_s^2} \end{gathered} \tag{6}$$

where $L_R = 0$ is the Rayleigh equation for a free halfspace (its single real root is $c = c_R$), and c_l, ρ and h_0 are taken as measurement units. Eqn. (6) has a single mode $c = c(q)$, which is real, if $c < c_s$.

If $q \to 0$, then $c \to c_R$: a plate of finite rigidity and mass does not influence long wave dispersion (more accurately: infinitely long wave). If $q > 0$, dispersion appears: c (linearly) decreases if q increases. The asymptotic behavior of c obtained from (6) is

$$\begin{gathered} c = c_R\left[1 - \alpha\rho_0 q + O\left(q^2\right)\right], \alpha = {}^1\!/_4\sqrt{1-c_R^2}\,(c_R/c_s)^2\,L_1^{-1} > 0, \\ L_1 = (1 + c_s^2 - 2c_R^2)\,[(1-c_R^2)\,(1 - c_R^2/c_s^2)]^2 - 2 + c_R^2/c_s^2 \end{gathered} \tag{7}$$

As can be seen from Eqn. (7), long wave dispersion strongly depends on the plate mass and, vice versa, is practically independent of its rigidity. In a middle spectrum, the decay of $c(q)$ stops, the dispersion curve reaches the minimum ($q_m = q_*, c_m = c_*$), then monotonically rises with increase in q up to $c = c_s$.

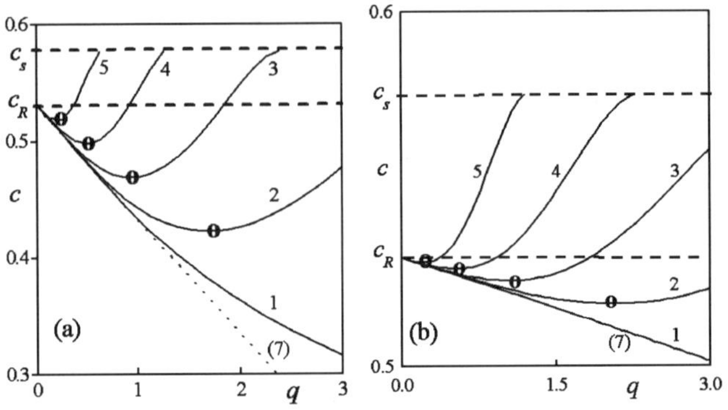

Figure 3. Dispersion curves for system: plate-halfspace: (a) "heavy" plate, $\rho_0 = 2$; (b) "light" plate, $\rho_0 = 0.5$. $1 - c_0 = 0, 2 - c_0 = 0.5, 3 - c_0 = 1, 4 - c_0 = 2, 5 - c_0 = .5$

Circles in Figure 3 are special points (minima), line $c = c_R$ is the Rayleigh dispersionless mode for a free halfspace. Asymptotic (7) coincides with a linear part of all curves. Note that here and in the subsequent calculations, the Poisson ratios are the same: $\nu_1 = \nu_2 = 0.25$, then $c_{js} = c_{jl}\big/\sqrt{3} \approx 0.5774,\ c_{jR} = 0.5318\ (j = 1, 2)$.

It can show that minimum points are inside the upper domains between two straight lines: asymptotic (7) and the plate bending mode $c = c_0 h_0 q/(12)^{1/2}$. Coordinates q_m, c_m can be obtained from the following formulas:

$$q_m = 2c_m/c_0,\ (\rho_0/c_0)\, c_m^5 \sqrt{1 - c_m^2} + \sqrt{3/8} L_R(c_m) = 0 \tag{8}$$

So, two critical velocities exist: $V_{cr} = c_R$, exciting a long wave spectrum of a low frequency $(\omega \to 0)$, and $V_{cr} = c_m$, exciting a medium wave spectrum $(q \sim q_m)$.

The less the dispersion in the vicinity q_m, c_m, i.e. the larger the index n in the approximating dependence $c \approx c_m + (q - q_m)^n$, the wider is the wavelength spectrum that shapes the resonance disturbances and the more intense is their growth in time, asymptotically $(t \to \infty)$ proportional to $t^{(n-1)/n}$ (see [3, 7]). Number $n (n \geq 2)$ is the first natural number for which $\partial^n c / \partial q^n \neq 0$. Therefore, in the case of a light and pliable plate (i. e. small ρ_0 and c_0) resonance regimes in the medium wave spectrum are to be suspected as more intensive. On the other hand, with the growth of c_0 the value of c_m approaches c_R, while q_m is removed into the long wave domain. This shows the possibility of superposition of surface waves in the halfspace $(V_{cr} = c_R, q \to 0)$ and bending waves in the plate $(V_{cr} = c_m, q = q_m)$, which will considerably strengthen disturbances if V is within (c_m, c_R). The question of which of the two mechanisms for the formation of disturbances growing in time will prevail can be solved in the analysis of the non-stationary problem.

3.2. THIN PLATE - LAYER - HALFSPACE

As distinct from the previous case, an infinite number of dispersion equation roots (modes) exist here - analogs of free wave modes propagating in a plane layer of finite thickness. If $q \to 0, c \to c_{R2}$ (the Rayleigh velocity for the halfspace) as in the previous case. The first (lower) mode has real roots, whereas the higher modes can be real or complex depending on the relation between the system parameters. A set of special points can exist in these modes.

Longwave ($q \to 0$) asymptotic of the dispersion equation (1) can be obtained as follows:

$$L \sim c_{2s}^4 L_{R2} + q \left[\rho_0 c^4 \sqrt{1 - c^2} + \rho_1 h_1 \Phi \left(c_{1l}; c_{1s}, c_{2s} \right) \right] \tag{9}$$

Here Φ is a finite function at $q \to 0$, measurement units are: ρ_2, c_{2l}, and h_0, also we assume that $v_0 = v_1 = v_2 = 0.25 (c_s/c_l = 0.5774, c_R = 0.9211 c_s)$.

A linear asymptotic $c(q)$ is proved by Eqn. (9) as in the previous case, Eqn. (6). The main distinction is that $c(q)$, depending on the expression in the square braces, can decrease or increase with growth in q. Obtaining analytical asymptotic for special points into middle-spectrum turns out problematic. Their analysis can be done on the basis of numerical solutions for dispersion equation (1). Some results are shown in Figure 4.

For a relatively rigid and heavy layer ($\rho_1 > \rho_2, c_{1l} > c_{2l}$) the first mode can receive maximum and minimum in the medium wave spectrum: see examples in Figures 4 (a) and (b). Curves 1 (Figure 4a) and 3 (Figure 4b) prove that in these cases the longwave dispersion is minimal. This is the most favorable situation for intense development of surface resonances.

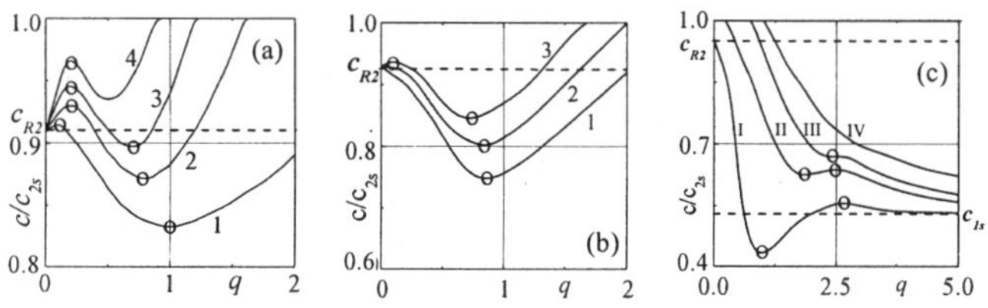

Figure 4. Dispersion in the system: plate-layer-halfspace:
(a) $\rho_0 = 2.5, h_1 = 10, \rho_1 = 2, c_0 = 1.05 c_{1l}$, *curves* 1 – 4 : $c_{1l} = 1.2, 1.6, 2.0, 2.4$;
(b) $\rho_0 = 2.5, \rho_1 = 2, c_0 = 1.92, c_{1l} = 1.6$, *curves* 1 – 3 : $h_1 = 2, 3, 5$;
(c) $\rho_0 = 2.0$, $\rho_1 = 0.6$, $h_1 = 5$, $c_0 = 1.92$, $c_{1l} = 0.6$, $I - IV$ are mode numbers.

In the example for a lighter ($\rho_1 < \rho_2$) and more pliable layer ($c_{1l} < c_l$) - Figure 4 (c), the first and second modes have maximum and minimum, while the third mode has an inflection point with a tangent parallel to the axis of q. The presence of diverse special points in the narrow spectrum of different modes testifies to the possibility of different oscillation forms with close wavelengths superposing in a narrow range of critical velocities.

4. Computer Simulation of the Unsteady Problem

The diversity of special points in a wide spectrum revealed above proves the development of a set of resonant disturbances if the surface load moves with critical velocities $V_{cr} = c_m$ and $V_{cr} = c_R$. Such processes were examined on the basis of direct numerical modelling of the problem considered. An explicit finite-difference scheme is used with a special method of mesh dispersion minimization initially developed in [11] and then realized, in particular, for numerical simulations of resonant processes (see [8,10]). The method enables long- and short-wave components to be calculated with the same accuracy on the basis of a static difference mesh.

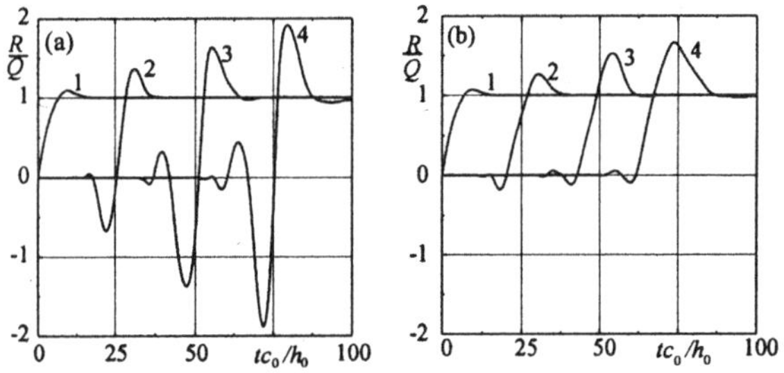

Figure 5. Formation of resonant waves in the system plate-halfspace: $\rho_2 = 0.4, c_{2l} = 1$. (a): $V = c_* = 0.48$, $(b) : V = c_* = 0.53$. *Curves* $14 : \ x = 0, 10, 20, 30$.

In Figure 5 an initial stage of resonant wave formation in the system plate-halfspace is shown for two critical velocities of the moving step load: $V = c_*$, and $V = c_R$. The depicted curves are relative normal reactions R/Q in cross-sections $y = 0, x = 0, 10, 20, 30$. Measurement units are h_0, c_0, ρ_0. A clear distinction can be seen between these two cases: (a) a strong rise and a fixed frequency of the flexural resonance corresponding to $V = c_*$, (b) a weak rise and decreasing frequency with time if $V = c_R$. Such peculiarities can be forecast by the analysis of dispersion roots in vicinities of special points: a weak dispersion is near the minimum $(q*, c*)$ and a strong dispersion is revealed in the longwave point $(0, c_R)$.

The same process in the system plate-layer-halfspace can be seen in Figure 6 for the case of a relatively rigid and heavy halfspace.

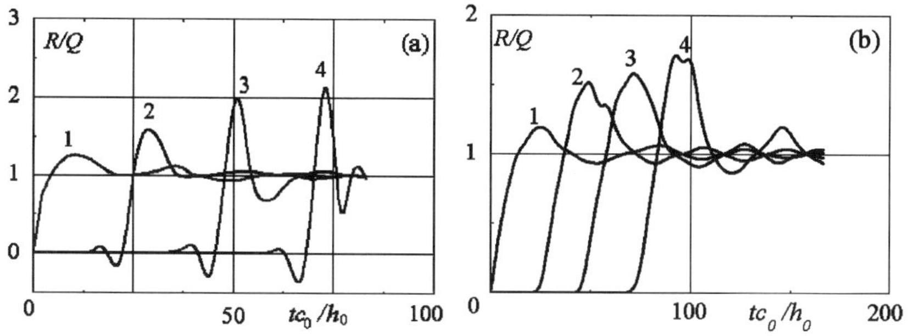

Figure 6. Formation of resonant waves in the system plate-layer-halfspace: $h_1 = 5, \rho_2 = 0.5, c_{2l} = 0.4, \rho_2 = 1.25$,
$c_{2l} = 1.12.(a) : V = c_* = 0.45, (b) : V = c_{R2} = 0.58$.
Curves 1 - 4: x = 0, 10, 20, 30.

5. Asymptotically-Equivalent System

Unfortunately, obtaining analytical solutions for resonant waves in the discussed systems is problematical. However, the problem can partially be solved on the basis of the known approach in which the original complex structure is changed by a simple structure possessing the same spectral properties in a fixes band. This approach is notably applicable in the case of resonances. We will use the simplest system for our analysis: plate on elastic foundation, which proves a single special point: the minimum in the middle spectrum. The rigidity of the foundation, the effective density and rigidity of the plate can be found on the basis of the best approximation (by the least square method, for example) of dispersion curves in the corresponding spectrum band. Let these parameters be found, then the system is described by Eqn. (2), in which the expression for R is expressed now as $R = wG$, where G is an effective rigidity of the foundation. The asymptotic solution ($t \to \infty$) for w can be written as follows (see [6]):

$$w(x,t) \sim \frac{Q}{\pi h q_*^2 c_* G}\left(\frac{t}{z}\right)^{1/2}\left[F_1(\kappa)\cos\eta q_* - F_2(\kappa)\sin\eta q_*\right],$$

$$z = \frac{1}{2}\left(\frac{dc_g}{dq}\right)_{q=q_*}, \quad \kappa = \frac{\eta}{(zt)^{1/2}}, \quad \eta = c_* t - x, \tag{10}$$

$$F_1(\kappa) = \int_0^\infty \frac{\sin y^2 \cos\kappa y}{y^2}, \quad F_2(\kappa) = \int_0^\infty \frac{(1-\cos y^2)\cos\kappa y}{y^2}.$$

So, resonant perturbations increase with time as $t^{1/2}$ in the vicinity of the loading front, $\kappa = 0$, also increased with time as $t^{1/2}$. Integrals in (10), envelopes for current frequency $\omega_* = c_* q_*$ are depicted in Figure 7.

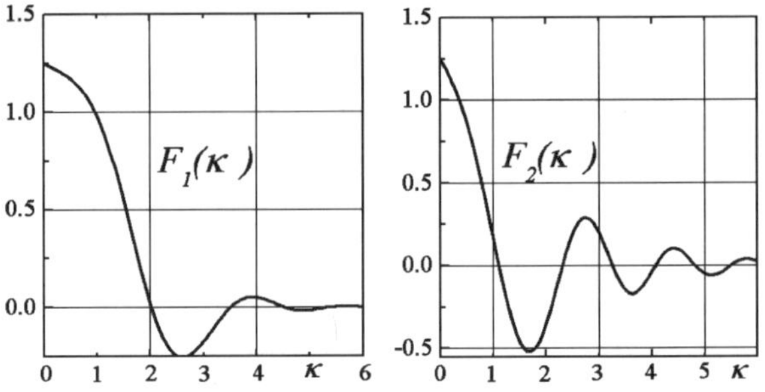

Figure 7. Quasi-steady state envelopes of resonant perturbations

The special point (minimum) is obtained from the dispersion equations of the model

$$(c/c_0)^2 = \left(h^2/12\right)q^2 + G_1 q^{-2}, G_1 = G/Eh$$

where E is the Young modulus of the plate. Its coordinates, q_* and c_* are

$$q_* = \left(12G_1/h^2\right)^{1/4}, c_* = c_0\left(h^2 G_1/3\right)^{1/4}$$

To prove the applicability of the asymptotic solution (10), a set of computer simulations has been performed (for the described system there are significantly simpler procedures than in the case of the original system). In Figure 8 three processes are compared calculated for cases when a step load moves with (a) subcritical, (b) critical and (c) supercritical velocities.

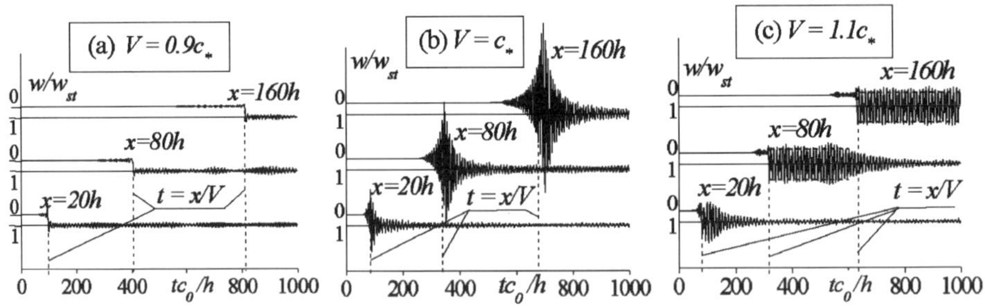

Figure 8. Unsteady response of the system plate-elastic foundation to a load moving with diverse velocities

In the figure the plate deflection w is normalized by its static value: $w_{st} = Q/G$, h is the effective plate thickness and c_0 is the effective sound velocity in the plate. Critical velocity $c_* = 0.2295c_0$. The comparison of asymptotic (10) with calculation results shows that solution (10) is practically accurate at $t > 20h/c_0$ (note that the static part of the solution is absent in formula (10), which serves for description of notably unsteady perturbations).

References

1. Goldshtein, R.V.: Rayleigh waves and resonance phenomena in elastic bodies, *Appl. Math. Mech.* 29 (1965), 3, 516-526.
2. Slepyan, L.I.: Resonance phenomena in plates and shells under moving loads, *Proc. VI All- Union Conf. on Theory of Plates and Shells*, Nauka, Moscow (1966), 690-696.
3. Ayzenberg, M.V. and Slepyan, L.I.: Resonance waves in a hollow cylinder immersed into a compressible liquid, *Proc. All-Union Symp. on Transient Processes in Plates and Shells*, Tallin (1967), 13-21.
4. Ayzenberg, M.V.: Resonance waves in a hollow cylinder, *Mechanics of Solids*, 4 (1969), 1, 84-90.
5. Ayzenberg, M.V.: Analysis of longitudinal resonant waves in a thin cylindrical shell immersed into fluid and subjected to moving load, *Proc. VII All-Union Conf. on Theory of Plates and Shells*, Nauka, Moscow (1970), 22-26.
6. Slepyan, L.I.: Time-Dependent Elastic Waves, Sudostroenie, Leningrad (1972).
7. Kadyrov, T., Stepanenko, M.V.: On particular features of elastic waves propagation in a layered halfspace, *J.Appl. Mech. Techn. Phys.* 18 (1988), 4, 109-115.
8. Alexandrova, N.I., Potashnikov, I.A. and Stepanenko, M.V.: Flexural resonant waves in a cylindrical shell under moving radial loading, *J. Appl. Mech. Techn. Phys.* 19 (1989), 3, 132- 137.
9. Ayzenberg-Stepanenko, M.V.: Resonance waves in elastic waveguides, *Proc. 25th Israel Conf. Mech. Engng.* (1994), Technion. Haifa
10. Slepyan, L.I. and Tsareva, O.V.: Energy flux for zero group velocity of the carrying wave, *Sov. Phys. Dokl.* 32 (1987), 6, 522-524.
11. Stepanenko, M.V.: On calculation of pulsed strain processes in elastic structures, *Sov. Mining Sci.* 12 (1976), 2, 53-57.

NUMERICAL ANALYSIS OF RAYLEIGH WAVES IN ANISOTROPIC MEDIA

A. V. KAPTSOV and S. V. KUZNETSOV
Institute for Problems in Mechanics
Prospect Vernadskogo, 101 Moscow 119526 Russia

Abstract. - Eigenvalues and eigenvectors of the Christoffel equation for the problem of Rayleigh waves propagating in media with arbitrary anisotropy are analyzed on the bases of three-dimensional complex formalism. Numerical algorithm for analysis of speed of Rayleigh waves is discussed.

Key words: nonlinear surface waves, elasticity, Rayleigh waves, scale invariance, nonlocality, spectral kernel, spatial kernel, Hilbert transform

Introduction

Spectral properties of the Christoffel equation are needed for construction of the exponentially decaying surface (Rayleigh) waves. The following analysis covers theorems of root characterization, structure of eigenvalues and corresponding eigenspaces of the Christoffel equation. The main results are obtained on the bases of three-dimensional complex formalism, which goes back to Rayleigh (1885).

For anisotropic media Rayleigh's approach was exploited by Farnell (1970) in numerical study of speed of Rayleigh waves, by Stoneley (1955), and later by Royer and Dieulesaint (1985) in analytical analysis of Rayleigh waves.

Six-dimensional formalism for Rayleigh wave analysis was proposed by Stroh (1962). This is based on similarity of solutions for moving line dislocations in unbounded media and propagation of Rayleigh waves. For Rayleigh wave analysis this approach was developed by Barnett and Lothe (1973, 1974), Lothe and Barnett (1976) and Chadwick et al. (1977, 1979). Under assumptions of this formalism theorems of uniqueness and existence for Rayleigh waves were proved.

R.V. Goldstein and G.A. Maugin (eds.),
Surface Waves in Anisotropic and Laminated Bodies and Defects Detection, 217–226.

Present analysis gives complete characterization of spectral properties of the Christoffel equation for Rayleigh waves propagating in media with arbitrary elastic anisotropy, and delivers a rational basis for constructing numerical algorithm for determination of energy and speed of genuine Rayleigh waves.

1. Basic notations

Equations of motion for anisotropic elastic medium can be written the form

$$\mathrm{div}\mathbf{C}\cdot\cdot\nabla\mathbf{u} - \rho\ddot{\mathbf{u}} = 0 \tag{1.1}$$

where $\mathbf{C}$ is the fourth-order elasticity tensor assumed to be positive definite, $\mathbf{u}$ is the displacement field, and ρ is density of a medium.

Rayleigh wave is composed of partial waves having simple exponential decay in depth (partial waves of different structure are not considered generally):

$$\begin{gathered}\mathbf{u}(\mathbf{x}) = \mathbf{m}e^{i(\gamma\boldsymbol{\nu}\cdot\mathbf{x}+\mathbf{n}'\cdot\mathbf{x}-ct)} = \mathbf{m}\,e^{-\alpha\boldsymbol{\nu}\cdot\mathbf{x}}e^{i(\beta\boldsymbol{\nu}\cdot\mathbf{x}+\mathbf{n}'\cdot\mathbf{x}-ct)}\\ \beta = \mathrm{Re}\,(\gamma), \qquad \alpha = \mathrm{Im}\,(\gamma)\end{gathered} \tag{1.2}$$

where $\mathbf{m}$ is the amplitude vector; $\boldsymbol{\nu}$ is the unit normal to a plane boundary Π_ν; $\gamma = \beta + i\alpha$ is the unknown complex parameter; $\mathbf{n}' \subset \Pi_\nu$ is the unit vector defining direction of propagation of Rayleigh wave; and c is the wave (phase) speed. Substitution of representation (1.2) into equations (1.1) produces the Christoffel equation:

$$\Big[(\gamma\boldsymbol{\nu}+\mathbf{n}')\cdot\mathbf{C}\cdot(\mathbf{n}'+\gamma\boldsymbol{\nu}) - \rho c^2\mathbf{I}\Big]\cdot\mathbf{m} = 0 \tag{1.3}$$

or equivalently

$$\det\Big[(\gamma\boldsymbol{\nu}+\mathbf{n}')\cdot\mathbf{C}\cdot(\mathbf{n}'+\gamma\boldsymbol{\nu}) - \rho c^2\mathbf{I}\Big] = 0 \tag{1.3$'$}$$

where it is assumed that the phase speed $c$is already specified. It is clear that left side of Eq. (1.3) represents polynomial of degree 6 on γ.

Simple analysis of (1.3) shows that if c is zero, then there are no real roots of (1.3), since matrix $(\gamma\boldsymbol{\nu}+\mathbf{n}')\cdot\mathbf{C}\cdot(\mathbf{n}'+\gamma\boldsymbol{\nu})$ is positive definite at real γ and its determinant cannot vanish. Similarly, asymptotic analysis produces

$$\gamma_i^2\ \frac{\rho c^2}{\lambda_i(\boldsymbol{\nu}\cdot\mathbf{C}\cdot\boldsymbol{\nu})}, c \to \infty \tag{1.4}$$

where $\lambda_i, i = 1, 2, 3$ are eigenvalues (necessarily positive) of the matrix $\boldsymbol{\nu}\cdot\mathbf{C}\cdot\boldsymbol{\nu}$. Expression (1.4) ensures that all roots of (1.3) are real at high values of c.

Suppose that γ is a real root corresponding to a certain value of the phase speed c. Introduction of a new parameter $\varphi = \arctan(\gamma)$ allows to rewrite Eq. (1.3) in terms of parameter φ:

$$\begin{aligned} &\det\left[(\tan(\varphi)\boldsymbol{\nu}+\mathbf{n}')\cdot\mathbf{C}\cdot(\mathbf{n}'+\tan(\varphi)\boldsymbol{\nu}) - \rho c^2\mathbf{I}\right] = \\ &= \det\left[\cos^{-2}(\varphi)\mathbf{w}\cdot\mathbf{C}\cdot\mathbf{w} - \rho c^2\mathbf{I}\right] = 0 \end{aligned} \tag{1.5}$$

where

$$\mathbf{w} = \sin(\varphi)\boldsymbol{\nu} + \cos(\varphi)\mathbf{n}' \tag{1.6}$$

it is clear that vector $\mathbf{w}$ in (1.6) is of the unit length.

DEFINITION 1.1. *Limiting speeds which correspond to the unit vectors $\boldsymbol{\nu}$ and $\mathbf{n}'$, are defined by*

$$c_i^{\lim} = \inf_{\varphi\in[-\frac{\pi}{2};\frac{\pi}{2}]}\left(\cos^{-1}(\varphi)\sqrt{\rho^{-1}\lambda_i(\mathbf{w}\cdot\mathbf{C}\cdot\mathbf{w})}\right),\ i=1,2,3 \tag{1.7}$$

where λ_i, i=1,2,3 are eigenvalues arranged in the descending mode.

Expression (1.7) gives:

$$c_i^{\lim} \leq \sqrt{\rho^{-1}\lambda_i(\mathbf{w}\cdot\mathbf{C}\cdot\mathbf{w})}, i = 1,2,3 \tag{1.8}$$

Analysis of (1.7) shows that there are three different cases, when complex roots exist: (i) if $c_2^{\lim} < c < c_1^{\lim}$, then Christoffel's equation has only one pair of complex-conjugate roots; (ii) if $c_3^{\lim} < c < c_2^{\lim}$, there are two pairs of complex-conjugate roots; and (iii) if $c < c_3^{\lim}$, there are three pairs of complex- conjugate roots. Since for existence of Rayleigh wave at least one pair of complex conjugate roots is needed, we arrive to

PROPOSITION 1.1. *Speed of the Rayleigh wave can not exceed $c_1^{\lim}$.*

REMARK 1.1. *a) Exponential decay in the lower half-space ($\boldsymbol{\nu}\cdot\mathbf{x} < 0$) is ensured by complex roots with negative imaginary part: $\alpha < 0$; b) precisely these roots will be chosen in the subsequent analysis;*

2. Properties of the Christoffel equation

Appropriate complex root γ being substituted in Christoffel's equation produces following equation for the unknown (complex) eigenvector:

$$(\mathbf{G} - \rho c^2\mathbf{I})\cdot\mathbf{m} = 0, \mathbf{G} \equiv \mathbf{A} + i\mathbf{B} \tag{2.1}$$

where Hermitian components $\mathbf{A}, \mathbf{B}$ of the matrix $\mathbf{G}$ have the form

$$\mathbf{A} \equiv \left[\mathbf{n}' \cdot \mathbf{C} \cdot \mathbf{n}' + (\beta^2 - \alpha^2)\,\boldsymbol{\nu} \cdot \mathbf{C} \cdot \boldsymbol{\nu} + \beta\,(\mathbf{n}' \cdot \mathbf{C} \cdot \boldsymbol{\nu} + \boldsymbol{\nu} \cdot \mathbf{C} \cdot \mathbf{n}')\right]$$

$$\mathbf{B} \equiv \alpha\,\left[2\beta\,\boldsymbol{\nu} \cdot \mathbf{C} \cdot \boldsymbol{\nu} + (\boldsymbol{\nu} \cdot \mathbf{C} \cdot \mathbf{n}' + \mathbf{n}' \cdot \mathbf{C} \cdot \boldsymbol{\nu})\right]$$

$$\mathbf{m}' \equiv \mathrm{Re}(\mathbf{m}), \mathbf{m}'' \equiv \mathrm{Im}(\mathbf{m})$$

LEMMA 2.1. *Matrices* $\mathbf{A}$ *and* $\mathbf{B}$ *both are non-zero and do not commute with each other.*

Following proposition flows out directly from the analysis of Eq. (2.1) and the lemma 2.1.

PROPOSITION 2.1.

a) If $\mathbf{m}'$, $\mathbf{m}'' \notin \ker(\mathbf{A} - \rho c^2\mathbf{I})$, *then* $\mathbf{m}'$, $\mathbf{m}'' \notin \ker(\mathbf{B})$ *and*

$$\left[\mathbf{B}^{-1} \circ (\mathbf{A} - \rho c^2\mathbf{I})\right]^2 = -\mathbf{I} \tag{2.2}$$

in the real (necessarily two-dimensional) space generated by vectors $\mathbf{m}'$ *and* $\mathbf{m}''$;

b) If $\mathbf{m}'$, $\mathbf{m}'' \in \ker(\mathbf{A} - \rho c^2\mathbf{I})$, *then* $\mathbf{m}'$, $\mathbf{m}'' \in \ker(\mathbf{B})$;

c) If $\mathbf{m}' \in \ker(\mathbf{A} - \rho c^2\mathbf{I})$ *and* $\mathbf{m}'' \notin \ker(\mathbf{A} - \rho c^2\mathbf{I})$, *then* $\mathbf{m}'' \in \ker \mathbf{B}$ *and* $\mathbf{m}'' \in \ker \mathbf{B}$ *and* $\mathbf{m}' \notin \ker \mathbf{B}$ *(and vice versa).*

It is obvious that proposition 2.1 covers all possibilities for Hermitian components $\mathbf{m}', \mathbf{m}''$ of the eigenvector $\mathbf{m}$ of matrix $\mathbf{G}$.

Corollary 1. Condition of the Proposition 2.1.a ensures:

a) Both matrices $\mathbf{B}$ and $(\mathbf{A} - \rho c^2\mathbf{I})$ are not of the fixed sign in the two dimensional subspace $Z \subset R^3$ generated by the components $\mathbf{m}'$ and $\mathbf{m}''$;

b) $\mathbf{B}(Z) \subset Z, (\mathbf{A} - \rho c^2\mathbf{I}) \subset Z$;

c) Unimodal matrix $\mathbf{B}^{-1} \circ (\mathbf{A} - \rho c^2\mathbf{I})$ in any basis in Z has the form

$$\mathbf{B}^{-1} \circ (\mathbf{A} - \rho c^2\mathbf{I}) = \begin{pmatrix} \delta & \chi \\ -\frac{1+\delta^2}{\chi} & -\delta \end{pmatrix} \tag{2.3}$$

where δ and χ are real, and $\chi \neq 0$; d) Matrices $\mathbf{A}$ and $\mathbf{B}$ do not commute with each other.

Corollary 2. Condition of the Proposition 2.1.c ensures

a) Vectors $\mathbf{m}'$ and $\mathbf{m}''$ both are non-collinear;

b) Tensors $\mathbf{A} - \rho c^2\mathbf{I}$ and $\mathbf{B}$ have following structure:

$$\begin{aligned} \mathbf{A} - \rho c^2\mathbf{I} &= 0\mathbf{m}' \otimes \mathbf{m}' + c_1\mathbf{v}_1 \otimes \mathbf{v}_1 - c_1\mathbf{v}_2 \otimes \mathbf{v}_2, \\ \mathbf{B} &= 0\mathbf{m}'' \otimes \mathbf{m}'' + c_2\mathbf{w}_1 \otimes \mathbf{w}_1 - c_2\mathbf{w}_2 \otimes \mathbf{w}_2 \end{aligned} \tag{2.4}$$

where c_1, c_2 are nonzero real numbers, vectors $\mathbf{m}', \mathbf{v}_1, \mathbf{v}_2 \in R^3$ are mutually orthogonal, vectors $\mathbf{m}'', \mathbf{w}_1, \mathbf{w}_2 \in R^3$ are also mutually orthogonal.
Corollary 3. Real part β of the complex root satisfies inequalities:

$$-\frac{\lambda_{\max}(\boldsymbol{\nu}\cdot\mathbf{C}\cdot\mathbf{n}'+\mathbf{n}'\cdot\mathbf{C}\cdot\boldsymbol{\nu})}{2\lambda_{\max}(\boldsymbol{\nu}\cdot\mathbf{C}\cdot\boldsymbol{\nu})} \leq \beta \leq -\frac{\lambda_{\min}(\boldsymbol{\nu}\cdot\mathbf{C}\cdot\mathbf{n}'+\mathbf{n}'\cdot\mathbf{C}\cdot\boldsymbol{\nu})}{2\lambda_{\min}(\boldsymbol{\nu}\cdot\mathbf{C}\cdot\boldsymbol{\nu})} \qquad (2.5)$$

Proof. Following inequalities flow out directly from the definition of matrix $\mathbf{B}$, which gives:

$$\underset{\mathbf{w}\in R^3;|\mathbf{w}|=1}{\forall\mathbf{w}} \quad \alpha(2\beta\lambda_{\min}(\boldsymbol{\nu}\cdot\mathbf{C}\cdot\boldsymbol{\nu})+\lambda_{\min}(\boldsymbol{\nu}\cdot\mathbf{C}\cdot\mathbf{n}'+\mathbf{n}'\cdot\mathbf{C}\cdot\boldsymbol{\nu})) \leq \mathbf{w}\cdot\mathbf{B}\cdot\mathbf{w} \leq$$

$$\leq \alpha(2\beta\,\lambda_{\max}(\boldsymbol{\nu}\cdot\mathbf{C}\cdot\boldsymbol{\nu})+\lambda_{\max}(\boldsymbol{\nu}\cdot\mathbf{C}\cdot\mathbf{n}'+\mathbf{n}'\cdot\mathbf{C}\cdot\boldsymbol{\nu}))$$

Corollary 1 ensures that in any case $\lambda_{\min}(\mathbf{B}) \leq 0$, so there exist such vectors $\mathbf{w} \in R^3; |\mathbf{w}| = 1$ that $\mathbf{w}\cdot\mathbf{B}\cdot\mathbf{w} = 0$. Taking into account that $\alpha > 0$ and $\lambda_{\frac{\max}{\min}}(\boldsymbol{\nu}\cdot\mathbf{C}\cdot\boldsymbol{\nu}) > 0$, we arrive to (2.5).

PROPOSITION 2.2. *For any complex root γ matrix $\mathbf{G}$ is not normal.*

Proof. of this proposition follows from consideration of the commutator $[\mathbf{A}, \mathbf{B}]$, which is not zero for complex γ, provided (1.9) holds.

REMARK 2.1. *At real γ matrix $\mathbf{G}$ is normal, that is due to its symmetric structure, which follows from (2.1) and the subsequent formulas for matrices $\mathbf{A}$ and $\mathbf{B}$.*

PROPOSITION 2.3. *Under the condition of proposition 2.1.b eigenspace in R^3 generated by vectors $\mathbf{m}', \mathbf{m}''$ is one dimensional.*

Proof. If this eigenspace is two- or three-dimensional, the symmetric matrices $\mathbf{A} - \rho c^2\mathbf{I}$ and $\mathbf{B}$ could be reduced to diagonal form by the same orthogonal transformation (in such a case $\mathbf{A} - \rho c^2\mathbf{I}$ and $\mathbf{B}$ have common eigenspace), this leads to there commutation, but these matrices can not commute since matrix $\mathbf{G}$ is not normal.

3. Spectral properties of the Christoffel equation

Let $W_{\mathbf{G}}$ be eigenspace of matrix $\mathbf{G}$ corresponding to the eigenvalue ρc^2. Propositions 2.1 - 2.3 produce

PROPOSITION 3.1. *a)* $\dim W_{\mathbf{G}} = 1$, *provided that both matrices $\mathbf{A}' \equiv \mathbf{A} - \rho c^2\mathbf{I}$ and $\mathbf{B}$ do not have zero eigenvalues, and relation (2.2) holds;*

b) $\dim W_{\mathbf{G}} = 2$, *provided that relation (2.2) holds, and both matrices $\mathbf{A}' \equiv \mathbf{A} - \rho c^2\mathbf{I}$ and $\mathbf{B}$ have zero eigenvalue;*

c) $\dim W_{\mathbf{G}} = 1$, *provided that relation (2.2) does not hold, and both matrices $\mathbf{A}' \equiv \mathbf{A} - \rho c^2\mathbf{I}$ and $\mathbf{B}$ have zero eigenvalue.*

Proof. a) Corollary 1 to proposition 2.1 shows that real space Z is two-dimensional. Let real vectors $\mathbf{e}_1, \mathbf{e}_2$ form an orthonormal basis in Z, then (complex) eigenvectors $\mathbf{m}', \mathbf{m}''$ admit representation

$$\mathbf{m}' = \mathbf{e}_1 + i(\mathbf{B}^{-1} \circ \mathbf{A}') \cdot \mathbf{e}_1, \mathbf{m}'' = \mathbf{e}_2 + i(\mathbf{B}^{-1} \circ \mathbf{A}') \cdot \mathbf{e}_2 \tag{3.1}$$

Substitution of relation (2.3) in (3.1) gives

$$\mathbf{m}' = \mathbf{e}_1 + i(\delta \mathbf{e}_1 - \tfrac{1+\delta^2}{\chi} \mathbf{e}_2), \mathbf{m}'' = \mathbf{e}_2 + i(\chi \mathbf{e}_1 - \delta \mathbf{e}_2) \tag{3.2}$$

Now, representation (3.2) shows that complex vectors $\mathbf{m}', \mathbf{m}''$ are linearly dependent: $\mathbf{m}'' = a\mathbf{m}'$ with complex constant $a = \frac{\chi\delta}{1+\delta^2} + i\frac{\chi}{1+\delta^2}$.

b) Linearly dependent complex vectors $\mathbf{m}', \mathbf{m}''$ satisfying (3.2) form one-dimensional space $K \subset C^3$. Now, eigenvector $\mathbf{m} = \mathbf{e}_3 + i\mathbf{0}$, where $\mathbf{e}_3$ is common eigenvector of matrices $\mathbf{A}'$ and $\mathbf{B}$ is orthogonal to K. This completes the proof.

c) Hermitian components of eigenvectors of the matrix $\mathbf{G}$ corresponding to ρc^2, belong to $\ker(\mathbf{A} - \rho c^2 \mathbf{I}) \cap \ker(\mathbf{B})$ due to proposition 2.1.b. If $\dim\left(\ker(\mathbf{A} - \rho c^2 \mathbf{I}) \cap \ker(\mathbf{B})\right) > 1$, then both matrices $\mathbf{A}$ and $\mathbf{B}$ have common eigenvectors, and therefore they commute with each other, but this contradicts to the proposition 2.2 (if Hermitian components $\mathbf{A}$ and $\mathbf{B}$ of the matrix G, commute, then the latter matrix becomes normal).

LEMMA 3.1. *If λ is a complex eigenvalue of matrix $\mathbf{G}$ and $\mathbf{m}$ is corresponding eigenvector with Hermitian components $\mathbf{m}', \mathbf{m}'' \in Z \subset R^3$, then $i\lambda$ is eigenvector of $\mathbf{G}$ with another eigenvector which Hermitian components also belong to Z.*

Proof is identical to proof of the proposition 2.1.a with the only difference that for the eigenvalue $i\lambda$ relation (2.2) changes to the following:

$$\left[\mathbf{A}^{-1} \circ (\mathbf{B} - \lambda \mathbf{I})\right]^{-2} = -\mathbf{I}$$

Similarly, values δ and χ from the expression (2.3) change.

PROPOSITION 3.2. *Algebraic multiplicity of the eigenvalue ρc^2 equals to its geometric multiplicity.*

Proof of the case corresponding to proposition 3.1.a. Let $Z \subset R^3$ is two-dimensional space from the Corollary 1 to proposition 2.1, then matrix $\mathbf{G}$, regarded as operator acting in C^3, can be decomposed in the direct sum:

$$\mathbf{G} = (\mathbf{A}'_Z + i\mathbf{B}_Z) \oplus (\mathbf{A}'_{CZ} + i\mathbf{B}_{CZ}) \tag{3.3}$$

where $\mathbf{A}'_Z, \mathbf{B}_Z, \mathbf{A}'_{CZ}, \mathbf{B}_{CZ}$ are restriction of the operators $\mathbf{A}', \mathbf{B}$ on Z, CZ, and $CZ = R^3 \backslash Z$. Since CZ is one-dimensional, its algebraic multiplicity coincides with algebraic multiplicity, and direct verification shows that corresponding eigenvalue does not equal to ρc^2 and $i\rho c^2$. Then, proposition 3.1.a shows that CZ is one-dimensional, if regarded as complex space. To complete the proof of this case it is sufficient to note that there is another eigenvector in Z, which corresponds to eigenvalue $i\rho c^2$. It should be noted that eigenvectors form Z are not orthogonal.

Proof of the cases corresponding to proposition 3.1.b, 3.1.c are mainly identical to the regarded case.

Proofs of the lemma 3.1 and proposition 3.2 ensure that for all regarded cases matrix $\mathbf{G}$ has 3 linearly independent complex eigenvectors and these eigenvectors form a basis in C^3. However, these eigenvectors are not orthogonal, otherwise matrix $\mathbf{G}$ would be normal.

Existence of three linearly independent eigenvectors allows to formulate *Corollary.* Matrix $\mathbf{G}$ is similar to the complex diagonal matrix.

PROPOSITION 3.3. *Let parameter ρc^2 be fixed, then (complex) eigenvectors corresponding to different complex roots with positive imaginary part, are linearly independent.*

Proof. Suppose that common eigenvector m corresponds to two different complex roots γ_1, γ_2, then

$$(\mathbf{G}(\gamma_1) - \mathbf{G}(\gamma_2)) \cdot \mathbf{m} = 0 \tag{3.4}$$

By the use of (1.3), (3.4) can be transformed to the following form

$$((\gamma_1 + \gamma_2)\boldsymbol{\nu} \cdot \mathbf{C} \cdot \boldsymbol{\nu} + (\boldsymbol{\nu} \cdot \mathbf{C} \cdot \mathbf{n}' + \mathbf{n}' \cdot \mathbf{C} \cdot \boldsymbol{\nu})) \cdot \mathbf{m} = 0 \tag{3.5}$$

Since $\mathrm{Im}(\gamma_1 + \gamma_2) \neq 0$, $\ker((\gamma_1 + \gamma_2)\boldsymbol{\nu} \cdot \mathbf{C} \cdot \boldsymbol{\nu} + (\boldsymbol{\nu} \cdot \mathbf{C} \cdot \mathbf{n}' + \mathbf{n}' \cdot \mathbf{C} \cdot \boldsymbol{\nu}))$ is nontrivial if and only if matrix $\boldsymbol{\nu} \cdot \mathbf{C} \cdot \boldsymbol{\nu}$ is degenerate or sign indefinite, but this is impossible.

REMARK 3.1. *Analysis of eigenspaces of the Christoffel equation shows that when the phase speed belongs to the interval $(0, c_3^{\lim})$ and hence, there are no multiple roots, there exist exactly three linearly independent complex eigenvectors forming partial waves in the Rayleigh wave representation (regular case). This flows out from propositions 3.2 and 3.3.*

The same propositions guarantee that degenerate case in the interval $(0, c_3^{\lim})$ can only occur when two conditions hold simultaneously: (i) there are multiple roots; and, (ii) multiplicity of the eigenvalue ρc^2 is one at any of the multiple roots.

It should be noted that in the isotropic case multiplicity of the parameter ρc^2 at multiple roots is two, and no degeneracy occurs.

4. Numerical analysis

Let elastic parameters of the cubic crystal be as follows:

$$\begin{aligned} &\eta = 1; \quad \lambda = 0.445\ 256\ 092\ 865\ 185\ 947\ 162\ 255\ 371\ 251\ 987...; \\ &\mu = 0.3; \qquad \rho = 1 \end{aligned} \tag{4.1}$$

The values for elastic constants in (4.1) ensure a kind of degeneracy, when no genuine Rayleigh wave of the type (1.2) exists. Direct verification shows, that at these values of the elastic parameters the elasticity tensor is positive definite.

Solution of dispersion equation gives the following value for the speed of Rayleigh wave:

$$c_R = 0.502\ 565\ 076\ 378\ 238\ 410\ 813\ 989\ 573\ 349\ 365 \cdots \tag{4.2}$$

Of course, the same value of the phase speed gives application of the Stroh sextic formalism.

Analysis of this situation within framework of the three-dimensional formalism is given in the following Remark.

REMARK 4.1. *For elastic parameters (4.1) computations based on representation (1.2), give the following complex roots γ_k of the Christoffel equation:*

$$\begin{aligned} &\gamma_1 = \gamma_2 = -0.586\ 302\ 258\ 870\ 134\ 534\ 743\ 187\ 664\ 288\ 300...i; \\ &\gamma_3 = -0.397\ 610\ 965\ 664\ 242\ 349\ 180\ 295\ 499\ 247\ 097...i \end{aligned} \tag{4.3}$$

Corresponding complex three-dimensional eigenvectors (amplitudes) of the Christoffel equation are as follows:

$$\begin{aligned} \mathbf{m}_1 = \mathbf{m}_2 = &(0.827\ 631\ 182\ 371\ 740\ 626\ 377\ 247\ 467\ 943\ 188...)\,\mathbf{n} \\ &+ i\,(0.561\ 272\ 327\ 810\ 622\ 254\ 491\ 318\ 496\ 469\ 724...)\,\boldsymbol{\nu} \\ &\mathbf{m}_3 = \mathbf{n} \times \boldsymbol{\nu} \end{aligned} \tag{4.4}$$

Substituting these roots and amplitude vectors into traction-free boundary conditions yields following (normalised) kernel eigenvector:

$$\vec{Z} = \left(\tfrac{1}{\sqrt{2}}, -\tfrac{1}{\sqrt{2}}, 0\right) \tag{4.5}$$

Bearing in mind (4.4), (4.5), representation for (genuine) Rayleigh wave takes the form:

$$\mathbf{u}(\mathbf{x}) = \frac{1}{\sqrt{2}}\mathbf{m}_1\, e^{ir(\gamma_1 \boldsymbol{\nu}\cdot\mathbf{x}+\mathbf{n}\cdot\mathbf{x}-c_R t)} - \frac{1}{\sqrt{2}}\mathbf{m}_1\, e^{i\,r(\gamma_1 \boldsymbol{\nu}\cdot\mathbf{x}+\mathbf{n}\cdot\mathbf{x}-c_R t)} = 0 \tag{4.6}$$

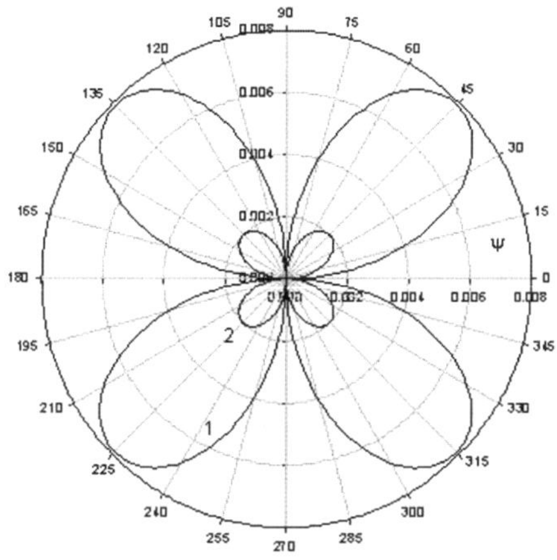

Figure 1. Variation of strain (1) and kinetic (2) energy functions vs. angle ψ

Now, for the regarded cubic crystal variation of strain and kinetic energy functions (plotted on the Π_ν-plane at $\boldsymbol{\nu}\cdot\mathbf{x} = 0$) due to variation of direction of propagation of genuine Rayleigh wave is analyzed (Fig.1). Both these energy functions are plotted at the wave number $r = 1$. Orientation of the vector $\mathbf{n}$ belonging to the Π_ν-plane is determined by the angle ψ, which is the angle between vector $\mathbf{n}$ and one of the crystallographic axes. These plots show that both strain and kinetic energy functions tend to zero as $\psi \to \pm 0 + {}^1\!/\!{}_2\pi n$ (n is integer). Thus, for the regarded cubic crystal "forbidden" directions for genuine Rayleigh waves are stable with respect to small variations of the angle ψ.

Acknowledgement

This paper study was supported by the Program for Supporting Outstanding Scientific Schools of the Russia (Grant 00-15-96066)

References

D. M. Barnett, and J. Lothe, Synthesis of the sextic and the integral formalism for dislocations, Greens functions, and surface waves in anisotropic elastic solids, *Phys. Norv.*, **7**, 1973, pp. 13-19.

D. M. Barnett, and J. Lothe, Consideration of the existence of surface wave (Rayleigh wave) solutions in anisotropic elastic crystals, *J. Phys., F: Metal Phys.*, **4**, 1974, pp. 671-686.

P. Chadwick and G. D. Smith, Foundations of the theory of surface waves in anisotropic elastic materials, *In: Advances in Applied Mechanics*, Acad. Press, N.Y., **17**, 1977, pp. 303-376.

P. Chadwick and D. A. Jarvis, Surface waves in a prestressed elastic body. *Proc. Roy. Soc. London.*, A**366**, 1979, pp. 517 - 536.

G. W. Farnell, Properties of elastic surface waves, *Phys. Acoust.*, 1970, **6**, pp. 109 - 166.

J. Lothe and D. M. Barnett, On the existence of surface wave solutions for anisotropic elastic half-spaces with free surface, *J. Appl. Phys.*, **47**, 1976, pp. 428-433.

D. Royer and E. Dieulesaint, Rayleigh wave velocity and displacement in orthorombic, tetragonal, and cubic crystals. *J. Acoust. Soc. Am.*, 1985, **76**, No. 5, pp. 1438 -1444.

R. Stoneley, The propagation of surface elastic waves in a cubic crystal. *Proc. Roy. Soc.*, 1955, A**232**, pp. 447 - 458.

A. N. Stroh, Steady state problems in anisotropic elasticity. *J. Math. Phys.*, 1962, **41**, pp. 77 - 103.

J. W. Strutt (Lord Rayleigh), On wave propagating along the plane surface of an elastic solid. *Proc. London Math. Soc.*, 1885, **17**, pp. 4-11.

GUIDED WAVES IN ANISOTROPIC MEDIA: APPLICATIONS

R. A. KLINE
Director San Diego Center for Materials Research San Diego State University
San Diego, CA, USA 92182

Abstract. This paper presents an overview of the use of guided waves for the characterization of anisotropic media.

Key words: nonlinear surface waves, elasticity, Rayleigh waves, scale invariance, nonlocality, spectral kernel, spatial kernel, Hilbert transform

Introduction

This paper presents an overview of the use of guided waves for the characterization of anisotropic media. For many years, isotropic (or more accurately statistically isotropic) materials dominated structural applications. With the advent of advanced composite materials and more recently, directionally solidified superalloys, anisotropic materials are being used with increasing frequency in structural critical applications. Hence, there is a great need for the development of nondestructive evaluation techniques for both basic property characterization in these materials as well defect identification to insure structural integrity. While bulk waves have received most of the research and development attention from the scientific community, there are several critical problems where guided waves are more suitable. For example, since laminated plate structures are commonly used in high performance applications, Lamb waves are often desirable. Further, since guided waves can propagate over long distances in structures without significant attenuation, they are advantageous for inspecting large scale structures.

R.V. Goldstein and G.A. Maugin (eds.),
Surface Waves in Anisotropic and Laminated Bodies and Defects Detection, 227–240.

1. Theory

The basic governing equation for guided wave propagation in anisotropic media, whether one is interested in Rayleigh, Lamb or Love waves, is based on the well-known Christoffel equation:

$$C_{ijkl}u_{k,jl} = \rho\ddot{u}_i \tag{1}$$

For many applications, it is advantageous to write this equation in a more compact form with differential operators as:

$$L_{ik}u_k = \delta_{ik}\rho\ddot{u}_k\,, \qquad L_{ik} \equiv C_{ijkl}\frac{\partial^2}{\partial x_j \partial x_l} \tag{2}$$

In order to obtain a non-trivial solution to this equation, it can be shown that

$$\det\, L_{ik} = 0 \tag{3}$$

One then usually assumes a solution in the form of a traveling wave of the form:

$$u(x,t) = A_0 a\, h(y)\, e^{i(kx\cos\theta + kz\sin\theta - \omega t)} \tag{4}$$

Substitution of this assumed form of the solution into the above determininantal equation for a non-trivial solution yields a sixth order differential equation

$$h(y) = e^{i\beta y} \tag{5}$$

The solution of this differential equation is governed by the boundary conditions as they determine the fundamental character of the solution and are fundamentally different for each type of guided wave.

1.1. SURFACE WAVES

For surface wave propagation, the free surface boundary condition requires that

$$\sigma_{21} = \sigma_{22} = \sigma_{23} = 0 \quad \text{at} \quad y = 0 \tag{6}$$

Further, we require that the displacements remain finite with increasing depth. This means that the imaginary part of β must be positive and we may omit the contributions from three of the roots. The nature of the solution has been investigated by a variety of investigators. Stonely [1] studied surface wave propagation in single crystals, but used a restrictive formulation based on isotropic media. This approach restricted solutions only to those where the amplitude decay was exponential with depth. This assumption led Stoneley to the conclusion that propagating solutions were not possible in certain directions. Synge [2] later introduced the concept

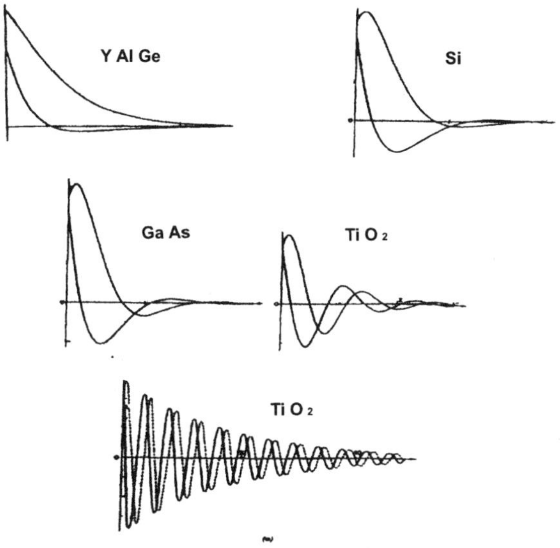

Figure 1. Rayleigh Wave Behavior in Anisotropic Media (after Royer and Dieulesaint [3])

of generalized Raylieh waves with complex (not pure imaginary) β values Complex values mean that the character of the solution is somewhat different than that observed for isotropic medium as the amplitude will exhibit not only the exponential decay with depth as in the isotropic case but will also show a harmonic variation with depth as well. Royer and Dieulesaint [3] published an extensive study of surface wave propagation in materials with several different crystal symmetries. Of particular interest is the variation in amplitude with depth as a function of anisotropy ratio A. These results are shown in Figure 1. Note that the solution closely resembles that for cases of weak anisotropy (A slightly greater than 1) and changes dramatically as A increases. For the more general case of surface wave propagation in materials of arbitrary symmetry the situation is quite similar to that found for materials with some degree of symmetry. Buchwald [4] investigated surface wave propagation for several different material symmetries. As long as the free surface coincided with a plane of material symmetry, Buchwald was able to obtain propagating wave solutions. However, when the free surface was not a plane of material symmetry, he found that there were directions in which no solutions could be obtained. Lim [5] later developed a formulation where solutions were obtained for the "forbidden" directions with the introduction of generalized Rayleigh waves.

1.2. PLATE WAVES IN ANISOTROPIC MEDIA

For plate wave propagation, the pertinent boundary conditions are that the top and bottom surfaces remain stress free. This requires

$$\sigma_{21} = \sigma_{22} = \sigma_{23} = 0 \quad \text{at} \quad y = \pm b \tag{7}$$

The solution to this problem has been thoroughly investigated for both isotropic and anisotropic materials [6-9]. Plates waves are multimode, with either symmetric or antisymmetric particle displacements and highly dispersive. For isotropic media, or for high symmetry directions, the equations of motion decouple. The boundary conditions may be satisfied with SH waves or with coupled P-SV waves. For arbitrary directions in anisotropic media, the modes of propagation are not purely longitudinal or transverse in character but rather quiasilongitudinal (QT) or quasitransverse (QT1,QT2). Further, the equations of motion are not separable and plate waves generally represent the coupled motion produced by the 3 possible waves (QL, QT1, QT2). Typical dispersion curves for a composite plate are shown below.

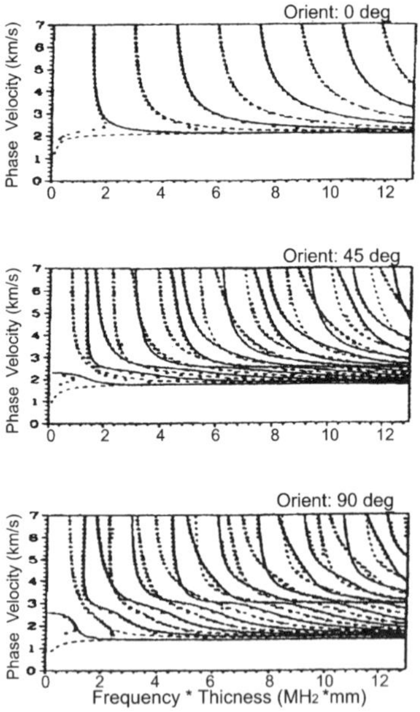

Figure 2. Dispersion curves for unidirectionally rainforced composites (after Mal and Bar-Cohen [16])

1.3. LOVE WAVES

A third type of guided wave in anisotropic media can be found substrates with a thin surface coating layer. These waves were first seen in seismic data records and originally investigated by A.E.H. Love[10], hence the name. Here, the boundary conditions are determined from the free surface at the top of the coating,

$$\sigma_{21} = \sigma_{22} = \sigma_{23} = 0 \qquad at \quad y = 0 \tag{8}$$

continuity of particle displacement at the interface between the coating and substrate,

$$\mathbf{u}_{COAT} = \mathbf{u}_{SUBST}, \ \mathbf{v}_{COAT} = \mathbf{v}_{SUBST}, \ \mathbf{w}_{COAT} = \mathbf{w}_{SUBST} \text{ at } y = H \tag{9}$$

continuity of traction at the interface

$$\begin{aligned} &\sigma_{21\ COAT} = \sigma_{21\ SUBST}, \quad \sigma_{22\ COAT} = \sigma_{22\ SUBST}, \\ &\sigma_{23\ COAT} = \sigma_{23\ SUBST} \ \text{ at } y = H \end{aligned} \tag{10}$$

and the requirement that the solutions remain finite with depth; hence only exponentially decaying solutions are allowed in the substrate. Research in this area has been fairly sparse. However, with the advent of carbon-carbon composites which require a protective coating layer (usually silicon carbide) and multifunctional micro electromechanical systems (MEMS) based devices produced by deposition on a single crystal silicon (cubic symmetry) substrate, interest in this area may be expected to increase in the near future. The most complete study of wave propagation in coated anisotropic materials may be found in the work of Farnell and Adler [11] at McGill University. They investigated Love wave propagation for isotropic coatings deposited on a cubic substrate, here gold on single crystal nickel. In their study attention was restricted to propagation in the high symmetry directions of the substrate. More recently, Bouden and Datta [12] extended their results to propagation in nonsymmetry directions, here in a transversely isotropic (unidirectionally reinforced composite) media with an isotropic coating layer. Like many other guided waves, these waves are found to be highly dispersive.

2. Experimental Considerations

Most of the experimental work to date with guided waves in anisotropic media has involved plate waves. As has been shown, these waves are multimode and highly dispersive. In most cases, this presents a problem as it difficult to isolate a particular mode of interest in the complex waveforms that are usually found in these experiments. Therefore, a great deal of research effort has been devoted to developing techniques of mode isolation for plate wave applications, both isotropic and anisotropic. These methods include:

Signal Processing Electromagnetic Acoustic Transducers (EMAT's) Laser Techniques Interdigital Transducers

The most common approach to mode isolation is through the use of advanced signal processing techniques. Spectral refinement using narrow band signals and Fourier processing techniques, for example, can be used to isolate a particular component of interest in a complex waveform. This is illustrated in Figure 3 from the work of Cawley and Alleyne [7].

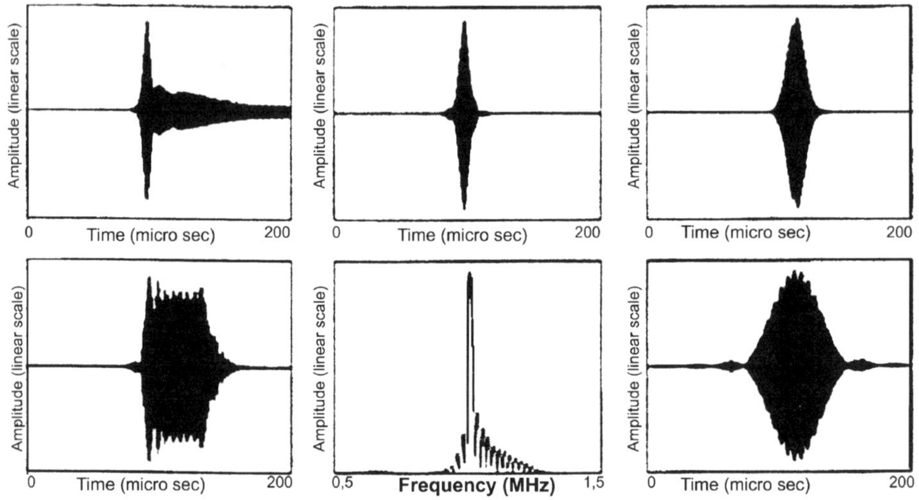

Figure 3. Special Analysis of Lamb Wave Propagation (after Alleyne and Cawley [13])

Alternatively, one can use specialized generation techniques, to produce a particular mode of interest by tailoring the source to match the desired characteristics of the mode of interest, Here, the goal is to excite a propagating disturbance of a single wavelength While the physical generation mechanism is diffetent, this approach is the basis for EMAT, laser and interdigital gemneration. Typical experimental geometry is shown in Figure 4 , in this case for laser generation. Here, a diffraction grating is used to produce multiple lines which are spaced according to the wavelength of the plate mode of interest. In this way a single mode can be readily isolated.

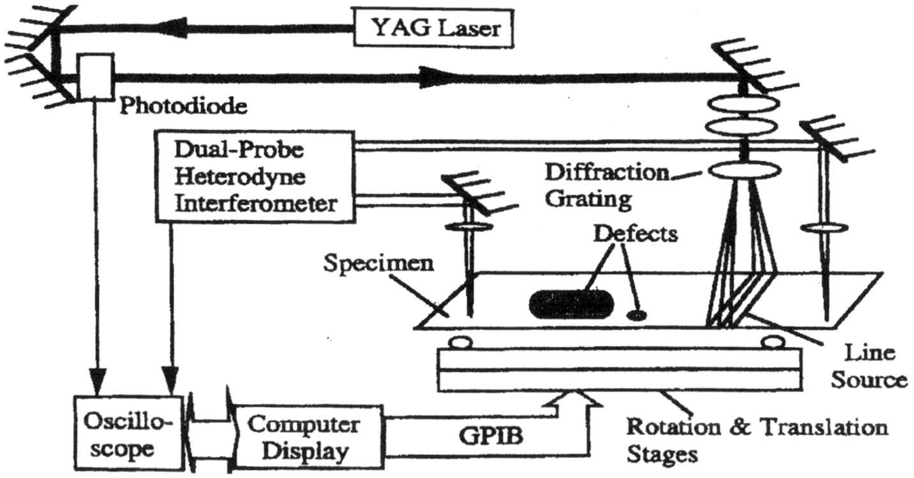

Figure 4. Laser Generation of Lamb Modes (after Niagata et. al. [14])

3. Material Property Reconstruction

While the dispersive nature of guided wave propagation in anisotropic media can be an experimental disadvantage, it also presents an opportunity which can be exploited for material characterization purposes. This approach is based on both the ability to accurately model the dispersive nature of guided wave propagation in anisotropic media as well as the ability to measure this behavior experimentally. Typically, this process is iterative. One begins with an initial estimate of the elastic properties of an unknown material and generates simulated dispersion curves for a material with these properties. The simulated data is then compared with the experimental measurements for the sample. Then, the initial property estimates are modified in a systematic, iterative fashion in order to bring the experimental and simulated curves into agreement with one another. Several algorithms have been developed for this purpose including genetic algorithms, simplex methods as well as several nonlinear numerical analysis approaches (e.g. Levenberg-Marquardt, Newton - Raphson).

Dispersion curves are usually measured experimentally in immersion via the so-called "leaky" Lamb wave technique. With this approach, a pitch catch arrangement in immersion is commonly utilized with the incident

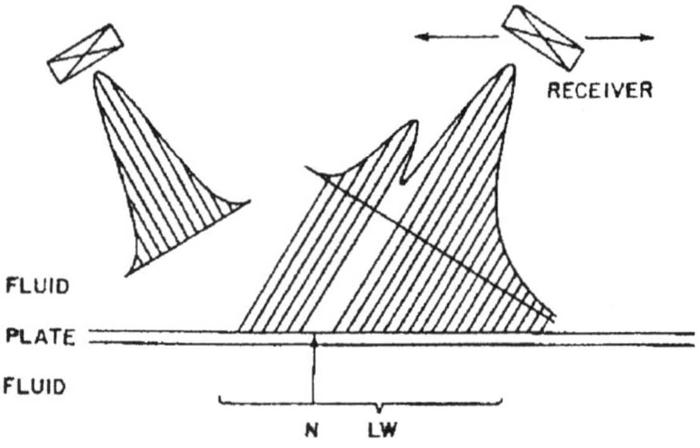

Figure 5. Leaky Lamb Wave Schematic (after Chimenti and Martin [15])

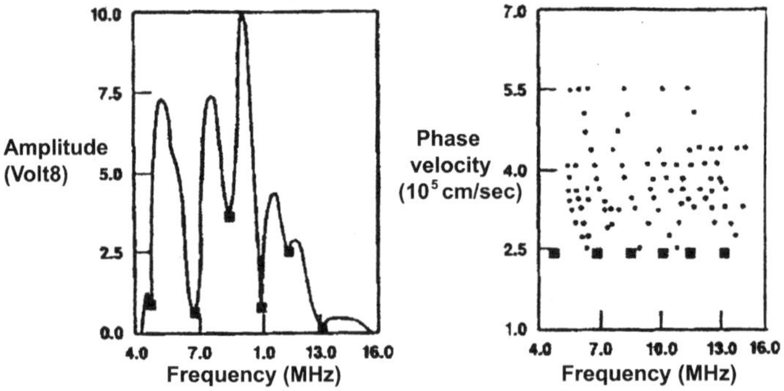

Figure 6. Leaky Lamb Wave Data Acquisition (after Bar Cohen and Mal [16])

pulse from the generating transducer producing both a specular reflection as well as exciting guided plate modes within the composite. The guided modes will lose energy into the surrounding immersion fluid hence the term leaky Lamb wave. The specular reflection and the leaky field will interfere to produce the pattern illustrated in Figure 5. Here, we find two acoustic regions separated by a null zone. The null zone is produced by the interference between the specular reflection and the leaky field and is a very sensitive feature which may be used to accurately characterize the guided modes generated within the composite. Since one wishes to measure the dispersion curves for each of the Lamb modes generated, it is necessary to perform these measurements over a wide frequency range. This can be achieved either by using a monochromatic tone burst and varying its frequency or by using a wide band pulse and analyzing its individual components in the frequency domain via Fourier transform techniques. Typical experimental results are shown in Figure 6. Each minimum in the amplitude-frequency curve corresponds to experimental conditions where a particular guided mode is excited in the composite and can be directly related to the wave number of the guided mode. This yields a point along the dispersion curve for each of the modes excited. This process is repeated for different experimental geometries until a complete picture of the dispersion is obtained. Elastic moduli are obtained numerically via an iteration scheme to obtain a best match between the predicted dispersion relations and those measured experimentally. One of the main issues in this elastic property reconstruction process, is the relative sensitivity of the approach to changes in a particular modulus valus. Generally speaking, this approach is quite sensitive in reconstructing the diagonal components of the stiffness matrix, but somewhat less sensitive to the off diagonal terms. Hence, higher experimental errors are to be expected for these terms.

4. Lamb Wave Tomography in Anisotropic Media

Tomographic imaging has been used in a variety of applications recently to yield full field property images. Typically, electromagnetic waves (i.e. X-rays and the associated CT scans) are used but the same principles apply to all types of waves, including guided waves. Since guided waves can propagate over relatively large distances without significant attenuation, acoustic tomography using guided waves is an ideal approach for inspecting large scale structures. The basic assumptions in tomography are detailed below:

Experimentally measured quantities are path integrals.

By dividing the domain into pixels with uniform properties, path integrals become summations.

With a sufficient number of measurements, one obtains a system of linear algebraic equations which can be solved for the unknown pixel properties. in this way, property distribution images may be obtained.

Acoustic tomography requires that one has an accurate forward propagation model to mathematically predict waveform characteristics for any source/receiver positions given the property distribution within the sample of interest. The solution approach is iterative. One begins with an initial estimate of the property distribution and then adjusts these values in a systematic way to bring the predictions and measurements into agreement.

To illustrate the process, consider the case of an ultrasonic velocity tomogram based on n sources and n receivers or n^2 source receiver combinations and an $n \times n$ pixel grid For simplicity here an isotropic medium is assumed as well as straight line ray paths. In reality for an anisotropic, non-homogeneous medium, one needs to account for ray bending, beam skew phenomena and the intrinsic variation in acoustic velocity with the propagation direction. Details for such corrections may be found in reference [20]. For this simplified discussion, the i-th ray ($i = 1$ *to* n^2) transit time is given by

$$\Delta T_i = \sum D_{ij} m_j = \sum D_{ij}/V(P_j) \tag{11}$$

where D_{ij} is the distance that the i-th ray travels in the j-th cell, m is the slowness, V is the acoustic velocity , and P_j is the reconstruction parameter of interest in the j-th cell. Then we have the following expressions for the k-th iterate transit time

$$\Delta T_i^k = F_i(P_1^k, P_2^k, ...) \tag{12}$$

and the $k+1$-th iterate transit time

$$\Delta T_i^{k+1} = F_i(P_1^k + \Delta P_1^k, P_2^k + \Delta P_2^k, ...) \tag{13}$$

Expanding in a Taylor series, we have

$$\Delta T_i^{k+1} = \Delta T_i^1 + \frac{\partial F_i}{\partial P_1^k} \Delta P_1^k + \frac{\partial F_i}{\partial P_2^k} \Delta P_2^k + ... + HOT \tag{14}$$

Here we seek corrections to the parameter estimates (ΔP_i) such that the predicted and measured time delays are brought into agreement with each other

$$\Delta T_i^{measured} - \Delta T_i^{k+1} = 0 \tag{15}$$

For an $n \times n$ array, this represents 1 equation in n^2 unknowns. For a solution, one generally adds the constraint that the sum of the squares of the correction terms

$$\sum_{j=1}^{N^2} (\Delta P_j^k)^2 \tag{16}$$

is a minimum. Then the optimal correction values can be uniquely determined to be

$$\Delta P_j^k = \frac{\partial F_i}{\partial P_j^k} \left[\Delta T_i - \Delta T_i^{k+1}\right] / \sum \left(\frac{\partial F_i}{\partial P_j^k}\right)^2 \tag{17}$$

Typically a damping factor λ is introduced at this point to account for experimental error in the measurement process

$$P_j^{k+1} = P_j^k + \lambda \Delta P_j^k, \quad 0 \leq \lambda \leq 1 \tag{18}$$

Experimentally, most of the work to date on Lamb wave tomography has been done with isotropic media. Achenbach and co-workers at Northwestern did most of the initial development work in this area [14] using laser based techniques. Subsequently, research groups at Batelle Northwest [17] and William and Mary [18] extended these results to more practical situations using piezoelectric transducer arrays. Typical array geometries and acoustic ray paths are illustrated in Figure 7.

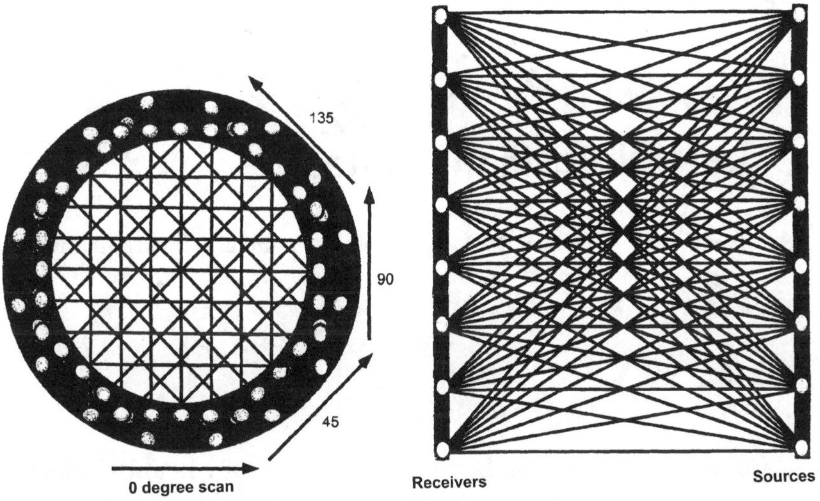

Figure 7. Geometry for Lamb Wave Tomography (after Malyranko and Hinders [18])

Here, the objective is to inspect large scale structures such as pressure vessels or aging aircraft assemblies for thickness changes due to corrosion as well as gross defects such as cracks. The approach also has the potential to isolate areas where to local stiffness has been degraded due to fatigue, environmental exposure, etc. Figure 8 depicts the ability of Lamb wave tomography to track material thickness variations.

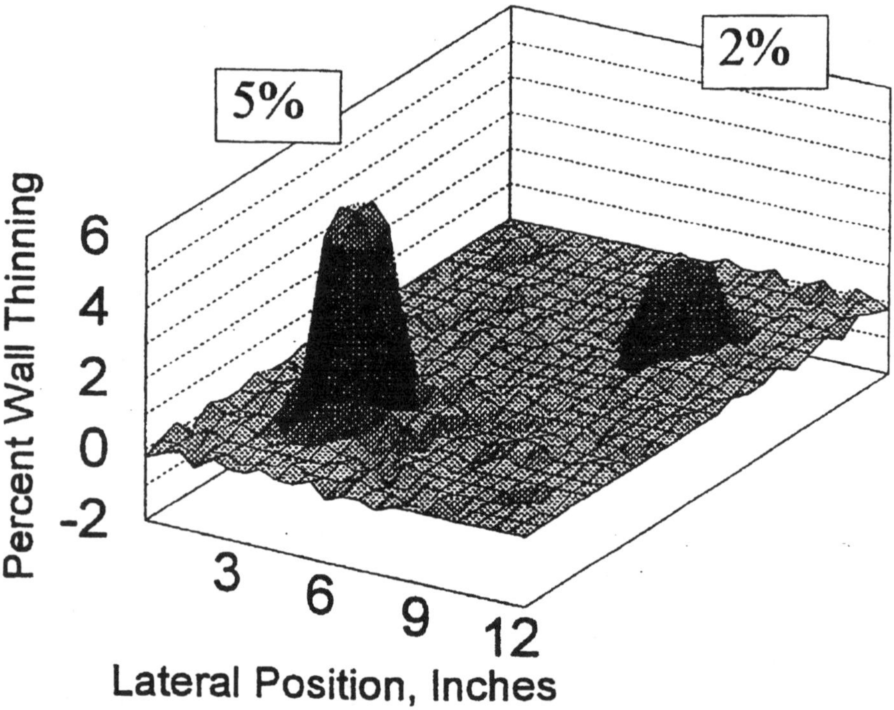

Figure 8. Lamb Wave Tomography Results - Wall Thinning (after Hilderbrand et. al. [17])

To date, very little work on Lamb wave tomography in anisotropic media has been conducted. Choi and Kline [19] have recently presented results of simulated experiments where the presence of residual strains introduces weak anisotropy into the material. Here, the acoustic velocity will be a known function of the principal strains in the material as well as the orientation of the principal axes. Using a multiparameter reconstruction formalism developed by Kline and Wang [20], it is possible to reconstruct both the local principal strains and local orientation of the principal axes. This is illustrated in Figures 9 for a plane strain state. Figure 9a shows the problem geometry along with the location of the residual strain region, the local residual strain values (principal strains) and local principal axis orientation for this two dimensional problem Results from the reconstruction are presented in Figures 9b-d.

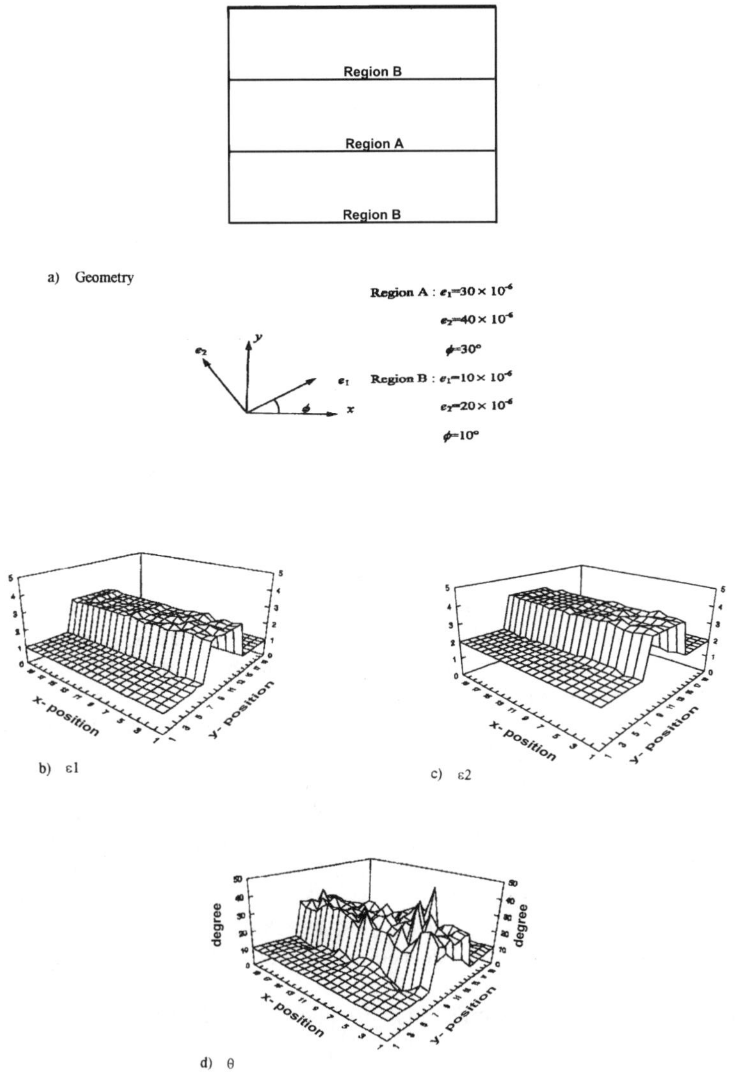

Figure 9. Residual Strain Field Reconstruction (after Kline and Choi [20]

5. Conclusion

Based upon these observations, we may conclude:

Guided wave propagation is well understood theoretically.

Practical applications of these techniques have been limited. The best developed application to this point has probably been materials characterization using the leaky Lamb wave approach.

Experimental advances have improved the ability to isolate individual modes of guided wave propagation.

Tomography offers several interesting and new possibilities for analyzing large scale structures using guided waves.

References

Stoneley, R., 1955, *Proc. Royal Soc.* London A, **A**2323:447-458.

Synge, R. 1957. *J. Math. Phys.*, **35**:323-328.

Royer, D. and E. Dieulesaint, 1984, *J. Acoust. Soc. Am.*, **87**:1438-1444.

Buchwald, V. 1961. *Quart. J. Mech. Appl. Math,* **14**:293-317.

Lim, T., PhD Thesis, McGill University (unpublished).

Kline, R., M. Doroudian and C. Hsiao, 1989, *J. Comp. Mat.*, **23**:505-533.

Chimenti, D. and A. Nayfeh, 1990, *J. Acoust. Soc. Am*, **87**: 1409-1415.

Green, W, 1982, *J. Mech. Appl. Math*, **XXXV**: 485-507.

Love, A., 1911, Some Problems of Geodynamics, Cambridge University Press.

Farnell, G. and E. Adler, 1972, *Physical Acoustics*, **9**:35-127.

Bouden, N. and S. Datta, 1989, *Review of Progress in QNDE*, **10**:1337-1344.

Karim, M., Mal, A. and Bar-Cohen, Y, 1990 *J. Acoust. Soc. Am.*,**88**:482-491

Alleyne, D. and P. Cawley (1994), *Review of Progress in QNDE*, **13**: 181-188.

Nagata, Y., J Huang, J. Achenbach and S. Krishnaswamy, 1996, *Review of Progress in QNDE*, **15**: 861- 868.

Chimenti, D. and Martin, R, 1991, *Ultrasonics*, **29**,:13-21.

Bar-Cohen, Y and A.Mal, 1989, in Proceedings, *Review of Progress in QNDE*, **9**: 1419-1424.

Hildebrand, B, T. Davis, G Posakony and J. Spanner, 1999 , *Review of Progress in QNDE*, **18**: 967 - 973.

E. Malyarenko and M. Hinders, 2001, *Ultrasonics* **39**: 269-281.

J. Choi and R. Kline, 2000, *KSME International Journal*, **14**:1-10.

Y.Q. Wang and R. A. Kline, 1994, *J. Acoustic Soc. Am.*, **95**: 2525-2532.

A GENERAL PURPOSE COMPUTER MODEL FOR CALCULATING ELASTIC WAVEGUIDE PROPERTIES, WITH APPLICATION TO NON-DESTRUCTIVE TESTING

M. J. S. LOWE and P. CAWLEY
Imperial College
Department of Mechanical Engineering
Exhibition Road
London SW7 2BX
UK

B. N. PAVLAKOVIC
Guided Ultrasonics Ltd
17 Doverbeck Close
Nottingham NG15 9ER
UK

Abstract.

Guided waves are increasingly used for Non-Destructive Testing (NDT) of structures; their attractions include speed of inspection and their ability to penetrate inaccessible regions. However developments of guided wave methods are relatively complex because of the existence of multiple frequency-dispersive modes. Successful development depends on a thorough knowledge of which modes may propagate in a structure, their velocity-frequency dispersion relations, and their profiles of stresses and displacements. The authors have developed a general purpose computer model which they have used for a wide variety of studies during the past ten years or so. The paper describes the function of the model in outline, and illustrates its use on some NDT applications.

Key words: nonlinear surface waves, elasticity, Rayleigh waves, scale invariance, nonlocality, spectral kernel, spatial kernel, Hilbert transform

1. Introduction

Elastic waves which are guided in a structure offer significant attractions for Non Destructive Testing (NDT), primarily in their potential for dramatic improvements in the speed of inspection. Whereas a conventional ultrasonic

R.V. Goldstein and G.A. Maugin (eds.),
Surface Waves in Anisotropic and Laminated Bodies and Defects Detection, 241–256.

inspection of an area of a structure requires a transducer to be scanned in two dimensions over the surface of the structure, collecting a series of single-point measurements, a guided wave inspects a line along the whole length or width of the structure with just one measurement. Thus only one dimension of scanning is required for a plate [Worlton, 1957; Wilcox et al., 2000], and only one point is required for the special case of inspecting a pipe [Thompson et al., 1972; Rose et al., 1994; Alleyne and Cawley, 1997]. The guided wave is launched at one location and travels along the structure. It is affected by interaction with defects (cracks or corrosion for example), and this is detected by receiving either a reflected signal at the same transducer location (pulse-echo) or a transmitted signal at a second, remote, transducer location (through-transmission). Another advantage of guided waves is that they can access regions of the structure which may be difficult to get to with conventional techniques; for example sections of pipelines which are buried under roadways [Alleyne et al., 2001]. Guided waves are also sensitive to the properties of the structure and can therefore be used to measure the elastic material properties of the structure or surface coatings [Chimenti, 1997].

Unfortunately, however, structure-guided waves are complicated, and this inhibits their exploitation. The complications include the possibility of exciting multiple different modes simultaneously, and the fact that most of the modes are frequency-dispersive: that is to say, their velocities vary with frequency. These issues are problematic even for the well-known modes, Lamb waves, which propagate in a simple plate of isotropic material. The complications are even greater for modes which travel in complex structures such as plates which are immersed in fluids, plates composed of multiple layers, anisotropic materials or pipes.

The authors therefore believe firmly that the most effective strategy for the development of new guided wave methods is first to devote considerable effort to predicting and examining the modes which may propagate in the waveguide; these are in fact properties of the waveguide system. It is the thorough understanding of these which enables the best choices to be made of the optimum modes for inspection, the transduction methods, and the interpretation of received signals once the practical development is under way. The choice of the optimum mode and frequency is decided not only from the sensitivity of the modes to the features to be detected, but also from their velocities and dispersion characteristics. Indeed with a good understanding, the complicated nature of the modes can become a benefit because potentially there are many features which can be sensitive to the property which is to be measured, if exploited with knowledge. Dispersion and the optimisation of the selection of modes and frequencies for guided wave inspection are discussed in [Alleyne and Cawley, 1992; Wilcox et al.,

1999; Wilcox et al., 2001].

Guided wave solutions for many specific structural forms are well-established in the literature; these include the simplest examples of the Rayleigh mode on a free surface of an isotropic material [Rayleigh, 1885], the Lamb modes in an unloaded isotropic plate [Lamb, 1917], and indeed very many specific cases of waves in multilayered isotropic or anisotropic structures. However, while the approach of deriving specific solutions for specific cases is necessary for fully understanding some complex new problems, it is not best suited to many applied researchers who are focused on finding new techniques for NDT. Their interest is in being able to make rapid assessments of the feasibility of new proposals for inspection strategies without becoming involved in the details of the equations. The authors have therefore taken the approach of creating a general-purpose model ("DISPERSE"), with the aim of covering a wide range of possibilities of structural forms in as flexible a manner as possible. The intention is to provide a tool which can be used straight away for the study of any new guided wave inspection concept. The program can model an arbitrary number of layers which can be flat or cylindrical. The layers may be elastic or damped isotropic solids, elastic anisotropic solids, or perfect or viscous fluid layers. Furthermore, the structure may be immersed in a fluid or embedded in a solid, in which case the leaky modes may be calculated. The primary output of the program is the dispersion curves, that is the frequency-velocity relationships of any modes which could travel in the structure. The program also predicts the attenuation of the modes, caused by leakage into surrounding materials or by material damping, and the profiles of the field quantities (mode shapes).

The authors' research group has specialised for the past decade on the development of guided wave inspection techniques, using DISPERSE as their core modelling tool. Studies which have been performed by the authors using this model include: development of a long range technique for detecting corrosion of petro-chemical plant pipework [Alleyne and Cawley, 1997; Lowe et al., 1998; Alleyne et al., 2001]; development of a guided wave technique for detecting defects in embedded bars such as reinforcing bars in concrete or the security bolts which are inserted in the rock in mines [Pavlakovic et al., 2001; Beard et al., 2001]; development of water leak detection techniques using guided waves in buried water pipes [Aristegui et al., 2001; Long et al., 2001]; development of a wire waveguide which can be used to monitor the properties of curing epoxy adhesives and a wide variety of liquids [Vogt et al., 2001]; and studies of the inspection of adhesively bonded joints [Lowe and Cawley, 1994; Lowe et al., 2000; Dalton et al., 2001].

2. Theoretical background

As indicated earlier, the purpose of the model is to make the analysis of guided wave problems more accessible to developers of NDT techniques, rather than to break new ground in waveguide theory. Accordingly the model embodies well-known theoretical methods and introduces very little novel theory itself. Nevertheless a great deal of effort and original thought has gone into the implementations, in order to solve the problems reliably and efficiently. In this section we outline the concepts of the operation of the model and indicate key references for more detailed information.

The earliest derivations for guided wave propagation were specific to particular kinds of modes and geometries: Rayleigh, Lamb, Love, Stonely and so on. However, by about the middle of the 20th century, researchers working on multi-layered geo-mechanical structures started to develop general forms of solution which could be applied to structures of arbitrary numbers of layers; notable for example are the model of Thomson [Thomson, 1950] and Haskell [Haskell, 1953], often called the "Transfer Matrices method". This has been used successfully in many forms but unfortunately suffers from a numerical instability at high values of the frequency-thickness product, the so-called "large f-d" problem [Dunkin, 1965]. Consequently numerous variants and alternatives have been pursued by many authors, and new approaches still continue to appear in the literature from time to time. The model developed by the authors is based on one such alternative, the "Global Matrix method" of Knopoff [Knopoff, 1964], later refined by Schmidt and Jensen [Schmidt and Jensen, 1985]. This avoids completely the instability problem and retains conceptual simplicity, although it is perhaps computationally less efficient than some others. Discussion of the instability and some of the alternative techniques, including the Global Matrix method is given in [Lowe, 1995]. Further details concerning the implementation in DISPERSE are given in [Lowe, 1995; Pavlakovic et al., 1997; Pavlakovic and Lowe, 1999; Lowe and Pavlakovic, 2001].

The Global Matrix method involves the construction of a single matrix equation which describe the displacement and stress fields associated with a harmonic wave propagating along a structure consisting of an arbitrary number of layers. The size of the matrix is determined by the number of layers and the possible form of the field within each layer. Within a layer any admissible field can be described without approximation by the summation in specific quantities of a set (of 2, 4 or 6, depending on the problem) of partial waves. The partial waves are simply bulk waves which could travel in an unbounded expanse of the material of the layer. For example, the field within a flat layer could be described by six partial waves in the following expression:

$$\begin{Bmatrix} u_1 \\ u_2 \\ u_3 \\ \sigma_{11} \\ \sigma_{12} \\ \sigma_{13} \end{Bmatrix} = [\mathbf{D}] \begin{Bmatrix} L_+ \\ L_- \\ SV_+ \\ SV_- \\ SH_+ \\ SH_- \end{Bmatrix} \tag{2.1}$$

where u_i are the particle displacements in three Cartesian coordinates, σ_{ij} are the stresses, and L, SV and SH ($\pm$) are the amplitudes of the partial waves. The matrix $[\mathbf{D}]$ is a function of the location in the layer, the material properties, the layer thickness, the frequency and the wavenumber. Only three of the possible stresses are entered in this relation because the remaining terms are not required for the solution. In this example the coordinate direction 1 is thus the normal to the plane of the layers.

Initially the amplitudes of the partial waves are not known, and they have to be found by considering the boundary conditions. An admissible set of fields for the whole collection of layers of the plate requires that the boundary conditions of compatibility and equilibrium be satisfied at all of the interfaces between the layers, and appropriate boundary conditions must also be satisfied at the extreme surfaces of the plate; in the common case of a plate which has free surfaces, the surface boundary condition is that the tractions must be zero. The calculations then consist of searches to find solutions when all of these boundary conditions can be satisfied simultaneously. Manipulation of the Global Matrix $[\mathbf{G}]$, incorporating these boundary conditions, results in the expression:

$$[\mathbf{G}]\{\mathbf{A}\} = 0 \tag{2.2}$$

where $\{\mathbf{A}\}$ is a vector of the partial wave amplitudes. Given that the materials and thicknesses remain constant or slowly changing (frequency-dependent material properties may be specified in the model) for the solution of any particular structure, the form of this equation is thus:

$$f(\omega, k) = 0 \tag{2.3}$$

where ω is the frequency and k is the wavenumber.

In practice the global matrix is assembled and its coefficients are populated using trial values of frequency and wavenumber. A valid solution for a guided wave is found when a pair of these values yield a singular global matrix. A root-tracing algorithm is then used to plot out sets of these solutions. These are the dispersion curves describing the variation of the wavenumber of a mode with frequency. The curves are best calculated in the frequency-wavenumber space [Lowe, 1995] but are readily transformed to

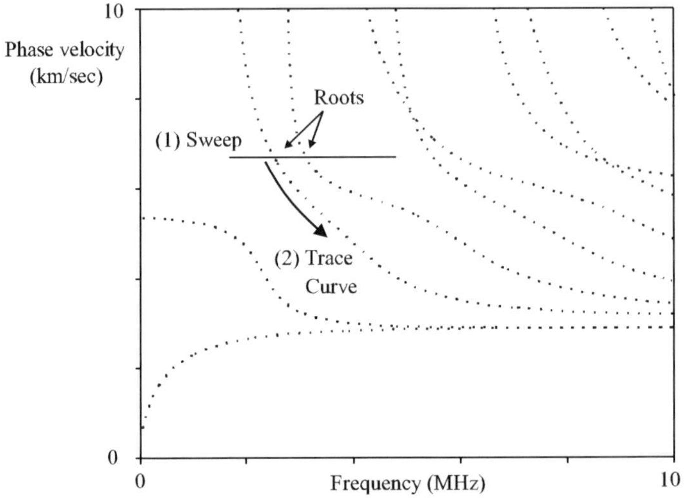

Figure 1. Procedure for tracing dispersion curves, shown for illustration using the phase velocity dispersion curves for Lamb waves. (1) a sweep is performed to find starting points at all positions where it crosses a dispersion curve; (2) a quadratic extrapolation algorithm is used to trace each curve from the starting points.

frequency-phase velocity. Dispersion curves for frequency-group velocity are easily calculated by numerical differentiation of the frequency-wavenumber curves, or alternatively the energy velocity curves can be calculated from the displacements and stresses, these curves being identical to the group velocity if the attenuation is zero [Bernard et al., 2001]. The procedure for tracing the dispersion curves is illustrated in Figure 1. First, a sweep is performed in which the function is evaluated along a line across the solution space. This reveals all locations where the line crosses dispersion curves. Then a quadratic extrapolation algorithm is used to trace each curve from the starting points; this has been found to be both efficient and stable. An iterative scheme is used to find the precise location of each root on the curve. The iterative scheme and the quadratic extrapolation method are described in [Lowe, 1995; Pavlakovic et al., 1997].

Finally, after finding the dispersion curves, the information about the fields in each layer can be used to plot their profiles through the layers (the mode shapes) at any location on a curve. A large number of options have been implemented, including displacements, stresses, strains, and energy densities. Additionally, the displacement mode shapes may be shown in a grid form, either static or animated, in which a deformed rectangular grid shows the shape of deformation of the structure.

An important variant of the basic model is the introduction of the possibility of embedding the structure in an infinite expanse of a solid or fluid. This allows the possibility of leaky waves in which energy is lost from the guided wave by radiation into the surrounding material. It also allows for subsonic waves in which the field extends into the surrounding material but radiation does not take place. In the case of leaky waves, the amplitude of the field in the waveguide must decrease as energy is lost from it, and this is described by an imaginary part of the wavenumber. The field equations are now described by complex quantities and the nature of the solution becomes:

$$f(\omega, k_{real}, k_{imag}) = 0 \tag{2.4}$$

where the real part of k describes the harmonic progression of the guided wave and the imaginary part describes its attenuation. Thus the attenuation is one of the unknowns which has to be found. This necessarily requires a significant increase in computational effort. The three-value search algorithm is described in [Lowe, 1995].

Another variant with a similar outcome is the option of material damping. The fundamental description in the model is of partial waves with complex wavenumbers. Thus the damping material properties are defined by the exponential decay of bulk waves in an infinite expanse of the material, in which the decay is a constant per wavelength of travel. This is is algebraically convenient, leading to exactly the same form of solution as was described for leaky waves in the preceding paragraph. However other forms of damping, such as viscous damping, are also permissible from this form, because the damping solution can be redefined at any frequency of calculation; this is handled automatically by the program.

Perfect fluids are represented straightforwardly by compressional partial waves only. Viscous fluids are represented by an equivalent solid material supporting both compressional and shear waves. The approach is described in [Nagy and Nayfeh, 1996] and the specific implementation in [Lowe and Pavlakovic, 2001]. Although this introduces some approximation, this approach is sensibly accurate for commonly occurring fluids.

Anisotropic materials may be modelled, but the permissible range of kinds of anisotropy is currently limited. Specifically the material may be orthotropic (9 elastic constants) but it must have a plane of symmetry which is parallel to the plane of the plate. The numerical solution is thus limited to cases in which the partial waves exist in pairs (each pair consists of an "upward" and a "downward" wave with the same properties), so that the eigen-solution for the partial waves is of 3rd order. The theoretical basis of the anisotropic model is given in [Nayfeh, 1995].

Wave propagation along pipes has been a very important topic of research for the authors, and necessitated the introduction of an option for cylindrical coordinates. The exponential wave functions are thus replaced by Bessel function. Initially the equations published by Gazis [Gazis, 1959a; Gazis, 1959b] for a single-layer pipe were implemented by the authors into the Global Matrix system, thus allowing the modelling of arbitrarily multi-layered structures. Following that, the complex wave field solution was developed so that the model can also predict leaky or damped guided waves in cylindrical structures [Pavlakovic, 1998; Lowe and Pavlakovic, 2001]. Arbitrary integer values of the circumferential component of the wavenumber are permissible.

Although the cylindrical model does not in general support anisotropic material properties, the specific case of transverse isotropy in which the plane of anisotropy is normal to the axis of the cylinder has been implemented [Mirsky, 1965; Berliner and Solecki, 1996; Nayfeh and Nagy, 1996; Pavlakovic and Lowe, 1999], again generalising the published results for arbitrary numbers of layers. This in fact is the most useful case of anisotropy for NDT applications, allowing modelling of fibres and drawn wires.

Finally, mixed boundary conditions between layers has been implemented as an option. The Global Matrix method lends itself very readily to manipulation in order to modify the conditions of continuity of displacement and traction. Thus the normal and shear contact conditions can be specified separately, for example to define perfect normal contact but free sliding. Each condition can be defined to be free, perfectly contact, or intermediate contact defined by a user-given spring stiffness [Lowe and Pavlakovic, 2001].

3. Examples of applications of the model

Figure 2 shows a typical example of the results of the model, applied to one of the simplest problem. Part (a) shows the well-known Lamb wave phase velocity dispersion curves for a steel plate 1mm in thickness; part (b) shows the deformed grid displacement mode shape of the s_0 and a_0 modes at the locations indicated by the arrows. Note that the elements of the grid have no significance other than as a means of presenting the shapes.

Figure 3 shows the effect of immersing the steel plate in water, in which case most of the waves are attenuated ("leaky Lamb waves") by radiation of bulk waves into the water. Part (a) shows the phase velocity dispersion curves, which are almost the same as those in the free plate case in the preceding figure, except for the additional branch at low velocity which is asymptotic at high frequency to the Scholte velocity. The upper grid in

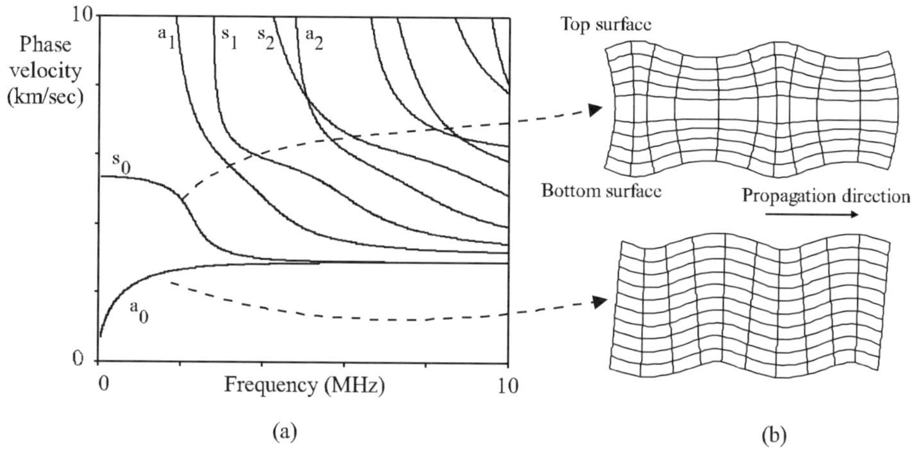

Figure 2. Lamb waves in a steel plate: (a) Phase velocity dispersion curves, (b) Mode shapes of s_0 and a_0

part (b) illustrates the radiation into the surrounding water; the lower grid illustrates the participation of the water, but without radiation of energy from the plate, in the sub-sonic antisymmetric mode.

The group velocity dispersion curves for a 3 inch schedule 40 steel pipe (internal diameter = 76mm, wall thickness = 5.5mm) are plotted in Figure 4. The inspection of petro-chemical pipelines has been a major area of research for the authors. The figure indicates a large number of modes in just a modest frequency range. Without careful study of these modes and careful development of instrumentation to target just one or two of them, any attempts to use guided waves for inspection would have resulted in the propagation and reception of large numbers of modes which would not have been possible to interpret. The logic of the mode selection and the commercialisation of a test instrument have been reported in [Lowe et al., 1998; Alleyne et al., 2001]. Later studies of the wave properties in pipes have included pipes with linings, pipes containing perfect or viscous fluids, pipes coated by highly attenuative bituminous materials, and pipes buried underground.

Turning to a yet more complicated example of the application of the model, Figure 5 shows the group velocity and attenuation dispersion curves for steel tendon embedded in cement grout. The axes have been scaled by the radius so that the results are applicable to any radius of bar. Similar results are found for anchor bolts embedded in rock, used in mining. The attenuation of the guided waves is caused almost entirely by leakage of energy from the bar into the surrounding material, although the model does also

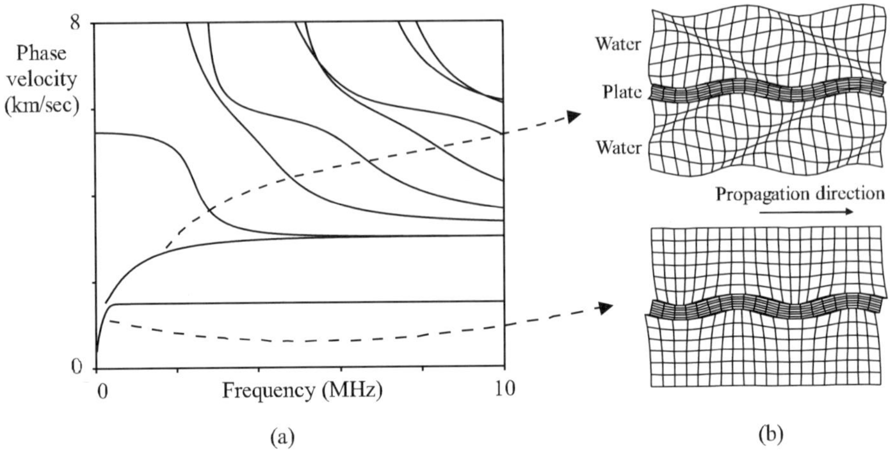

Figure 3. Leaky Lamb waves in a steel plate immersed in water: (a) Phase velocity dispersion curves, (b) Mode shapes of fundamental flexural mode, showing leaky branch (above) and non-leaky branch (below). A finite region of water is shown, but model assumes water extends to infinity above and below plate.

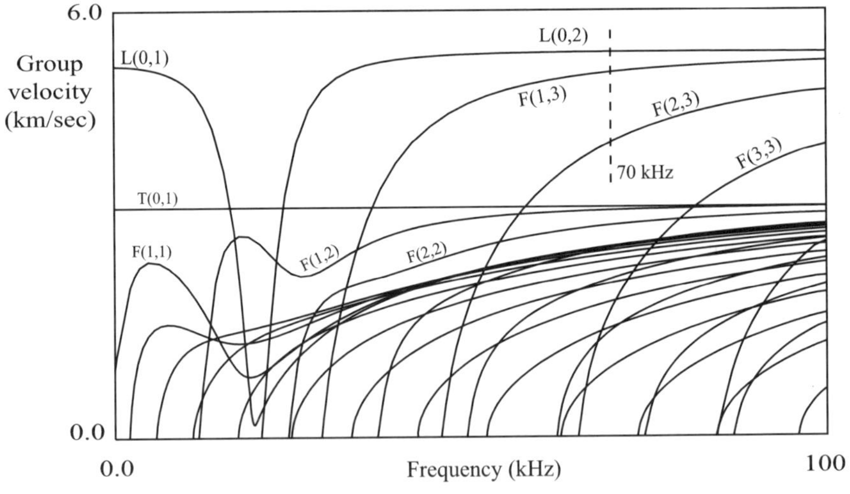

Figure 4. Group velocity dispersion curves for a 3 inch schedule 40 steel pipe (internal diameter = 76mm, wall thickness = 5.5mm), showing typical test frequency when using the extensional mode L(0,2).

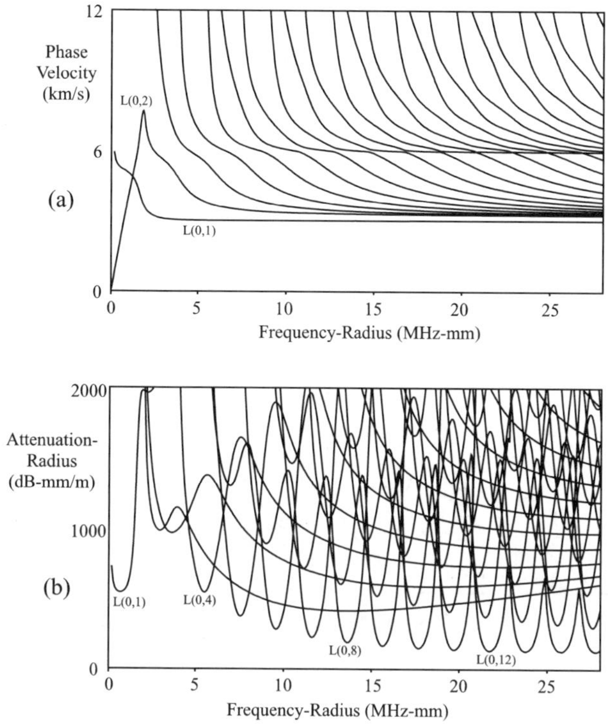

Figure 5. Dispersion curves for longitudinal modes (circumferential wavenumber=0) in a steel tendon embedded in cement grout; (a) Phase velocity; (b) attenuation due to leakage of energy into the grout.

include damping properties of the materials, and so damping contributes also to a minor extent. The attenuation is, for the most part, extremely high. However, a very interesting outcome of modelling this system is that it can be seen that there are certain modes, for example those labelled L(0,8), L(0,12), which have relatively low attenuation at very specific high frequencies. A detailed study and explanation of this phenomenon is given in [Pavlakovic et al., 2001]. This phenomenon would not naturally have been anticipated and this example serves well to illustrate the value of performing such model studies. In fact it has been found that waves of this kind have sufficiently low attenuation to propagate from the exposed end of an anchor bolt (3m long, 22mm diameter), to reflect at the embedded end and then return to be detected at the exposed end, and so may be used to inspect the bolts [Beard et al., 2001].

Figure 6 shows phase velocity and attenuation dispersion curves for a

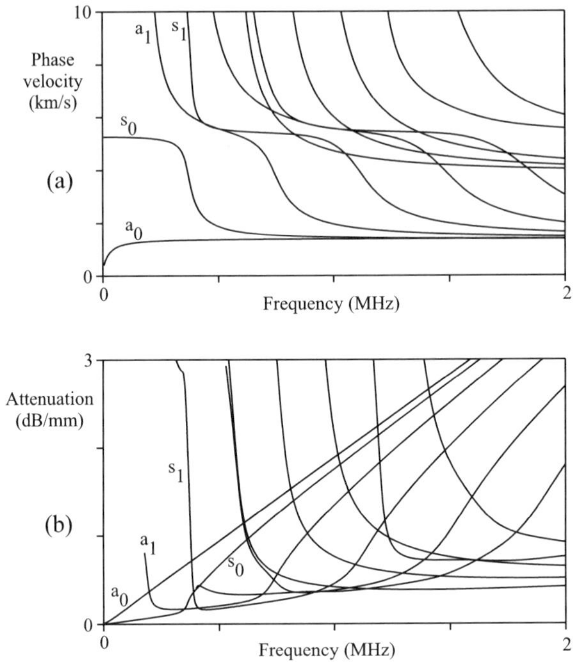

Figure 6. Dispersion curves for modes propagating along a principal axis of a [0/90/+45/-45] symmetric carbon-epoxy plate: (a) Phase velocity, (b) attenuation due to material damping.

cross-ply carbon-epoxy composite [Kwun et al., 2002]. The material lay-up is [0/90/+45/-45] symmetric and the plate thickness is 3.6mm. The predictions are for propagation along one of the principal axes. The attenuation is caused by material damping which has been included in the model. The velocity dispersion curves for composite materials have been calculated by other researchers in the past, for example [Nayfeh and Chimenti, 1989; Chimenti and Nayfeh, 1990; Nayfeh, 1995], therefore phase velocity predictions such as these are not novel. However the prediction of the attenuation due to damping is unusual and valuable. The elastic and the damping properties of the material were measured at the University of Bordeaux, using their rig and measurement technique [Hosten and Castings, 1993]. The attenuation curves show the s_0 mode at low frequency to have the lowest attenuation. However it is also clear that all values of attenuation are high and that they generally increase with frequency. The practical outcome of this in an aircraft structure, for example, is that any waves

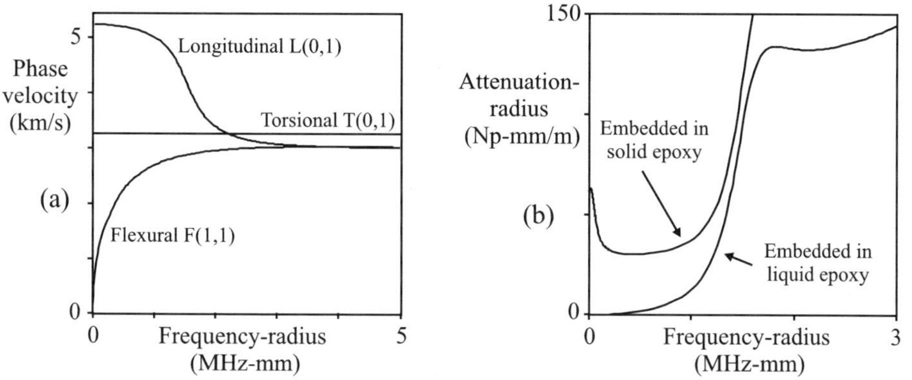

Figure 7. Dispersion curves for guided waves in a steel wire immersed in epoxy adhesive: (a) Phase velocity of fundamental modes, (b) attenuation of L(0,1) mode.

which propagate significant distances must have low frequencies, typically of a few hundred kHz or lower. A further limitation is that these predictions are for a continuous plate; if stiffeners or other structural joints are present then the long range propagation will be reduced even more. This issue is discussed in the context of stiffened metallic aircraft structures in [Dalton et al., 2001].

A final example, illustrating the use of a wire waveguide to measure the properties of a material [Vogt et al., 2001], is shown in figure 7. The dispersion curves are for a steel wire which is immersed in an epoxy adhesive. Part (a) of the figure shows the three fundamental modes: the simple longitudinal (extensional) mode L(0,1), the torsional mode T(0,1) and the flexural mode F(1,1). Part (b) shows the attenuation of the L(0,1) mode which is caused by leakage of energy from the wire into the surrounding adhesive. Two different curves are shown, one for the early stage of cure when the epoxy is still a liquid, and the other for a later stage of cure when the epoxy has become a solid. Essentially the difference between these states is the development of cross-links between the molecular chains of the polymer, resulting in the development of its shear stiffness. There is a clear difference between the attenuation curves, showing the significance of the shear stiffness of the epoxy in increasing the leakage. Thus a configuration of the wire waveguide in which the attenuation is continuously measured can be used to monitor the development of the shear stiffness during the cure cycle. This technique can also be used in many other applications to measure material properties.

4. Conclusions

This paper has presented a brief summary of the general purpose predictive model Disperse, which has been developed by the authors in order to be able to study and implement guided wave techniques for inspection and monitoring. Several examples of its use have been shown, illustrating the value of such modelling for understanding guided wave behaviour as a pre-requisite to the optimum development of guided wave techniques.

References

Alleyne, D. and P. Cawley: 1992, 'Optimization of Lamb wave inspection techniques'. *NDT & E Int.* **25**, 11–22.

Alleyne, D. and P. Cawley: 1997, 'Long range propagation of Lamb waves in chemical plant pipework'. *Mater. Eval.* **55**, 504–508.

Alleyne, D., B. Pavlakovic, M. Lowe, and P. Cawley: 2001, 'Rapid, long range inspection of chemical plant pipework using guided waves'. *Insight* **43**, 93–96,101.

Aristegui, C., M. Lowe, and P. Cawley: 2001, 'Guided waves in fluid-filled pipes surriunded by different fluids'. *Ultrasonics* **39**, 367–375.

Beard, M., M. Lowe, and P. Cawley: 2001, 'Inspection of rock bolts using guided ultrasonic waves'. In: D. Thompson and D. Chimenti (eds.): *Review of Progress in Quantitative NDE*, Vol. 20. pp. 1156–1163.

Berliner, M. and R. Solecki: 1996, 'Wave propagation in fulid-loaded, transversely isotropic cylinders. Part I. Analytical formulation'. *J. Acoust. Soc. am.* **99**, 1841–1847.

Bernard, A., M. Lowe, and M. Deschamps: 2001, 'Guided waves energy velocity in absorbing and non-absorbing plates'. *J. Acoust. Soc. am.* **110**, 186–196.

Chimenti, D.: 1997, 'Guided waves in plates and their use in materials characterization'. *Applied Mechanical Review* **50**, 247–284.

Chimenti, D. and A. Nayfeh: 1990, 'Ultrasonic reflection and guided wave propagation in biaxially laminated composite plates'. *J. Acoust. Soc. am.* **87**, 1409–1415.

Dalton, R., P. Cawley, and M. Lowe: 2001, 'The potential of guided waves for monitoring large areas of metallic aircraft fuselage structure'. *J. Nondestruct. Eval* **20**, 29–46.

Dunkin, J.: 1965, 'Computation of modal solutions in layered, elastic media at high frequencies'. *Bulletin of the Seismological Society* **55**, 335–358.

Gazis, D.: 1959a, 'Three dimensional investigation of the propagation of waves in hollow circular cylinders. I. Analytical foundation'. *J. Acoust. Soc. am.* **31**, 568–578.

Gazis, D. C.: 1959b, 'Three-Dimensional Investigation of the Propagation of Waves in Hollow Circular Cylinders. II. Numerical Results'. *J. Acoust. Soc. am.* **31**(5), 573–578.

Haskell, N.: 1953, 'The dispersion of surface waves on multi-layered media'. *Bulletin of the American Seismological Society* **43**, 17–34.

Hosten, B. and M. Castings: 1993, 'An acoustic method to predict the effective elastic constants of orthotropic and symmetric laminates'. In: D. Thompson and D. Chimenti (eds.): *Review of Progress in Quantitative NDE.* pp. 1201–1207.

Knopoff, L.: 1964, 'A matrix method for elastic wave problems'. *Bulletin of the Seismological Society of America* **54**, 431–438.

Kwun, H., G. Burkhardt, and C. Teller: 2002, 'Propagation of Lamb waves in anisotropic and absorbing plates: theoretical derivation and experiments'. In: D. Thompson and D. Chimenti (eds.): *Review of Progress in Quantitative NDE*, Vol. 21. p. in press.

Lamb, H.: 1917, 'On waves in an elastic plate'. In: *Proc. R. Soc. London, Ser. A*. pp. 114–128.

Long, R., K. Vine, M. Lowe, and P. Cawley: 2001, 'Monitoring acoustic wave propagation in buried cast iron water pipes'. In: D. Thompson and D. Chimenti (eds.): *Review of Progress in Quantitative NDE*, Vol. 20. pp. 1202–1209.

Lowe, M.: 1995, 'Matrix techniques for modeling ultrasonic waves in mutilayered media'. *IEEE Trans. Ultrason. Ferroelectr. Freq. Control* **42**, 525–542.

Lowe, M., D. Alleyne, and P. Cawley: 1998, 'Defect detection in pipes using guided waves'. *Ultrasonics* **36**, 147–154.

Lowe, M. and P. Cawley: 1994, 'The applicability of plate wave techniques for the inspection of adhesive and diffusion bonded joints'. *J. Nondestruct. Eval* **13**, 185–199.

Lowe, M., R. Challis, and C. Chan: 2000, 'The transmission of Lamb waves across adhesivley bonded lap joints'. *J. Acoust. Soc. Am.* **111**, 1333–1345.

Lowe, M. and B. Pavlakovic: 2001, 'DISPERSE User Manual, version 2.0.11d'. Technical report, Imperial College of Science, Technology and Medicine, London, UK; www.me.ic.ac.uk\dynamics.

Mirsky, I.: 1965, 'Wave propagation in transversely isotropic circular cylinders Part I: Theory'. *J. Acoust. Soc. am.* **37**, 1016–1021.

Nagy, P. B. and A. Nayfeh: 1996, 'Viscosity-induced attenuation of longitudinal guided waves in fluid-loaded rods'. *J. Acoust. Soc. am.* **100**, 1501–1508.

Nayfeh, A.: 1995, *Wave propagation in layered anisotropic media with application to composites*. Elsevier.

Nayfeh, A. and D. Chimenti: 1989, 'Free wave propagation in plates of general anisotropic media'. *J. Appl. Mech* **556**, 881–886.

Nayfeh, A. and P. Nagy: 1996, 'General study of axisymmetric waves in layered anisotropic fibers and their composites'. *J. Acoust. Soc. am.* **99**, 931–941.

Pavlakovic, B.: 1998, 'Leaky guided ultrasonic waves in NDT'. Ph.D. thesis, University of London, UK.

Pavlakovic, B. and M. Lowe: 1999, 'A general purpose approach to calculating the longitudinal and flexural modes of multi-layered, embedded, transversely isotropic cylinders'. In: D. Thompson and D. Chimenti (eds.): *Review of Progress in Quantitative NDE*, Vol. 18. pp. 239–246.

Pavlakovic, B., M. Lowe, D. Alleyne, and P. Cawley: 1997, 'DISPERSE: A general purpose program for creating dispersion curves'. In: D. Thompson and D. Chimenti (eds.): *Review of Progress in Quantitative NDE*, Vol. 16. pp. 185–192.

Pavlakovic, B., M. Lowe, and P. Cawley: 2001, 'High frequency low loss ultrasonic modes in embedded bars'. *J. Appl. Mech* **68**, 67–75.

Rayleigh, L.: 1885, 'On waves propagating along the plane of an elastic solid'. *Proc. London Math. Soc.* **17**.

Rose, J., J. Ditri, A. Pilarski, K. Rajana, and F. Carr: 1994, 'A guided wave inspection technique for nuclear steam generator tubing'. *NDT & E Int.* **27**, 307–310.

Schmidt, H. and F. Jensen: 1985, 'A full wave solution for propagation in multilayered viscoelastic media with application to Gaussian beam reflection at liquid-solid interfaces'. *J. Acoust. Soc. am.* **77**, 813–825.

Thompson, R., G. Alers, and M. Tennison: 1972, 'Application of direct electromagnetic Lamb wave generation to gas pipeline inspection'. In: *Proceedings of the 1971 IEEE Ultrasonic Symposium*. pp. 91–94.

Thomson, W.: 1950, 'Transmission of elastic waves through a stratified solid medium'. *J. Appl. Phys.* **21**, 89–93.

Vogt, T., M. Lowe, and P. Cawley: 2001, 'Cure monitoring using ultrasonic guided waves in wires'. In: D. Thompson and D. Chimenti (eds.): *Review of Progress in Quantitative NDE*, Vol. 20. pp. 1642–1649.

Wilcox, P., M. Lowe, and P. Cawley: 1999, 'Long range Lamb wave inspection: the effect of dispersion and modal selectivity'. In: D. Thompson and D. Chimenti (eds.): *Review of Progress in Quantitative NDE*, Vol. 18. pp. 151–158.

Wilcox, P., M. Lowe, and P. Cawley: 2000, 'Lamb and SH wave transducer arrays for the inspection of large areas of thick plates'. In: D. Thompson and D. Chimenti (eds.): *Review of Progress in Quantitative NDE*, Vol. 19. pp. 1049–1056.

Wilcox, P., M. Lowe, and P. Cawley: 2001, 'The effect of dispersion on long range inspection using ultrasonic guided waves'. *NDT & E Int.* **34**, 1–9.

Worlton, D.: 1957, 'Ultrasonic testing with Lamb waves'. *NonDestr. Test.* **15**, 218–222.

THE INFLUENCE OF THE INITIAL STRESSES ON THE DYNAMIC INSTABILITY OF AN ANISOTROPIC CONE

Y. A. ROSSIKHIN and M. V. SHITIKOVA
Voronezh State University of Architecture and Civil Engineering, ul.Kirova 3-75, Voronezh 394018, Russia

Abstract. The problem of surface instability of a right circular cone with an arbitrary opening made of a hexagonal single crystal is investigated. The surface of the cone is free from normal and tangential stresses, but in the layer near the surface initial constant tensile or compressive stresses act in the hoop direction and in the direction of the cone's generators. Surface instability is analyzed by the use of weak nonstationary disturbances which propagate along the surface of the cone in the form of the nonstationary surface wave of the "whispering gallery" type polarized perpendicular to the sagittal plane. The weak nonstationary surface wave is interpreted as the line of discontinuity (diverging circle) on which partial derivatives of the stress and strain tensor components with respect to coordinates and time have a discontinuity, but the components of these tensors are continuous. The line of discontinuity propagates with a constant normal velocity along the cone's surface in the direction of its generators and is obtained as a result of the exit onto the cone surface of a real conic wave surface of weak discontinuity (wave of the "whispering gallery" type). The analysis is carried out within the framework of the theory of discontinuities based on the kinematic, geometric and dynamic conditions of compatibility, using which the velocity of the surface wave propagation and its intensity have been found. It has been shown that the surface wave velocity is dependent only on the initial stress acting in the direction of the propagation of a surface disturbance, whereas the damping coefficients for the intensity of the surface waves is dependent not only on this stress but on the initial stress acting in the hoop direction as well. The relationship for the critical magnitude of the force compressive in the hoop direction has been obtained, and it has been shown that under the hoop compressive forces in excess of this magnitude the intensity of the surface wave of the "whispering gallery" type begins to increase without bounds during its propagation, i.e. the surface of the cone loses stability with respect to weak nonstationary disturbance.

Key words: nonlinear surface waves, elasticity, Rayleigh waves, scale invariance, nonlocality, spectral kernel, spatial kernel, Hilbert transform

R.V. Goldstein and G.A. Maugin (eds.),
Surface Waves in Anisotropic and Laminated Bodies and Defects Detection, 257–270.

1. Introduction

The propagation of harmonic surface waves in an initially deformed isotropic body was investigated by Flavin [1] and Hayes and Rivlin [2]. The propagation of surface waves in the material with the Mooney potential was studied in [1]. The equation defining the velocity of the Rayleigh wave propagation was obtained, in so doing it was noted that the roots of this equation could be obtained only by the use of numerical methods. Hayes and Rivlin [2] based on the data presented in Green *et al.* [3] discussed the questions of the propagation of the surface waves of the Rayleigh and Love types in the half-space subjected to large homogeneous deformations. The characteristic equation for the Rayleigh waves was derived and its complete analysis was carried out. The condition for the occurrence of the Love surface waves were obtained in a general form. A thermoelastic surface wave was investigated in Flavin [4]. Makhort [5] considered the influence of large and small deformations on the propagation velocity of harmonic Rayleigh waves in a nonlinearly elastic, initially isotropic body with an arbitrary form of the elastic potential. For three variants of the problem formulation, the characteristic equations were obtained and numerical results for variations in the surface wave velocity as a function of the magnitude of the initial stress were presented. Investigations of nonstationary surface waves in nonlinearly elastic bodies were carried out by Bestuzheva *et al.* [6], wherein the question on surface instability of the boundary of a prestressed nonlinearly elastic half-space was touched on.

In the present paper, investigations are undertaken into surface instability of an elastic right circular cone made of a hexagonal single crystal with respect to weak nonstationary disturbances resulting in the propagation of the lines of weak discontinuity (diverging circumferences) along the cone generators: the wave of the "whispering gallery" type. The statement of the problem about surface instability is similar to the definition of stability problems for beams, plates, and shells, i.e. it is assumed that constant compressive stresses act within a thin layer near the surface of the elastic cone. This is justified by the fact that the nonstationary surface waves (lines of discontinuity) are completely localized within the thin layer at the cone's surface, and their characteristics are very sensitive to the stressed state of the given layer. Differentiating from the enumerated papers, wherein only surface wave velocities were investigated, in the present paper, the influence of initial stresses both on surface wave velocities and their intensities is investigated.

2. Problem Formulation

Let us consider the dynamic stability of an infinite right circular cone with a free surface (normal and tangential stresses on the conic surface are equal to zero) under the compression. Note that the loss of stability takes place in the immediate vicinity of the free surface and attenuates with distance from the surface. All phenomena occurring in the zone nearby the free surface are called as "skin effects" [7].

Weak (sound) nonstationary surface waves of the "whispering gallery" type are used as small disturbances. Note that Biot [8] was the first to investigate the stability of the free surface of a half-space.

Since the discussion is about the problem of elastic stability, then one can limit himself to the case of small deformations, but large displacements. To be exact, when developing the equations of stability, one can ignore the differences between lengths, cross-sectional areas, and volumes before and after deformation takes place. This assumption evidently falls inside the scope of applied theories [9].

Furthermore we shall consider the cone with the opening 2α made of a hexagonal monocrystal, where a cone axis coincides with a crystal's axis of isotropy (Fig. 1). Undisturbed equilibrium of the cone in the Cartesian coordinate system x_1, x_2, x_3 with the origin at the cone vertex and the x_2-axis directed along the axis of isotropy is characterized by the components of the displacement vector u_i^0, the stress tensor σ_{ij}^0, and the vectors of volume X_i^0. The characteristics of the undisturbed equilibrium satisfy the following set of equations:

$$[\sigma_{jk}^0(\delta_{ik} + u_{i,k}^0)]_{,j} + X_i^0 = 0 \tag{1}$$

subjected to the boundary conditions

$$[\sigma_{jk}^0(\delta_{ik} + u_{i,k}^0)]N_j = 0, \tag{2}$$

where $\mathbf{N}$ ($N_1 = \cos\alpha\cos\varphi$, $N_2 = \cos\alpha\sin\varphi$, $N_3 = \sin\alpha$) is the unit normal vector to the undeformed surface of the cone, φ is the angular cylindrical coordinate (Fig. 1), δ_{ij} is the Kronecker's symbol, and an index after a comma labels the partial derivative with respect to the corresponding spatial Cartesian coordinates x_1, x_2, x_3.

Let us give the cone such small deviations from the undisturbed equilibrium state which result in the propagation of the weak nonstationary surface wave of the "diverging circle" type along the free surface of the cone in the direction of its generators (Fig.1), in so doing the wave is polarized perpendicular to the sagittal plane. We shall refer such a wave as the nonstationary wave of the "whispering gallery" type. The surface wave represents the line of weak discontinuity (diverging circumference) on which partial derivatives of the stress and strain tensor components with respect to coordinates and time experience a discontinuity, but the components of these tensors are continuous.

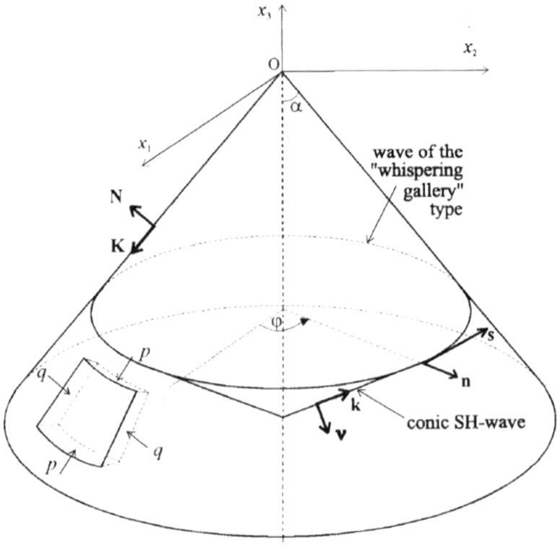

Figure 1. A scheme of the wave fron location of the surface wave of the "whispering gallery" type in a hexagonal cone with initial homogeneous stresses

The components of the characteristics of disturbed motion (we shall mark them by the sign $\sim$) have the form

$$\widetilde{u}_i = u_i^0 + u_i, \tag{3.1}$$

$$\widetilde{\sigma}_{ij} = \sigma_{ij}^0 + \sigma_{ij}, \tag{3.2}$$

$$\widetilde{X}_i^0 = X_i^0 + X_i, \tag{3.3}$$

where σ_{ij}, u_i and X_i are the characteristics of the disturbance itself, ρ is the density, and an overdot denotes the time-derivative. Further let us consider that at the static equilibrium the volume forces X_i are absent, and $X_i^0 = -\rho\ddot{u}_i$.

The components of disturbed motion should also satisfy Eq.(1) and the boundary conditions (2). Substituting (3) in (1) and (2), and linearizing the resulting relationships, we obtain the equations in variations and the corresponding boundary conditions [9], [10]:

$$[\sigma_{jk}(\delta_{ik} + u_{i,k}^0) + \sigma_{jk}^0 u_{i,k}]_{,j} = \rho\ddot{u}_i, \tag{4.1}$$

$$[\sigma_{jk}(\delta_{ik} + u_{i,k}^0) + \sigma_{jk}^0 u_{i,k}]N_j = 0. \tag{4.2}$$

Assume that the initial deformed state can be described by the geometrically linear theory of elasticity. This assumption is in compliance with ignoring the values $|u^0_{i,j}|$ as compared with unity. Fulfilling the corresponding assumptions and considering that the boundary surface of the cone is free from normal and tangential stresses both in undisturbed and disturbed states, from (4.1) and (4.2) we obtain the equations of motion and the boundary conditions

$$\sigma_{ij,j} + (\sigma^0_{jk} u_{i,k})_{,j} = \rho \ddot{u}_i, \tag{5.1}$$

$$\sigma_{ij} N_j = 0, \tag{5.2}$$

$$\sigma^0_{ij} N_j = 0, \tag{5.3}$$

which refer to the equations of stability for the second variant of the theory of small subcritical deformations (Biot [8], Guz' [10]).

The relationship

$$\dot{\sigma}_{ij} = \lambda_{ijkl} \dot{u}_{k,l} \tag{6}$$

should be added to Eqs.(5), where the components λ_{ijkl} of the tensor of elasticity are defined by the matrix

$$\left\| \begin{matrix} \lambda_{1111} & \lambda_{1122} & \lambda_{1133} & 0 & 0 & 0 \\ & \lambda_{1111} & \lambda_{1133} & 0 & 0 & 0 \\ & & \lambda_{3333} & 0 & 0 & 0 \\ & & & \lambda_{1313} & 0 & 0 \\ & & & & \lambda_{1313} & 0 \\ & & & & & \lambda_{1212} \end{matrix} \right\|$$

wherein $\lambda_{1212} = \frac{1}{2}(\lambda_{1111} - \lambda_{1122})$.

Note that since the influence of the thin initially deformed layer near the surface of the elastic cone on the behaviour of the surface wave characteristics is investigated in the present paper, and specifically the influence of stability or instability of this layer upon the changes in surface wave intensities during its propagation along the cone's generators, then when considering the behaviour of the near surface layer by the use of Eqs.(5) and (6) one can put, as it is done during the analysis of the problems of shell stability, $p = \sigma^0_{ij} K_i K_j = \; const$ and $q = \sigma^0_{ij} s_i s_j = const$, where $\mathbf{K}$ ($K_1 = \sin\alpha\cos\varphi$, $K_2 = \sin\alpha\sin\varphi$, $K_3 = -\cos\alpha$) is the unit vector directed along the generator of the hexagonal cone, and $\mathbf{s}$ ($s_1 = -\sin\varphi$, $s_2 = \cos\varphi$, $s_3 = 0$) is the unit vector directed tangentially to the cone's guide. The conditions $p = const$ and $q = const$ are realized at $\sigma^0_{ij} = 0$ $(i \neq j)$ and $\sigma^0_{11} = \sigma^0_{22} = const$, $\sigma^0_{33} = const$, i.e. the components of the tensor of the initial stresses can be carried out from the sign of the derivative in Eqs.(5.1).

Similar problem on stability of the initially stressed layer near the surface of a hexagonal cone with respect to nonstationary sound surface waves of the "diverging circle" type can arise during investigation of a blanket of snow on mountain tops with the aim of revealing the possibilities for coming off of snow avalanches. As this takes place, the sound surface waves can act as the source of information during control over the state of snow blankets in avalanche-dangerous mountain regions.

3. The Wave of the "Whispering Gallery" Type

Let us consider the nonstationary volume conic SH-wave with the opening $180° - 2\beta$ propagating with the normal velocity G (Fig. 2). For definiteness, we shall assume that $\beta < \alpha$, however, all further reasonings are valid in the case $\beta > \alpha$ as well.

The conic SH-wave Σ intersecting with a free surface of a hexagonal cone generates a surface wave of the "whispering gallery" type (a diverging circumference) which propagates along the generators of the material cone with the velocity $G_S > G$ and is polarized along the unit vector $\mathbf{s}$ directed tangentially to this circumference.

Considering that

$$K_i\nu_i = GG_S^{-1} = \cos(\alpha - \beta),$$

$$N_i\nu_i = -n = -\sin(\alpha - \beta), \quad n = \sqrt{1 - G^2G_S^{-2}}, \tag{7}$$

for the components ν_i of the unit vector ν normal to the wave surface Σ we obtain

$$\nu_i = K_i\cos(\alpha - \beta) - N_i\sin(\alpha - \beta) \tag{8.1}$$

or

$$\nu_1 = \sin\beta\cos\varphi, \ \nu_2 = \sin\beta\sin\varphi, \ \nu_3 = -\cos\beta. \tag{8.2}$$

To find the equation of the surface Σ in Cartesian coordinates, we use the following relationships (Thomas [11]):

$$\delta x_i/\delta t = G\nu_i, \tag{9.1}$$

$$x_{i,1} = \partial x_i/\partial u^1 = k_i, \tag{9.2}$$

where $\delta/\delta t$ is the Thomas δ-derivative [11], $\mathbf{k}$ ($k_1 = \cos\beta\cos\varphi$, $k_2 = \cos\beta\sin\varphi$, $k_3 = \sin\beta$) is the unit vector directed along the generator of the wave surface Σ, u^1 is the coordinate on the surface Σ directed along its generator and measured from the point of intersection of the normal leaving the vertex of the hexagonal cone with the surface Σ, and positive

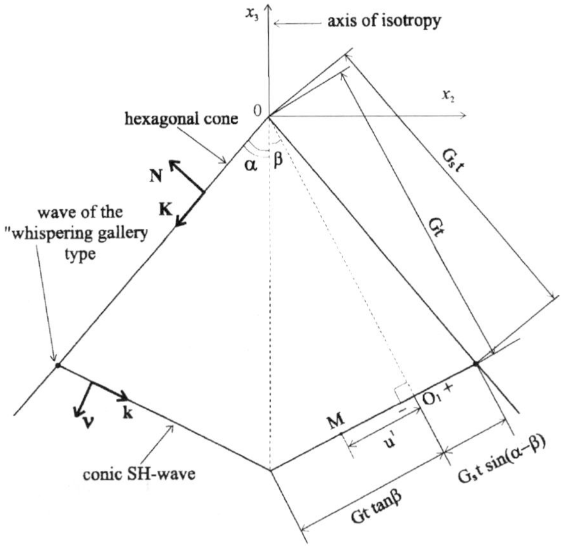

Figure 2. A scheme of the wave front location in the $x_2 - x_3$ section of a hexagonal cone

and negative directions of a reading of the value u^1 are indicated by the signs "+" and "-", respectively.

Considering that $\delta u^1/\delta t = \delta u^2/\delta t = 0$, where $u^2 = \varphi$, from formulas (9) we find

$$x_i = Gt\nu_i + u^1 k_i. \tag{10}$$

Using Eq.(10) and the relationships

$$g_{\alpha\beta} = x_{i,\alpha}x_{i,\beta}, \;\; b_{\alpha\beta} = x_{i,\alpha\beta}\nu_i = \frac{\partial^2 x_i}{\partial u^\alpha \partial u^\beta}\nu_i,$$
$$\nu_{i,\alpha} = -g^{\beta\gamma}b_{\beta\alpha}x_{i,\gamma}, \;\; g^{\beta\gamma}g_{\beta\alpha} = \delta^\gamma_\alpha, \tag{11}$$

where $g_{\alpha\beta}$ and $g^{\alpha\beta}$ are the covariant and contravariant components of the metric tensor of the wave surface Σ, respectively, δ^γ_α is the Kronecker's symbol, $b_{\alpha\beta}$ are the coefficients of the second quadratic form of this surface, a Greek symbol after a comma denotes a covariant derivative with respect to the corresponding coordinate u^1 or u^2 on the surface Σ, and Greek indices take on the values 1 and 2, we calculate the geometric characteristics to be used in further discussion.

As a result we obtain

$$
\begin{gathered}
x_{i,2} = \partial x_i / \partial u^2 = r s_i, \\
r = \sqrt{x_1^2 + x_2^2} = Gt \sin\beta + u^1 \cos\beta, \\
\nu_{i,1} = 0, \quad \nu_{i,2} = s_i \sin\beta, \\
g_{11} = 1, \quad g_{12} = 0, \quad g_{22} = r^2, \\
b_{11} = 0, \quad b_{12} = 0, \quad b_{22} = -r \sin\beta.
\end{gathered} \tag{12}
$$

Using Eqs.(5)-(12), one can determine the velocity and intensity of the conic wave surface Σ, in so doing during the transition across this surface the first-order partial derivatives of the stress tensor components and the second order partial derivatives of the displacement vector components with respect to the time t and the coordinates x_i experience discontinuities. Writing (5.1) and (6) on the different sides of the surface Σ and taking their difference yeilds

$$
[\sigma_{ij,j}] + \sigma^0_{jk}[u_{i,kj}] = \rho[\dot{v}_i], \tag{13.1}
$$

$$
[\dot{\sigma}_{ij}] = \lambda_{ijkl}[v_{k,l}], \tag{13.2}
$$

where $v_i = \dot{u}_i$ are the displacement velocity vector components, $[Z] = Z^+ - Z^-$, and Z^+ and Z^- are the magnitudes of a certain function Z calculated immediately ahead of and behind the surface Σ, respectively.

Applying the first order conditions of compatibility [11]

$$
\begin{gathered}
[\sigma_{ij,l}] = b_{ij}\nu_l, \qquad [\dot{\sigma}_{ij}] = -Gb_{ij}, \qquad b_{ij} = [\sigma_{ij,l}]\nu_l, \\
[v_{i,j}] = \omega_i \nu_j, \ [\dot{v}_i] = -G\omega_i, \ G[u_{i,kj}] = -\omega_i \nu_k \nu_j, \ \omega_i = [v_{i,l}]\nu_l,
\end{gathered} \tag{14}
$$

from Eqs.(13) we find

$$
Gb_{ij}\nu_j - \sigma^0_{jk}\omega_i \nu_k \nu_j = -\rho G^2 \omega_i, \tag{15.1}
$$

$$
-Gb_{ij} = \lambda_{ijkl}\omega_k \nu_l. \tag{15.2}
$$

Eliminating the values b_{ij} from (15.1) and (15.2) and considering that on the SH-wave

$$
\omega_i = \omega s_i, \quad \omega = \sqrt{\omega_i \omega_i}, \tag{16}
$$

we are led to the equation

$$
\left\{ \lambda_{ijkl}\nu_j \nu_l s_k - (\rho G^2 - \sigma^0_{jk}\nu_j \nu_k) s_i \right\} \omega = 0, \tag{17}
$$

from which we obtain at $\omega \neq 0$

$$
\rho G^2 = \lambda_{1212} \sin^2\beta + \lambda_{1313} \cos^2\beta + \sigma^0_{jl}\nu_j \nu_l. \tag{18}
$$

If we write the formula (18) on the free surface of the hexagonal cone and take the relationships (7), (8.1), and (5.3) into account, then we obtain

$$\rho G^2 = \rho G_0^2 + p\cos^2(\alpha - \beta), \tag{19.1}$$

$$\rho G_S^2 = [\rho G_0^2 + p\cos^2(\alpha - \beta)]\cos^{-2}(\alpha - \beta), \tag{19.2}$$

$$\rho G G_S = [\rho G_0^2 + p\cos^2(\alpha - \beta)]\cos^{-1}(\alpha - \beta), \tag{19.3}$$

where $\rho G_0^2 = \lambda_{1212}\sin^2\beta + \lambda_{1313}\cos^2\beta$, and G_0 is the velocity of the SH-wave propagation in an unstressed cone.

To find the angle β, we use the boundary conditions (5.2) which after differentiating with respect to time, eliminating the values $\dot{\sigma}_{ij}$ by the use of (6), and evaluating on the line of discontinuity take the form

$$\lambda_{ijkl}N_j[v_{k,l}] = 0. \tag{20}$$

From Eq.(20) with the help of the conditions of compatibility we find the relationship

$$\lambda_{ijkl}s_k\nu_l N_j = 0, \tag{21}$$

from which we obtain the equation for determining the angle β

$$\tan\beta = \lambda_{1313}\lambda_{1212}^{-1}\tan\alpha. \tag{22}$$

To obtain the equation satisfied by the amplitude of the volume conic wave Σ when it exits onto the surface of the hexagonal cone, we differentiate Eqs.(5.1) and (6) with respect to the time t and write the resulting relationships on the front of this wave. As a result we have

$$[\dot{\sigma}_{ij,j}] + \sigma_{jk}^0[v_{i,kj}] = \rho[\ddot{v}_i], \tag{23.1}$$

$$[\ddot{\sigma}_{ij}] = \lambda_{ijkl}[\dot{v}_{k,l}]. \tag{23.2}$$

Using the second order conditions of compatibility [11]

$$\begin{aligned}
&[v_{i,kj}] = L_i\nu_k\nu_j + g^{\alpha\beta}\omega_{i,\alpha}(\nu_j x_{k,\beta} + \nu_k x_{j,\beta}) - \omega_i g^{\alpha\beta}g^{\sigma\tau}b_{\alpha\sigma}x_{k,\beta}x_{j,\tau},\\
&[\dot{v}_{k,l}] = -GL_k\nu_l + \frac{\delta\omega_k}{\delta t}\nu_l - Gg^{\alpha\beta}\omega_{k,\alpha}x_{l,\beta},\\
&[\ddot{v}_i] = G^2L_i - 2G\frac{\delta\omega_i}{\delta t}, \qquad L_i = [v_{i,jl}]\nu_j\nu_l,\\
&[\dot{\sigma}_{ij,l}] = -GB_{ij}\nu_l + \frac{\delta b_{ij}}{\delta t}\nu_l - Gg^{\alpha\beta}b_{ij,\alpha}x_{l,\beta},\\
&[\ddot{\sigma}_{ij}] = G^2B_{ij} - 2G\frac{\delta b_{ij}}{\delta t}, \qquad B_{ij} = [\sigma_{ij,kl}]\nu_k\nu_l,
\end{aligned} \tag{24}$$

from Eqs.(23.1) and (23.2) we find

$$-B_{ij}\nu_j G+\frac{\delta b_{ij}}{\delta t}\nu_j-Gg^{\alpha\beta}b_{ij,\alpha}x_{j,\beta}+\overset{0}{\sigma}_{jk}\Big\{L_i\nu_k\nu_j \quad +\omega_{i,\alpha}g^{\alpha\beta}(\nu_j x_{k,\beta}+\nu_k x_{j,\beta})$$

$$-\omega_i g^{\alpha\beta}g^{\sigma\tau}b_{\alpha\sigma}x_{k,\beta}x_{j,\tau}\Big\}=\rho\left(G^2L_i-2G\frac{\delta\omega_i}{\delta t}\right), \qquad (25.1)$$

$$G^2B_{ij}-2G\frac{\delta b_{ij}}{\delta t}=\lambda_{ijkl}\left(-GL_k\nu_l+\frac{\delta\omega_k}{\delta t}\nu_l-Gg^{\alpha\beta}\omega_{k,\alpha}x_{l,\beta}\right). \qquad (25.2)$$

Eliminating the values B_{ij} from (25.1) and (25.2) and regarding to (15) yields

$$\Big\{(\rho G^2-\overset{0}{\sigma}_{jl}\nu_j\nu_l)\delta_{ik}-\lambda_{ijkl}\nu_j\nu_l\Big\}L_k=2\rho G\frac{\delta\omega_i}{\delta t}$$

$$+\lambda_{ijkl}\left\{(\omega_k\nu_l)_{,\alpha}x_{j,\beta}+\omega_{k,\alpha}x_{l,\beta}\nu_j\right\}g^{\alpha\beta}$$

$$+\ \overset{0}{\sigma}_{jk}\left\{\omega_{i,\alpha}(\nu_j x_{k,\beta}+\nu_k x_{j,\beta})-\omega_i g^{\sigma\tau}b_{\alpha\sigma}x_{k,\beta}x_{j,\tau}\right\}g^{\alpha\beta}. \qquad (26)$$

Putting $L_i = Ls_i$ and considering the relationships (8.1), (9.2), (12), and (16), we are led to the equation

$$2\rho G\frac{\delta\omega}{\delta t}+\omega_{,1}\left\{(\lambda_{1212}-\lambda_{1313})\sin 2\beta+p\sin 2(\alpha-\beta)\right\}+$$
$$+(\lambda_{1212}+q)\frac{1}{r}\omega\sin\beta=0, \qquad (27)$$

which defines the value $\delta\omega/\delta t$ on the conic SH-wave.

Write the boundary condition (5.2) in the form

$$\lambda_{ijkl}[\dot v_{k,l}]N_j=0 \qquad (28)$$

and apply the conditions of compatibility (24). Having regard to the expression (21), as a result we obtain

$$\lambda_{ijkl}N_j\omega_{k,\alpha}x_{l,\beta}g^{\alpha\beta}=0. \qquad (29)$$

From (29) we find

$$\omega_{,1}=\omega\frac{æ}{r}, \qquad æ=\frac{\lambda_{1212}\cos\alpha}{\lambda_{1212}\cos\alpha\cos\beta+\lambda_{1313}\sin\alpha\sin\beta}. \qquad (30)$$

The wave surface Σ coming out on the free surface of the cone at $u^1 = s\sin(\alpha-\beta)$ ($s = G_S t$ is the distance traveled by the surface wave along the generator of the hexagonal cone during the time t) generates the circumference of the radius $r = r_S = s\sin\alpha$ on which the relationships (27) and (30) are valid at a time and, moreover, the derivatives $\delta/\delta t$ and $\partial/\partial u^1$ appear to be connected in terms of the derivative d/ds along the direction of the vector $\mathbf{K}$ by the following formula:

$$\frac{d}{ds} = \frac{1}{G_S}\frac{\delta}{\delta t} + n\,\frac{\partial}{\partial u^1}\,. \tag{31}$$

Arranging the relationships (27), (30), and (31) on the surface wave of the "whispering gallery" type, we are led to the equation for determining the intensity of the surface wave

$$\frac{d\omega}{ds} + \gamma\frac{1}{s}\omega = 0, \tag{32.1}$$

$$\gamma = \frac{1}{2\rho G G_S}\left[(\lambda_{1212}+q)\frac{\sin\beta}{\sin\alpha} + \chi\ \cot\alpha\right], \tag{32.2}$$

where

$$\chi = \frac{[(\lambda_{1212}-\lambda_{1313})\sin 2\beta + p\sin 2(\alpha-\beta) - n2\rho G G_S]\lambda_{1212}}{\lambda_{1212}\cos\alpha\cos\beta + \lambda_{1313}\sin\alpha\sin\beta}.$$

It can be shown that the value χ is equal to zero, so

$$\gamma = \frac{1}{2\rho G G_S}(\lambda_{1212}+q)\frac{\sin\beta}{\sin\alpha}, \tag{32.3}$$

or after using the formula (19.3)

$$\gamma = \frac{(\lambda_{1212}+q)\sin\beta}{2(\rho G_S^{0\,2}+p)\cos(\alpha-\beta)\sin\alpha}, \tag{32.4}$$

where $\rho G_S^{0\,2} = \rho G_0^2\cos^{-2}(\alpha-\beta)$, G_S^0 is the propagation velocity of the wave of the "whispering gallery" type in an unstressed cone, and $\rho G_S^{0\,2}+p>0$.

Integrating Eq.(32.1) yields

$$\omega = Cs^{-\gamma}, \tag{33.1}$$

where C is an arbitrary constant dependent on the initial conditions.

If $\alpha=\pi/2$, i.e. an infinite cone becomes a half-space, then the solution to Eq.(32.1) has the form (33.1), wherein

$$\gamma = \frac{\lambda_{1212}+q}{2(\rho G_S^{0\,2}+p)}, \qquad \rho G_S^{0\,2} = \rho G_0^2. \tag{33.2}$$

If $\alpha=0$ and in the limit $s\sin\alpha = R$, where R is a certain constant value, then an infinite cone goes over into an infinite cylinder, and the solution to Eq.(32.1) takes the form

$$\omega = C, \quad \rho G_S^{0\,2} = \rho G_0^2, \tag{33.3}$$

where C is an arbitrary constant.

Formulas (19) and (33) define the main characteristics of the surface wave of the "whispering gallery" type propagating along the free surface of the hexagonal cone with the homogeneous initial stresses acting near its surface.

From these formulae it is seen that the velocity of the surface wave of the "whispering gallery" type is dependent on the tensor σ_{ij}^0 components acting in the vicinity of the surface of the cone in the direction of its generators, whereas the damping coefficient of the wave intensity is dependent on the components of this tensor acting both in the direction of the cone generators and in the hoop direction.

Under the action of the compressive stresses along the generators of the hexagonal cone ($p < 0$), the velocity of the surface wave of the "whispering gallery" type decreases, but the damping coefficient of its intensity γ increases. Under the action of the compressive stresses in the hoop directions ($q < 0$), the velocity of the surface wave does not change, but the damping coefficient of its intensity decreases.

Under the hoop compressive stresses $q < -\lambda_{1212}$, the intensity of the surface wave of the "whispering gallery" type begins to increase without bounds during its motion along the free surface of the hexagonal cone, i.e. at $q = q_{\text{crit}} = -\lambda_{1212}$ the surface of the cone becomes unstable with respect to the nonstationary dynamic excitation of the SH-wave type. Similar result can be obtained if one investigates the surface instability of an orthotropic half-space under static loads (Guz' [10]).

These conclusions are not in conflict with the experimental investigations carried out by Crecraft [12] and Smith [13], from which it follows that the magnitude of the velocity of the harmonic shear wave polarized perpendicular to the action of the stress decreases, but the velocity of the harmonic shear wave polarized along the action of the stresses increases with increase in the compressive force.

Once the velocity G_S and the damping coefficient γ of the pure shear surface wave are found from the experimental investigations, then for determining the unknown values p and q from the formulas (19) and (32.2) the following relationships are obtained

$$\begin{aligned} p &= \rho G_S^2 - \left(\lambda_{1212}\sin^2\beta + \lambda_{1313}\cos^2\beta\right)\cos^{-2}(\alpha-\beta), \\ q &= 2\gamma\rho G_S^2\cos(\alpha-\beta)\sin\alpha\sin^{-1}\beta - \lambda_{1212}. \end{aligned} \tag{34}$$

Note that the corresponding characteristics for an unstressed cone were obtained by Rossikhin [14].

For an isotropic cone, it is necessary to put $\lambda_{1212} = \lambda_{1313} = \mu$ in

Eqs.(19), (22), and (32.4). As a result we obtain

$$\alpha = \beta, \quad \rho G_S^2 = \mu + p, \quad \omega = Cs^{-\frac{\mu+q}{2(\mu+p)}} . \tag{35}$$

If we set $p = q = 0$ in (35), then we are led to the known relationships

$$\rho G_S^2 = \mu, \quad \omega = Cs^{-\frac{1}{2}}, \tag{36}$$

which are also valid for the surface SH-wave propagating along the free boundary of an unstressed elastic isotropic half-space.

4. Conclusion

Results of the investigations carried out show that the characteristics of the nonstationary surface wave of the "whispering gallery" type are very sensitive to the level of the tensile or compressive stresses in the near-surface layer of an elastic isotropic or anisotropic cone. As this takes place, the compressive forces acting in the direction of the surface wave propagation as if "close" these waves, resulting in decrease in velocity of this wave and increase in the damping coefficients of its intensity. In contrast, the compressive forces acting in the direction perpendicular to the direction of wave propagation, may result in an infinite increase in intensity of this wave during its propagation. These properties allow one to use this type of surface waves for the analysis of surface instability of elastic bodies, as well as for nondestructive testing of residual stresses in the vicinity of surfaces of different bodies.

Another application of the results obtained can be found in seismology and seismic survey. Knowing the magnitudes of the surface wave velocity and its damping coefficient, one can determine the level of stress concentration in the Earth's crust and predict the potentiality of earthquakes.

References

Flavin, J.N. (1963) Surface waves in pre-stressed Mooney material, *Quart. J. Mech. Appl. Math.* **16**, 441–449.

Hayes, M. and Rivlin, R.S. (1961) Surface waves in deformed elastic materials, *Arch. Rat. Mech. Anal.* **8**, 358–380.

Green, A.E., Rivlin, R.S. and Shield, R.T. (1952) General theory of a small elastic deformations superposed on finite elastic deformations, *Proc. Roy. Soc.* **211 A**, 128–154.

Flavin, J.N. (1962) Thermo-elastic Rayleigh waves in a prestressed medium, *Proc. Cambridge Phil. Soc.* **58**, 532–538.

Makhort, F.G. (1971) To the theory of surface wave propagation in an elastic body with initial deformations (in Russian), *Prikl. Mekh.* **7**, 34–40.

Bestuzheva, N.P., Bukovtsev, G.I. and Durova, V.N. (1981) Investigation of nonstationary surface waves in nonlinear elastic media (in Russian), *Prikladnaja Mekhanika*, **Vol.17**, 27–33.

Biot, M.A. (1966) Fundamental skin effect in anisotropic solid mechanics, *J. Solid Struct.* **2**, 645–663.

Biot, M.A. (1965) *Mechanics of Incremental Deformations*. Wiley, New York.

Bolotin, V.V. (1963) *Non-conservative Problems of the Theory of Elastic Stability*. Pergamon, Oxford.

Guz', A.N. (1971) *Stability of Three-dimensional Deformable Bodies* (in Russian). Naukova Dumka, Kiev.

Thomas, T.Y. (1961) *Plastic Flow and Fracture in Solids*. Academic Press, New York.

Crecraft, D.J. (1962) Ultrasonic wave velocities in stressed nickel steel, *Nature* **195**, 1193–1194.

Smith, R.T. (1963) Stress-induced anisotropy in solids - the acoustoelastic effect, *Ultrasonics* **1**(3), 135–147.

Rossikhin, Yu.A. (1992) Non-stationary surface waves of "diverging circle" type on conic surfaces of hexagonal crystals, *Acta Mechanica* **92**, 183–192.

EMBEDDING THEOREM AND MUTUAL RELATION FOR THE INTERFACE AND SHEAR WAVESPEEDS

I. V. SIMONOV
Institute for Problems in Mechanics
Russian Academy of Science,
pr. Vernadskogo, 101-1, 119526 Moscow, Russia

Key words: nonlinear surface waves, elasticity, Rayleigh waves, scale invariance, nonlocality, spectral kernel, spatial kernel, Hilbert transform

1. Introduction

Since *Rayleigh and Stoneley* pioneering works many efforts have been taken to achieve progress in the surface wave analyses. The present paper is targeted to clear up the following points:

- Generalization of some well-known and new obtained results on the interface waves: existence, comparison of the existence domains in the phase space of moduli of elasticity for various contact conditions in bimaterials, and analysis of the associated wavespeeds.
- Eliminating *the contradiction.*
- Clarifying some *"blank spaces"*.

Herein these points will be analyzed in detail, whereas previously they were published only in the form of the thesis [1].

The results are relevant in tentative analyses of the fastly moving interface load, crack, dislocation, *etc.* In the case of the problem solving by both numerical and analytical method, it is important to know in advance what kind of the interface waves can be radiated and what is the relation between their wavespeeds.

R.V. Goldstein and G.A. Maugin (eds.),
Surface Waves in Anisotropic and Laminated Bodies and Defects Detection, 271–276.

2. Physical assumptions and necessary relations

Let two homogeneous (transversely-)isotropic materials 1 and 2 occupy the half-planes $y > 0$ and $y < 0$ of orthogonal coordinate system x, y. We will consider the four traction free uniform boundary conditions over the whole interface $y = 0$, belonging to four classical types: open, full, slip and anti-slip contact. The interface waves analyzed mean free vibrations localized in the vicinity of the interface, i.e., they obey the well-known decay condition of the type of Rayleigh wave as $|y| \to \infty$. Linear theory of elasticity yields the following algebraic characteristic equations for the corresponding wavespeeds c [1-5]:

$$R_j(c) = 0, \qquad S(c) = 0, \qquad P(c) = 0, \qquad Q(c) = 0,$$

$$S = \frac{D^2 - PQ}{R_1 R_2} = \sum_1^3 S_i, \tag{1}$$

$$S_m = \mu^{2(k-1)}(\beta_{1k}\beta_{2k} - 1)R_m, \quad m \neq k = 1, 2,$$

$$S_3 = -\mu\{2\gamma_1\gamma_2 + (1-\beta_1)(1-\beta_2)(\beta_{11}\beta_{22} + \beta_{12}\beta_{21}),$$

$$D = \gamma_1 R_2 - \mu\gamma_2 R_1, \qquad P = \mu\beta_{12}(\beta_2 - 1)R_1 + \beta_{11}(\beta_1 - 1)R_2,$$

$$Q = \mu\beta_{22}(\beta_2 - 1)R_1 + \beta_{21}(\beta_1 - 1)R_2,$$

$$\beta_{mj} = (1 - \frac{c^2}{c_{mj}^2})^{1/2}, \qquad \beta_j = \frac{1}{2}(1 + \beta_{2j}^2), \qquad \mu = \frac{\mu_1}{\mu_2},$$

$$R_j = \beta_{1j}\beta_{2j} - \beta_j^2, \qquad \gamma_j = \beta_{1j}\beta_{2j} - \beta_j.$$

Here c_{1j} and c_{2j} are the compression and shear wavespeeds along the $x-$direction for the material $j = 1, 2$ with the shear modulus μ_j; R_j and S are the modified Rayleigh and Stoneley functions; in turn, P and Q correspond to the cases of slip and anti-slip contact conditions. Any of the above mentioned problems has a non-trivial solution, if the corresponding equation in (1) has a positive root, c_R, c_S, c_P, or c_Q; if there is no root, there is no wave.

Below it will be used the following inequalities, asymptotics, and notations:

$$0 < \beta_{mj} < 1, \quad 0 < |c| < c_{mj}; \qquad \beta_{mj} = 0, 1, \quad |c| = c_{mj}, 0,$$

$$\frac{1}{2} \leq \beta_j \leq 1, \quad 0 \leq \beta_j(1 - \beta_j) \leq \frac{1}{4}, \quad -\frac{1}{2} \leq \gamma_j \leq 0, \qquad |c| \leq c_{2j},$$

$$\beta_{1j} > \beta_{2j}, \quad \gamma_j < \beta_{2j} - \beta_j < 0, \quad \beta_j > \beta_2 j, \qquad 0 < c \leq c_{2j},$$

$$c_R^{max} = \max\{c_{Rj}\}, \qquad c_R^{min} = \min\{c_{Rj}\}, \tag{2}$$

$$R_j(c) > 0, \quad c < c_{Rj}; \quad R_j(c) < 0, \quad c > c_{Rj}; \quad R_j(c_{Rj}) = 0,$$

$$R_j(c) \sim \frac{c^2}{2}\left(\frac{1}{c_{2j}^2} - \frac{1}{c_{1j}^2}\right), \qquad \gamma_j \sim -\frac{c^2}{2c_{1j}^2}, \qquad c \to 0,$$

$$P(c) \sim A_P c^4, \quad Q(c) \sim A_Q c^4, \quad c \to 0; \qquad A_P,\ A_Q < 0.$$

Suppose that the cuts for the radicals β_{mj} join the branch points $\pm c_{mj}$ throughout the infinity along the Re c-axis in the complex c-plane, and the condition $\beta_{mj}(0) = +1$ specifies the branches of complex functions $\beta_{mj}(c)$. Then the following properties of any function F of the set $P,\ Q,\ R,\ S$ to be even and continuable through the Re c-axis take place:

$$F(c) = F(-c), \qquad F^*(c^*) = F(c). \tag{3}$$

We use the symbol * to denote a complex conjugation.

Some other limits are also useful. For the case of identical materials, $c_{m1} = c_{m2}$, the necessary limits are

$$D = 0, \qquad P = 2\beta_{11}(\beta_1 - 1)R_1\ , \qquad Q = 2\beta_{21}(\beta_1 - 1)R_1\ . \tag{4}$$

In contrast, if medium 2 asymptotically tends to a rigid one in the sense $\mu \to 0, \quad c_{m1}/c_{m2} \to 0$ we get

$$D/R_2 \to \gamma_1\ , \qquad P/R_2 \to \beta_{11}(\beta_1 - 1)\ , \qquad Q/R_2 \to \beta_{21}(\beta_1 - 1). \tag{5}$$

3. The theorems of existence and embedding

The necessary and sufficient conditions for existence of the unique positive root c^2 of any equation (1) is the nonnegativity of the function $F = P,\ Q,\ R,\ S$ at the point $c = c_2$

$$F(c_2) > 0, \qquad c_2 = \min\{c_{2j}\}. \tag{6}$$

In the Rayleigh case, it is well-known that this condition is justified for any element $\pi \in \Pi^E$, where Π^E is the whole phase space of elastic parameters for both materials. By applying the argument principle and the residual theorem, statements (6) were proved for the functions $P(c)$ and $S(c)$ in [2,3]. The same result for $Q(c)$ was presented in [4], but this paper was concerned only with real roots. In addition, similarly to [2,3], the absence of complex roots of the equation $Q(c) = 0$ was verified in [1].

Inequalities (6) express the essence of the existence theorem and also specify the existence domains , $\Pi^F \subset \Pi^E$, with the boundaries $F(c_2) = 0$. Now we will prove the embedding theorem for these domains

$$\Pi^S \subset \Pi^P \subset \Pi^Q \equiv \Pi^R \equiv \Pi^E. \tag{7}$$

All but the first inclusion in (7) were already known [2-4]. As to the first one, we note that due to $P(c_2) > 0$ for the case $c_R^{max} \leq c_2$, which readily follows from the expression for $P(c)$ in (1), i.e., the root $c_P > 0$ exists, it is sufficient to consider the case $c_R^{max} > c_2$.

Thus, suppose $\pi \in \Pi^S$ when $c_R^{max} > c_2$; then putting $c = c_2$, from definitions (1) and inequalities (2) we get

$$S > 0, \quad R_1 R_2 < 0, \quad Q > 0, \quad D^2 - PQ < 0, \quad c = c_2 \, . \tag{8}$$

Now, from (8) and the first expression for $S(c)$ in (1) it follows at once that $P(c_2) > 0$. Consequently, in this case, by theorem (6), there exists a non-trivial root c_P . With the above mentioned note, this means that the root c_P exists everywhere when so does the root c_S. Everything taken together completes the proof of theorem (7).

In addition, the relation

$$c_P < c_S \tag{9}$$

takes place if $\pi \in \Pi^S$, i.e., when c_S and c_P exist. In fact, by putting $c = c_p$ in the expression for $S(c)$ in (1), the following inequality can be proved:

$$S(c_P) = D^2 (R_1 R_2)^{-1} (c_P) < 0.$$

It is true since $D(c_P) \neq 0$, and $R_1(c_P)$ and $R_2(c_P)$ are opposite in sign because the root c_P always ranges over $c_R^{min} < c_P < c_R^{max}$ as it can be easily shown. On the other hand, by assumption $\pi \in \Pi^S$ and (6), $S(c_2) > 0$. Thus, $S(c)$ changes sign when moving from c_P to c_2 and has a single zero within the range $c_P < c < c_2$, which justifies inequality (9).

Interestingly enough, the values of c_S for 900 combinations of metals were examined, and it occurs that only 30 of which are suitable with respect to the existence of Stoneley wave [5], so that Π^S is a relatively small domain.

4. Contradiction

Let us consider one contradiction associated with restrictions for the Stoneley wavespeed, existing in the literature. Thus, in the paper by *Koppe, H., 1948* [6], next in the work by *Owen, T.E., 1964* [7] with reference to *Koppe* and at last in the book by *Grinchenko and Meleshko, 1981* [5] with reference to *Owen*, the restrictions for the c_S variation were stated as

$$c_R^{max} \; < \; c_S \; < \; c_2. \tag{10}$$

On the other hand, *Gol'dstein R.V., 1966* [8] has pointed out that for the materials being closely related by waveproperties, i.e., as $c_{j1} \to c_{j2}$, it is possible the inequality

$$c_R^{min} \; < \; c_S \; < \; c_R^{max} \; < c_2 \tag{11}$$

which contradicts (10).

What actually happens is that the c_S is varied within the range

$$c_R^{min} < c_S < c_2 \tag{12}$$

which is the sum of (10) and (11). This eliminates the contradiction.

The idea of a cumbersome proof of (12) consists in analysis of new quantity $G = S/\mu$ as a function only of μ at all bulk wavespeeds c_{mj} being frozen (in this case the values β_j, β_{mj}, R_j, γ_j become constants). By using the definition for $S(c)$ in (1) and the properties (2), due to the relatively simple dependence of G on $\mu \in (0, \infty)$, the bounds (12) can be proved.

5. Generalization of mutual relations and limits for wavespeeds

In this section, the general conclusions about the interface wavespeed restrictions, and also their various asymptotics will be made. They become evident from the previous works [1-6], results (2-8), and some additional reasoning. Let us start with disposition of the wavespeeds.

1. Suppose $\pi \in \Pi^S$; Then

$$c_P < c_S < c_2, \qquad c_R^{min} < c_Q < c_P < c_R^{max},$$

$$c_P = c_Q = c_{R1}, \qquad c_{R1} < c_S < c_2, \qquad c_{R1} = c_{R2}.$$

2. Suppose $\pi \in \Pi^P - \Pi^S$; Then

$$c_R^{min} \leq c_Q \leq c_P \leq c_R^{max}.$$

3. Suppose $\pi \in \Pi^E - \Pi^P$; Then

$$c_R^{min} < c_Q < c_2 < c_R^{max}.$$

Asymptotic behavior of the wavespeeds is as follows.

1. Suppose

$$\pi \in \Pi^F \cup \pi \to \partial\Pi^F, \qquad F = S,\ P,$$

i.e., the point π approaches the boundary of the corresponding existence domain; Then $c_F \to c_2$.

2. Suppose $c_{m1} \to c_{m2}$, $\mu \to 1$ or the mismatch parameters die out; Then

$$c_P,\ c_Q \to c_2, \quad \pi \in \Pi^P; \qquad c_S \to c_2, \quad \pi \in \Pi^S.$$

3. Suppose media 2 becomes rigid in the sense $c_{m1}/c_{m2} \to 0, \quad \mu \to 0$; Then

$$\pi \in \Pi^E - \Pi^P, \qquad c_{R1} < c_Q \to c_{21}.$$

6. Conclusion

In the current study, the general questions concerning the interface wave existence, embedding existence domains, and mutual relations of the associated wavespeeds as well as their restrictions and asymptotics, are resolved for homogeneous (transversely-)isotropic elastic bimaterials.

References

1. Simonov I.V. (1986) Propagation of cracks along the interface between two elastic media, *Thesis*, Institute for problems in mechanics AS, Moscow.
2. Berberian L.G. (1941) To question on wave propagation along the interface between two various media slipping one over another, *Proceedings AS Gruzinskoy SSR* **2**, 321–326.
3. Gogoladze V.G. (1947) Reflection and refraction of elastic waves. General theory of the surface Rayleigh waves, *Proceedings of the Institute of Seismology AS* **125**, 1–43.
4. Gol'dstein, R.V. (1967) On interface waves in joined elastic materials and their relation to the crack propagation along the interface, *PMM [J. Appl. Math. Mech.]* **31**, 468–475.
5. Grinchenko, V.T. and Meleshko V.V. (1981) *Harmonic vibrations and waves*, Naukova Dumka, Kiev.
6. Koppe H. (1948) Über Reyleigh-Wellen an der Oberfläche zweier Medien, *Z. angew. Math. Mech.* **28**, 355–360.
7. Owen T.E. (1964) Surface wave phenomena in ultrasonics, *Progr. Appl. Mater. Res.* **6**, 69–87.
8. Gol'dstein, R.V. (1966) On steady motion of a crack along the rectilinear interface between two dissimilar materials, *Inzhenerniy Zhurnal, MTT [J. Engng. Mech. Solids]* **5**, 93–102.

Part IV

Applications of surface waves to analysis of fracture and damage

THE NON-UNIQUENESS OF CONSTANT VELOCITY CRACK PROPAGATION

K. B. BROBERG
Department of Mathematical Physics,
University College Dublin, Belfield, Dublin 4, Ireland

Key words: nonlinear surface waves, elasticity, Rayleigh waves, scale invariance, nonlocality, spectral kernel, spatial kernel, Hilbert transform

1. Introduction

The theory of the mechanism for fast crack propagation has undergone several revisions during the latter half of the 20th century, and there are still unresolved questions, which seem to require more experimental work. Originally, the classical Griffith-Orowan-Irwin theory of fracture in a brittle material [Griffith 1920, Orowan 1952, Irwin 1957] was adopted, i.e., the energy dissipation per unit of crack growth and crack front length (the specific energy dissipation) was assumed to be a velocity independent material constant. However, this view predicted that a crack expanding in a sufficiently large plate, subjected to remote loading, would accelerate toward the Rayleigh velocity [Broberg, 1960], but although experiments showed that a crack accelerated to a constant terminal velocity, this was distinctly lower than the Rayleigh velocity (e.g. Schardin 1959).

The idea of a velocity independent part of the energy dissipation, such as the surface energy, had to be abandoned after Paxson and Lucas (1973) recorded a specific energy dissipation in PMMA that was about 50 times larger at the highest crack velocity measured than at slow crack growth. Such a high energy dissipation could hardly be attributed to increased plastic flow outside a process region of velocity-independent size. A considerable increase in the process region size had to be assumed [Broberg, 1979]. This would govern a corresponding size increase of the plastic region.

R.V. Goldstein and G.A. Maugin (eds.),
Surface Waves in Anisotropic and Laminated Bodies and Defects Detection, 279–288.

For explaining the increase of the process region size with crack velocity, the cell model of materials [Broberg, 1979] was developed. A cell may be viewed as the smallest material unit that contains reasonably sufficient information about the fracture behaviour of the material [Broberg 1996, 1999]. At a low crack velocity, the process region is essentially confined to a single central layer of cells, because cells that have reached the decohesive state impart unloading on offside cells, thereby preventing them to reach the decohesive state.

At high crack velocities, the information about unloading in the central layer arrives too late to prevent cells in adjacent layers to reach the decohesive state. The process region may therefore extend across several layers, and for each added layer, its size becomes less and less dependent on the cell size. This is effectively a loss of an intrinsic length parameter [Broberg, 1979]. For dimensional reasons, the presence of such a parameter is known to explain why a relation between static fracture toughness and crack length exists. For dynamic crack propagation, the effective absence of such a parameter would predict the absence of a relation between the dynamic fracture toughness and crack velocity.

With hindsight it is realized that the absence of a unique relation between the dynamic fracture toughness and crack velocity followed from experiments such as those by Kobayashi and Dally (1977), who showed increase of the stress intensity factor during constant velocity propagation. However, the most compelling results were given by Ravi-Chandar (1982). He found that the constant crack velocity was dependent on the way by which the load was applied (he used crack face loading of different magnitudes). Thus, the constant velocity, that had been obtained in a great number of experiments since the 1930s, could not be considered as a material constant, or as the maximum attainable velocity in the material.

It may be noted that constant crack velocity in mode I has been observed in laboratory experiments for two different geometrical configurations. One is an edge crack in a long strip subjected to fixed grip loading, e.g. [Paxson and Lucas, 1973]. Another one is a central crack in a large plate subjected to constant remote loading, including equivalent configurations, such as the one used by Ravi-Chandar (1982).

Constant velocity crack propagation in modes II or III has been observed either as earthquake slip or (at least nearly) in laboratory experiments; cf. e.g. [Kalthoff, 1990] and [Rosakis *et al.*, 1999]. To prevent kinking, these modes require a sufficiently high compressive load or weak planes in the body.

In addition, steady state constant velocity crack propagation may occur in mode I by wedging, either in a strip or in a large plate. However, it might be difficult to obtain the high velocities usually associated with dynamic crack propagation by wedge loading.

Steady state uni-directional slip in mode II, self-similar crack propagation in mode I and steady state crack propagation in a long strip will be considered here. Also, possible implications for constant velocity wedge loading will be discussed.

2. Steady state uni-directional Slip in Mode II

For very slow uni-directional slip under Coulomb friction along the plane $y = 0$ in a large body, subjected to a remote shear stress, τ_{xy}^{∞}, and a remote normal stress, σ_y^{∞}, the length $2a$ of the sliding region and the deposited slip Δ on each side are (e.g. Broberg 1999, page 347)

$$2a = \frac{K_{II}^2}{2\pi(\tau_{xy}^{\infty} - \tau_f)^2} \tag{2.1}$$

$$\Delta = \frac{K_{II}^2}{2(1-k^2)\mu(\tau_{xy}^{\infty} - \tau_f)} \tag{2.2}$$

where K_{II} is the stress intensity factor at the leading edge, μ is the modulus of rigidity, k is the ratio c_S/c_P between the S and P wave velocities, and $\tau_f = \mu_d \sigma_y^{\infty}$, where μ_d is the friction coefficient. Energy-neutral healing is assumed, i.e. no energy is released at the trailing edge, and it is also assumed that smooth slip can take place, i.e. that no stick-slip effect is present. Note that both crack length and deposited slip are uniquely determined.

If the crack is running with a velocity $V < c_R$, where c_R is the Rayleigh wave velocity, then

$$2a = \frac{[K_{II}(\beta)]^2}{2\pi[\tau_{xy}^{\infty} - \tau_f(\beta)]^2} \tag{2.3}$$

$$\Delta = \frac{[K_{II}(\beta)]^2 Y_{II}(\beta)}{2(1-k^2)\mu[\tau_{xy}^{\infty} - \tau_f(\beta)]} \tag{2.4}$$

where $\beta = V/c_P$ and

$$Y_{II}(\beta) = \frac{2k(1-k^2)\beta^2\sqrt{k^2-\beta^2}}{R(\beta)} \tag{2.5}$$

where $R(\beta) = 4k^3\sqrt{(1-\beta^2)(k^2-\beta^2)} - (2k^2-\beta^2)^2$ is the Rayleigh function.

Obviously, in the dynamic case, only a relation between crack length and velocity is obtained. Thus, the velocity has to be specified in advance in order to find the crack length, or *vice versa*. For each velocity, a unique crack length is found (but the opposite is not generally true). If K_{II} and

μ_d are independent of β, then any crack velocity (less than c_R) is possible, and the crack length is independent of the velocity.

3. Self-similar Propagation in Mode I

3.1. CONTINUUM VERSUS MICRO-STRUCTURAL SCALING

Compare geometrically similar three point bend specimens for fracture toughness testing with specimens without a crack but otherwise identical. The plastic collapse load for the specimens without crack scales in proportion to W^2, where W is the length of the beam. Assuming that the specimens with crack are large enough to ensure small scale yielding, the fracture collapse load scales in proportion to $W^{3/2}$. The difference, which may be perceived as the difference between continuum scaling and micro-structural scaling, depends on the fact that the size of the fracture process region is independent of the scale.

Similarly, during slow crack motion, the process region size stays approximately constant, and continuum scaling, which would imply that the process region size increases in proportion to the crack length, does not work. However, for a fast running crack, continuum scaling becomes again applicable, in spite of the micro-scopic events, such as formation and growth of micro-separations, in the process region. This is a consequence of the effective loss of an intrinsic length parameter, according to the cell model, i.e., the process region behaves as a continuum (Broberg 1979, 1999). Thus, an expanding crack in a large plate, subjected to remote loading, will obey continuum scaling laws, and self-similar motion will be approximated after constant crack velocity has been reached. In particular, the specific energy dissipation scales in proportion to the crack length. This kind of continuum scaling will be called (linear) *length scaling.*

3.2. LENGTH SCALING AND VELOCITY SCALING

It is known that the angular stress distribution in the elastic field near the edge of a running crack is approximately the same for all velocities below about half the Rayleigh wave velocity. This fact enables continuum scaling with respect to the crack velocity, (linear) *velocity scaling* as distinct from length scaling. Thus, with a moving coordinate system with the z axis along the crack edge, a point (x_1, y_1) near the edge of a crack travelling with velocity V_1 will experience the same stress-strain history as a point (x_2, y_2), $x_2 = V_2 x_1/V_1$, $y_2 = V_2 y_1/V_1$, near the edge of a crack travelling with velocity V_2. This is the "similarity argument" discussed in [Broberg, 1979]. In particular, the specific energy dissipation scales in proportion to the (constant) crack velocity.

3.3. MODE I CRACK EXPANDING WITH CONSTANT VELOCITY

Consider a central crack $|x| < a$ in a large plate subjected to a remote tensile stress, σ_y^∞. It is assumed that the constant velocity phase has been reached and that small scale yielding prevails. The velocity is assumed to be sufficiently high for continuum scaling to apply, i.e., self-similarity prevails due to the linear length scaling. If the velocity is not too high, then linear velocity scaling also prevails, for comparison between cases of crack expansions under different velocities. Thus, the specific energy dissipation,

$$\Gamma \propto Va \tag{3.6}$$

Now, cf. [Broberg, 1999], the energy flux into the dissipative region,

$$\mathcal{G} = \frac{\pi(\sigma_y^\infty)^2 a \cdot k^2 \sqrt{(1-\beta^2)^3}\, R(\beta)}{2\mu\beta^2[g_1(\beta)]^2} \tag{3.7}$$

where

$$\begin{aligned} g_1(\beta) = {} & [(1-4k^2)\beta^2 + 4k^4]\mathbf{K}\left(\sqrt{1-\beta^2}\right) \\ & - \beta^{-2}[\beta^4 - 4k^2(1+k^2)\beta^2 + 8k^4]\mathbf{E}\left(\sqrt{1-\beta^2}\right) \\ & - 4k^2(1-\beta^2)\mathbf{K}\left(\sqrt{1-\beta^2/k^2}\right) \\ & + 8k^4\beta^{-2}(1-\beta^2)\mathbf{E}\left(\sqrt{1-\beta^2/k^2}\right) \end{aligned}$$

where $\mathbf{K}$ and $\mathbf{E}$ are the complete elliptic integrals of the first and second kind. For $\beta \to 0$, the function $R(\beta) \to 2k^2(1-2k^2)\beta^2$ and $g_1(\beta) \to 2k^2(1-k^2)$.

Because $\mathcal{G} = \Gamma$, relations (3.7) and (3.6) combine to a relation between the constant crack velocity and the remote stress,

$$\sigma_y^\infty = \sigma_0 \cdot \frac{g_1(\beta)}{(1/\beta^2 - 1)^{3/4}\sqrt{R(\beta)}} \tag{3.8}$$

where σ_0 is undetermined. This relation, assuming, temporarily, that σ_0 is independent of β, is shown by the full-drawn curve in Figure 1.

Recall that only a certain velocity interval has been considered. Thus, the full-drawn curve in Fig. 1 is not reliable above about half the Rayleigh wave velocity. It is neither applicable nor meaningful for velocities below the region for continuum scaling. For these velocities, the "minimum energy argument" (Broberg 1979, 1999) is applicable, i.e. complete decohesion must occur in at least one layer of cells. Thus, rather than approaching

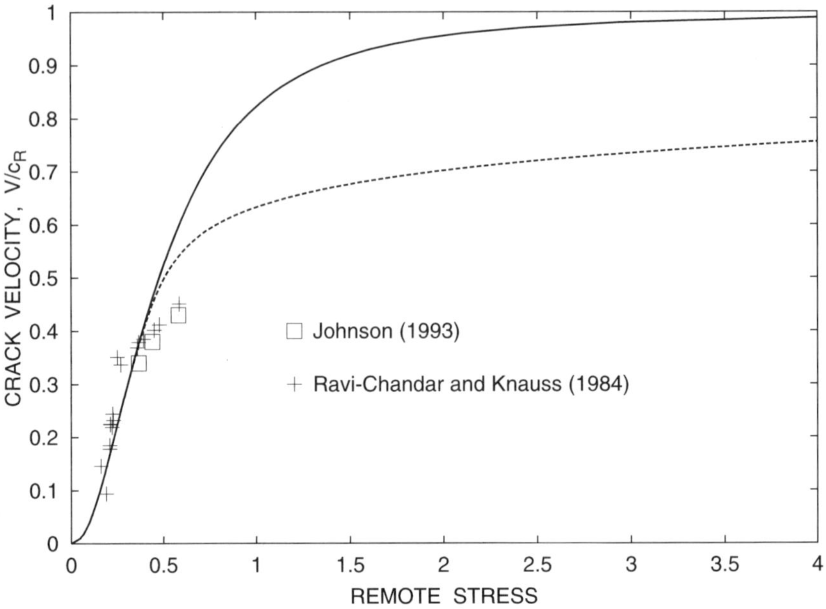

Figure 1. Edge velocity, V/c_R, of an expanding crack, after the constant velocity phase has been reached, versus remote load, σ_y^∞/σ_0, calculated under the assumption that the similarity argument is valid for all velocities (full-drawn curve). The dashed curve shows an attempt to consider the deviation from the similarity argument at high velocities.

zero for low velocities, as suggested by (3.6), the specific energy dissipation approaches a constant; cf. Figure 2, which indicates that continuum scaling requires a crack velocity above about a quarter of the Rayleigh wave velocity.

The dashed line in Figure 2, shows linear velocity scaling. The deviation from this line for high velocities appears to be due to dramatically changed angular stress distribution near the crack edge, particularly in the ratio σ_x/σ_y [Broberg, 1979]. It is not obvious how consideration of this change would modify the relation shown by the fulldrawn line in Figure 1, except that it would lower the crack velocity. Therefore, an attempt is made to use the experimental results for PMMA in Figure 2, by assuming the same ratio between the specific energy dissipation (fulldrawn curve) and the dissipation according to linear velocity scaling (dashed line) for the expanding crack. This results in the dashed curve in Figure 1.

Figure 1 also shows experimental data for Homalite-100 by Ravi-Chandar and Knauss (1984) and results from numerical simulations, using the cell model, by Johnson (1993). These data were obtained for crack face loading, but may be re-interpreted for remote loading. The undetermined stress σ_0,

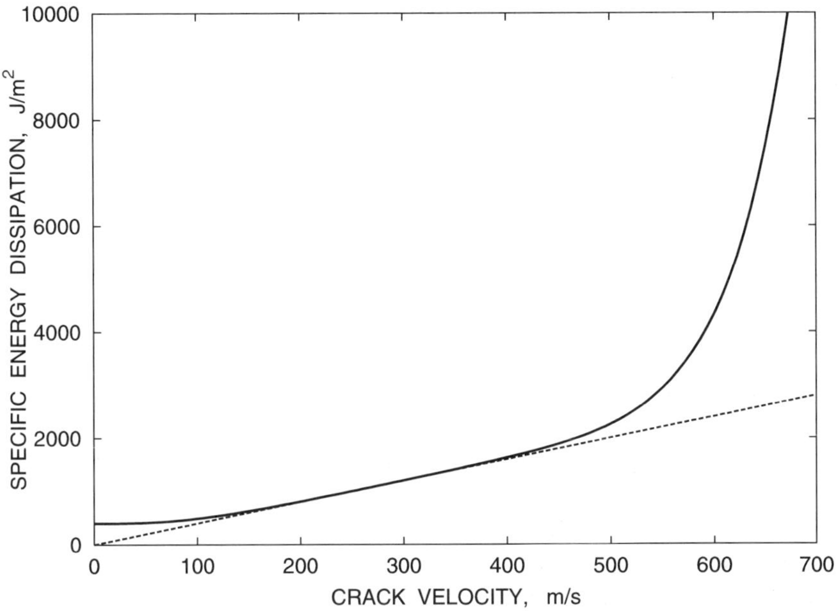

Figure 2. Specific energy dissipation as function of the crack velocity for the long strip configuration. The curve is based on results for mode I steady state crack propagation by Shioya and Zhou (1995) for PMMA, extrapolated toward zero velocity by assuming the specific energy dissipation during slow crack growth to be 400 J/m^2. The similarity argument approximation is shown by the dashed line. The Rayleigh velocity is probably slightly higher than 900 m/s.

assumed to be velocity independent, has been chosen to give a good fit in the velocity region where continuum scaling is applicable. Some remarkable numerical simulations by Abraham *et al.* (1998) may also be mentioned. They used atomistic molecular dynamics (which may be envisaged as a special case of the cell model) for crack expansion in crystalline silicon. Application of an extremely high load resulted in acceleration to a constant velocity of about $0.85c_R$ in a few picoseconds. This would correspond to a point far to the right of (or beyond) Figure 4 (but recall that the dashed curve was obtained for PMMA and it might not be representative for crystalline silicon).

Because small scale yielding is assumed, an obvious condition is that the applied remote load must be significantly smaller than the yield stress or, in a brittle solid, the cohesive stress. This implies that the curves in Figure 1 should be cut off beyond some maximum value of the remote stress, say around a third of the yield stress or the cohesive stress.

4. The Non-uniqueness for Steady State and Self-similarity

The reason why no unique solution was obtained for steady state uni-directional slip or for self-similar crack expansion is obviously associated with the fact that steady state and self-similarity were assumed *a priori*. But how were they introduced? The complete problem has to consider the introductory phase.

For the self-similar case, the introductory phase may simply consist of an acceleration from a starter crack. The cell model suggests that different acceleration history (e.g. due to different lengths of blunted starter cracks) would lead to different process region size at the same distance from the origin, after self-similarity has been approached. Relation (3.8) will still be valid, but the constant σ_0 is likely to be different for different acceleration histories. Obviously, this constant is then also different between cases with different remote load, except, perhaps, by the coincidence that the acceleration histories happen to produce the same σ_0. Such coincidence might, perhaps, occur approximately in laboratory tests with the same setup, but different load magnitudes, for instance those by Ravi-Chandar (1982).

The same non-uniqueness is also likely to characterize self-similar crack propagation in modes II and III. However, the relative ease by which velocities close to the Rayleigh wave velocity is obtained for mode II, cf. Rosakis *et al* (1999), indicates that the size increase of the process region with velocity is much less pronounced for the shearing modes. For mode II, even intersonic velocities can be reached.

The introductory phase for uni-directional slip is more complicated. Slip may start by expansion, i.e. with two leading edges travelling in opposite directions. If one of the edges is arrested at an obstacle, it will reverse direction, and uni-directional slip will result, cf. Johnson (1990). Quite obviously, such an introductory phase may produce widely different crack lengths and crack velocities when steady state is approached. The relations (2.3) and (2.4) will still be valid, but, by analogy with the self-similar case, it seems likely that the dynamic stress intensity factor K_{II} depends on the history during the introductory phase, as well as on the steady state velocity. Thus, not even the relation between crack length and velocity seems to be unique.

5. Other Steady State Cases

Two cases of steady state crack propagation other than uni-directional slip may be mentioned. One is the long strip configuration, in mode I, as used by several investigators, e.g. Paxson and Lucas (1973) and Shioya and Zhou (1995). In this case, the specific energy dissipation is accurately known, and it can be related directly to the crack velocity. However, there is inevitably an introductory phase. In view of the discussion about the self-similar case, it is likely that the specific energy dissipation is dependent on the history leading up to constant velocity propagation as well as upon the crack velocity. Different acceleration histories may be realized by triggering blunted starter cracks of different lengths. However, to the author's knowledge, no systematic investigation of this kind has been published.

Another case of steady state crack propagation consists of wedging of a crack in a long strip or in a large plate. It might be difficult to obtain constant velocity crack propagation for high velocities, and to prevent stick-slip effects or crack kinking. The theory for high velocity wedging in a large plate is given by Broberg (1975). The length of the crack in front of the wedge depends both on the wedge velocity and on the specific energy dissipation. Then, it is obvious that different histories of the introduction to constant velocity crack propagation (if such is possible) would be expected to lead to different relations between crack length and velocity, and therefore also between specific energy dissipation and velocity.

6. Conclusions

Experiments have shown that constant velocity appears in small scale yielding cases of approximate self-similar crack expansion. This fact contradicts theoretical evidence if the specific energy dissipation is assumed to be a velocity independent material constant. Moreover, the constant velocity obtained is not a material constant, but it depends on the remote load.

Theoretical analyses of constant velocity propagation under steady state or self-similarity fail to produce a unique solution, because the introductory phase is not taken into account. Consideration of this phase can only be done with full knowledge of the material behaviour, in particular that of the process region. A material model like the cell model strongly indicates a non-continuum behaviour for low velocities, gradually developing into a continuum behaviour. For crack propagation in simple atomic lattices, atomistic molecular dynamics appears to work well, but so far only for extremely small crack lengths.

A history dependence on the development of the process region during the introductory phase leading to constant velocity crack propagation is suggested by the cell model, but it is difficult to estimate how strong it is. It might be less strong in some materials than in others, and it might be less strong in the shearing modes than in the opening mode.

References

Abraham, F.F., Broughton, J.Q., Bernstein, N. and Kaxiras, E. (1998) Spanning the length scales in dynamic simulation. *Computers in Physics* **12**, 538-546.

Broberg, K.B. (1960) The propagation of a brittle crack. *Arkiv för Fysik* **18**, 159-192.

Broberg, K.B. (1975) On the theory of wedging. *Reports of the Tohoku Research Institute for Strength and Fracture of Materials*, Tohoku University, Vol. II, 1-27.

Broberg, K.B. (1979) On the behaviour of the process region at a fast running crack tip. In *High Velocity Deformation of Solids*, edited by K. Kawata and J. Shioiri, Springer-Verlag, Berlin Heidelberg, 182-194.

Broberg, K.B. (1996) The cell model of materials. *Computational Mechanics* **19**, 447-452.

Broberg, K.B. (1999) *Cracks and Fracture*, Academic Press, London.

Griffith, A.A. (1920) The phenomena of rupture and flow in solids, *Phil. Trans. Roy. Soc. (London)* **A221**, 163-198.

Irwin, G.R. (1957) Analysis of stresses and strains near the end of a crack traversing a plate. *J. Appl. Mech.* **24**, 361-364.

Johnson, E. (1990) On the initiation of unidirectional slip. *Geophys. J. Int.* **101**, 125-132.

Johnson, E. (1993) Process region influence on energy release rate and crack tip velocity during rapid crack propagation. *Int. J. Fract.* **61**, 183-187.

Kalthoff, J.F. (1990) Transition in the failure behaviour of dynamically shear loaded cracks. *Appl. Mech. Rev.* **43**, S247-S250.

Kobayashi, T. and Dally, J.W. (1977) Relation between crack velocity and the stress intensity factor in birefringent polymers. In *Fast Fracture and Crack Arrest, ASTM STP 627*, edited by G.T. Hahn and M.F. Kanninen. American Society for Testing and Materials, 257-273.

Orowan, E. (1952) Fundamentals of brittle behaviour in metals. In *Fatigue and Fracture of Metals*, edited by W.M. Murray. John Wiley, New York, 139-154.

Paxson, T.L. and Lucas, R.A. (1973) An experimental investigation of the velocity characteristics of a fixed boundary fracture model. In *Proceedings of an International Conference on Dynamic Crack Propagation*, edited by G.C. Sih. Noordhoff International Publishing, Leyden, 415-426.

Ravi-Chandar, K. (1982) An experimental investigation into the mechanics of dynamic fracture. *Ph.D. Thesis* California Institute of Technology, Pasadena, California

Ravi-Chandar, K. and Knauss, W.G. (1984) An experimental investigation into dynamic fracture: III On steady-state crack propagation and crack branching, *Int. J. Fract.* **26**, 141-154.

Rosakis, A.J., Samudrala, O. and Coker, D. (1999) Cracks faster than the shear wave speed. *Science* **284**, 1337-1340.

Schardin, H. (1959) Velocity effects in fracture. In *Fracture*, edited by B.L. Averbach, D.K. Felbeck, G.T. Hahn and D.A. Thomas, John Wiley & Sons, New York, 297-329.

Shioya, T. and Zhou, F. (1995) Dynamic fracture toughness and crack propagation in brittle material. In *Constitutive Relation in High/Very High Strain Rates*, edited by K. Kawata and J. Shioiri, Springer-Verlag, Tokyo, 105-112.

EMBEDDING FORMULAE FOR PLANAR CRACKS

R. V. CRASTER
Department of Mathematics, Imperial College of Science, Technology and Medicine, London SW7 2BZ, U.K.

A. V. SHANIN
Department of Physics (Acoustics Division), Moscow State University, 119992, Leninskie Gory, Moscow, Russia

Abstract.

Embedding formulae allow one to decompose scattering problems apparently dependant upon several angular variables (angles of incidence and observation) into those dependant upon fewer angular variables. In terms of facilitating rapid computations across considerable parameter regimes this is a considerable advantage. In this short article we concentrate on embedding formulae for a typical problem from acoustics in three dimensions.

Key words: Embedding, integral equations, acoustics, asymptotics, ray theory

1. Introduction

In three dimensions, the solution of a diffraction problem is usually represented as a function of four angular variables: two of them specify the direction of the wave vector of the incident plane wave illuminating the obstacle, and the other two are the direction of the scattered wave. The far-field diffraction pattern is a function of these directions. For numerical work, it can be time consuming to perform a parametric study - all angular variables must be independently varied and the numerical routine rerun for each value.

Fortunately, for many practically important cases there exists an elegant mathematical theory, unfortunately it is little known and not often utilized, that enables one to reduce the dimension of the problem. The essence of this theory is the following: instead of directly solving the main diffraction problem with the desired plane-wave incidence, a set of different auxiliary problems are solved. For example, if the obstacle is a planar crack in the

R.V. Goldstein and G.A. Maugin (eds.),
Surface Waves in Anisotropic and Laminated Bodies and Defects Detection, 289–299.

medium, the auxiliary problems are associated with the excitation of the field by a point source located asymptotically close to the edge of the crack. We could also interpret these auxiliary solutions as unphysically singular eigensolutions of the problem, in the sense that they no longer have the usual local square root dependence of the acoustic potential at the edge (i.e. having the dependence of the form $\sim r^{1/2}$) but are square root singular there (i.e. have the form $\sim r^{-1/2}$).

The solution of the auxiliary diffraction problem (in 3D) depends on only three variables: the position of the crack edge, where the source is located and the two angles which determine the direction of the scattering. The solution of the original diffraction problem is represented as the integral of the solutions of the auxiliary problems. Such a representation is called an embedding formula.

Embedding formulae have been derived by several previous authors for various diffraction problems, these have used a different set of auxiliary problems, or have used theories based explicitly upon integral equations beginning with [Williams, 1980], and this triggered further applications to cracks in elastic solids, [Martin and Wickham, 1983], and recently the method has been embraced by [Biggs et al., 2000; Biggs and Porter, 2001; Biggs and Porter, 2002]. However, except for the article by [Williams, 1980] who uses grazing incidence to generate the auxiliary solutions, the derivation of embedding formulae is typically through complicated manipulations of integral equations which obscures the final structure of the formula.

One purpose of this article is to demonstrate an easy way to derive embedding formulae, which has a physical interpretation and can be easily implemented, that is, we use a set of auxiliary solutions that have immediate interpretations. Here we shall consider incident fields that consist of plane waves and this is important for the success of the embedding technique as we utilize this in an operator we apply.

2. Formulation

We suppose that the Helmholtz equation

$$\nabla^2\phi + k_0^2\phi = 0 \tag{2.1}$$

holds in 3D where Cartesian coordinates (x, y, z) are utilized, so $\phi(x, y, z)$ and the crack occupies area S in the (x, y) plane.

For definiteness we take the Neumann boundary condition $\partial_z\phi = 0$ on the faces of the planar defect/crack; the approach remains valid for Dirichlet or in electromagnetic theory for impedance boundary conditions.

The total field ϕ is the sum of an incident field $\phi^{(\mathrm{in})}$ and a scattered field $\phi^{(\mathrm{sc})}$. The incident field is assumed to be a plane wave

$$\phi^{(\mathrm{in})} = \exp[-i(\mathbf{k}^{(\mathrm{in})} \cdot \mathbf{x} + \sqrt{k_0^2 - |\mathbf{k}^{(\mathrm{in})}|^2} z)] \tag{2.2}$$

where $\mathbf{k}^{(\mathrm{in})} = (k_x^{(\mathrm{in})}, k_y^{(\mathrm{in})})$ and $\mathbf{x} = (x, y)$.

We also require for physically meaningful solutions that suitable edge conditions (Meixner's) are fulfilled, which for our planar problem means that the field near the edge of the crack has the asymptotic behaviour that

$$\phi \sim K r^{1/2} \cos \varphi/2, \tag{2.3}$$

where r is the distance between the observation point and the edge of the crack/defect; φ is the angle in the local cylindrical coordinates taken such that φ lies along the crack face on $z = 0_+$.

We utilize uniqueness, that is, we consider only the scattered field, i.e. $\phi = \phi^{(\mathrm{sc})}$ and assume that the Helmholtz equation, boundary, radiation and edge conditions be fulfilled. Then $\phi = 0$ identically. We assume that the theorem of uniqueness is satisfied by all diffraction problems considered here.

Using the spectral language, we assume that the parameter k_0 of equation (2.1) does not belong to the spectrum of the problem, that is, we cannot have any trapped modes. Note that the method proposed below is applicable even if k_0 belongs to the spectrum and has finite degeneration. However, in this case the method should be modified.

2.1. AUXILIARY SOLUTIONS OF THE DIFFRACTION PROBLEMS

We require the solutions of auxiliary problems, namely diffraction problems with point source incidence (or a line source for a 2D problem). The scatterer is assumed to have the same geometry and (homogeneous) boundary conditions as the scatterer of the initial diffraction problem, and the source is located near the edge of the crack. For our present purpose one cannot simply place the source near the edge of the crack, since the Neumann condition is fulfilled on the crack faces, and we still assume the physically meaningful, Meixner's condition, is fulfilled at the edge. It is also assumed that the radiation condition at infinity holds.

We now consider a limiting procedure, that is, we quantify how near the source is to the edge, in terms of which the auxiliary functions will be treated. We introduce a coordinate l along the edge of the crack, and take a point lying in the (x, y) plane a small distance, ϵ, from a position $l_0 = (x_0, y_0, 0)$ on the contour Γ (see Fig. 1). We consider a diffraction problem with a pair of point sources, strength $-\pi\epsilon^{-1/2}/2$, above and below

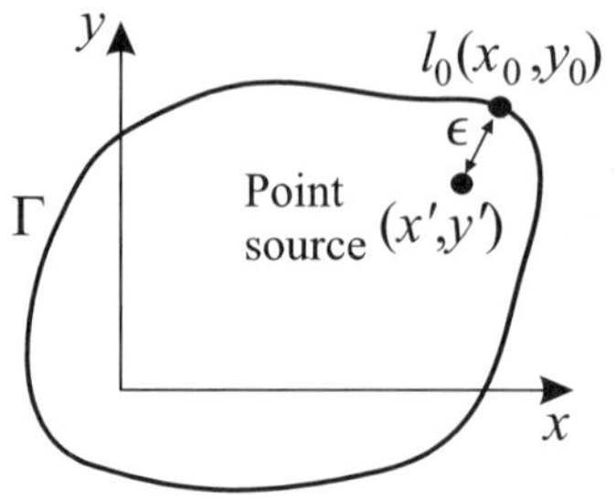

Figure 1. Location of a point source

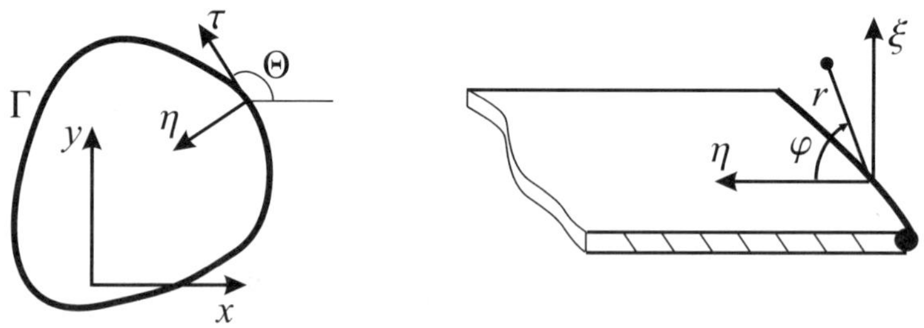

Figure 2. To the geometry of the problem

the crack; we solve the inhomogeneous Helmholtz equation for the function $\hat{\phi}_\epsilon(x, y, z; l_0)$:

$$\nabla^2 \hat{\phi}_\epsilon + k_0^2 \hat{\phi}_\epsilon = -\frac{1}{2}\pi\epsilon^{-1/2}\delta(x - x')\delta(y - y')\delta(z - 0) +$$

$$\frac{1}{2}\pi\epsilon^{-1/2}\delta(x - x')\delta(y - y')\delta(z + 0), \tag{2.4}$$

where δ is the delta-function,

$$x' = x_0 - \epsilon \sin\Theta, \qquad y' = y_0 + \epsilon \cos\Theta.$$

Here Θ is the angle between the vector dl tangential to Γ and the x-axis (see Fig. 2).

A detailed study shows that for each point (x, y, z), with the exception of the point l_0 on Γ, there exists a finite limit

$$\hat{\phi}(x, y, z; l_0) = \lim_{\epsilon \to 0} \hat{\phi}_\epsilon(x, y, z; l_0). \tag{2.5}$$

The function $\hat{\phi}(x, y, z; l)$ is the auxiliary solution.

The auxiliary problem has one important property: it depends on fewer variables than the physical diffraction problem. The function $\phi^{(\mathrm{sc})}$ depends explicitly on 3 variables (the spatial coordinates) and implicitly on 2 variables: the parameters k_x and k_y of the incident wave, i.e., the total number of variables is 5. The number of the arguments of $\hat{\phi}$ is 4. We assume that the function $\hat{\phi}(x, y, z; l)$ is known and express the solution of the initial diffraction problem, with plane-wave incidence, in terms of this function.

Below we shall use the asymptotics of the auxiliary solution at the edge. Consider the integral

$$\phi^*(x, y, z) = \int_\Gamma \rho(l)\hat{\phi}_\epsilon(x, y, z; l)dl, \tag{2.6}$$

where the sources of the field are concentrated along the contour Γ and have line density $\rho(l)$. We assume this density to be a continuous function having period equal to the length of the contour Γ. One can show that for fixed r

$$\lim_{\epsilon \to 0} \phi \sim \frac{\rho(l)\cos(\varphi/2)}{r^{1/2}} + O(r^{1/2}). \tag{2.7}$$

One can see that the source near the edge of the crack leads to a field with edge asymptotics stronger, than is allowed by the usual Meixner's conditions. This property will be used below.

We see that there are two equivalent ways to introduce the auxiliary solution. The first one is to introduce a point source near the the edge and the corresponding limiting procedure. The other is to formally introduce a solution having the edge asymptotics that is stronger than it is allowed by the edge conditions. However, the second way is a bit cumbersome in the 3D case, where it is necessary to provide the oversingular behaviour at a single point of the edge.

2.2. DIRECTIVITY OF THE FIELD

In the far field zone the leading term of the scattered field can be written as the modulated spherical wave:

$$\phi^{(sc)}(x, y, z) \sim -\frac{e^{ik_0 R}}{2\pi R} D(\theta_x, \theta_y; \theta_x^{(in)}, \theta_y^{(in)}), \tag{2.8}$$

where $R = \sqrt{x^2 + y^2 + y^2}$; $\theta_x = \arccos(x/R)$; $\theta_y = \arccos(y/R)$; $\theta_x^{(in)} = \arccos(k_x^{(in)}/k_0)$; $\theta_y^{(in)} = \arccos(k_y^{(in)}/k_0)$; D is the directivity of the field. Utilizing the Green's formula, one can express the directivity as the Fourier-transform of normal derivative of the scattered field on the crack:

$$D(\theta_x, \theta_y; \theta_x^{(in)}, \theta_y^{(in)}) = ik_0\sqrt{1 - (\cos^2\theta_x + \cos^2\theta_y)} \times$$

$$\iint_S \phi^{(\mathrm{sc})}(x, y, +0)\, e^{-\mathrm{i}k_0(x\cos\theta_x + y\cos\theta_y)} dx\, dy. \tag{2.9}$$

Analogously, the auxiliary solution can also be represented using its directivity:

$$\hat{\phi}^{(\mathrm{sc})}(x, y, z; l) \sim -\frac{e^{\mathrm{i}k_0 R}}{2\pi R} \hat{D}(\theta_x, \theta_y; l), \tag{2.10}$$

and the directivity can be calculated as

$$\hat{D}(\theta_x, \theta_y; l) = \mathrm{i}k_0 \sqrt{1 - (\cos^2\theta_x + \cos^2\theta_y)} \times$$

$$\iint_S \hat{\phi}^{(\mathrm{sc})}(x, y, +0; l)\, e^{-\mathrm{i}k_0(x\cos\theta_x + y\cos\theta_y)} dx\, dy. \tag{2.11}$$

The embedding formula, which will be derived below, expresses the function $D(\theta_x, \theta_y; \theta_x^{(\mathrm{in})}, \theta_y^{(\mathrm{in})})$ in terms of $\hat{D}(\theta_x, \theta_y; l)$.

2.3. DERIVATION OF THE EMBEDDING FORMULA

We are going to derive the embedding formula in three steps: beginning with applying operators to the total field, followed by an application of the uniqueness theorem and the reciprocity principle.

Consider

$$\mathbf{H} = (H_x, H_y) = [\nabla + \mathrm{i}\mathbf{k}^{(\mathrm{in})}]\phi = (\partial_x + \mathrm{i}k_x^{(\mathrm{in})}, \partial_y + \mathrm{i}k_y^{(\mathrm{in})}) \tag{2.12}$$

Apply one of this operators (say, H_x) to the total field ϕ related to the initial diffraction problem with a plane wave incidence. The function

$$\overline{\phi}(x, y, z) = H_x[\phi(x, y, z)]$$

has the following properties: it satisfies the Helmholtz equation (2.1), contains no incoming waves from infinity or growth at infinity (note that $H_x[\phi^{in}] \equiv 0$), and $\overline{\phi} = 0$ on the crack surfaces. The conditions of the uniqueness theorem are satisfied, except for the edge condition. If the local asymptotics of the field, ϕ, near the edge are

$$\phi \sim K(l) r^{\frac{1}{2}} \cos\left(\frac{\varphi}{2}\right) + O(r^{\frac{3}{2}}),$$

then

$$\overline{\phi} \sim -\frac{1}{2} K(l) r^{-\frac{1}{2}} \sin\Theta \cos\left(\frac{\varphi}{2}\right) + O(r^{\frac{3}{2}}), \tag{2.13}$$

where θ is the angle between the x-axis and the unit vector dl tangential to the contour Γ. That is, $\overline{\phi}$ has an overly singular behaviour at the edge.

Comparing the edge asymptotics of the function $\overline{\phi}$ in (2.13) and the integral of the auxiliary functions (2.7), one finds that the combination

$$w(x,y,z) = \overline{\phi}(x,y,z) + \frac{1}{2}\int_{\Gamma} K(l)\sin\Theta(l)\,\hat{\phi}(x,y,z;l)dl \qquad (2.14)$$

obeys the usual Meixner's condition at the edge. Furthermore this function obeys the Helmholtz equation, the radiation condition and the Dirichlet boundary condition. Therefore, we apply uniqueness to this combination and thus $w(x,y,z) \equiv 0$, and

$$H_x[\phi] = -\frac{1}{2}\int_{\Gamma} K(l)\sin\Theta(l)\,\hat{\phi}(x,y,z;l)dl. \qquad (2.15)$$

This is a *weak form* of the embedding formula.

The function $K(l)$ in (2.15) remains unknown, to generate the complete embedding formula we must express $K(l)$ in terms of $\hat{\phi}(x,y,z;l)$. Instead of having an incident plane wave let us take a point source of the unit strength located at a point (X,Y,Z), such that

$$X = R\frac{k_x}{k_0}, \qquad Y = R\frac{k_y}{k_0}, \qquad Z = \sqrt{R^2 - X^2 - Y^2}$$

and the lengthscale R is much greater than both the size of the scattering region and the wavelength (being more accurate, we assume that the point (X,Y,Z) is located in the far field zone). The incident field from the source is asymptotically a plane wave having the form (2.2) multiplied by the factor $-(4\pi R)^{-1}e^{ik_0R}$. To find $K(l)$ we take the observation point, in the (x,y) plane, to be at a small distance ϵ from the point l on the edge contour Γ. We multiply the value of the field at the observation point by $\epsilon^{-1/2}$ and take the simultaneous limits that $R \to \infty$ and $\epsilon \to 0$. The result is $K(l)$ from the formula (2.15) multiplied by $-(4\pi R)^{-1}e^{ik_0R}$.

We now use the reciprocity principle [Junger and Feit, 1986] and interchange the source and observation point in the limit procedure described above, that is, the source is now near the edge, and the observation point is at (X,Y,Z). From the Helmholtz reciprocity principle, the value of the field for this interchanged problem is the same as that of the original problem. The diffraction problem with the point source located near the edge is the auxiliary problem, the solution under the appropriate limit is $\hat{\phi}(x,y,z;l)$. Hence

$$K(l) = 4\lim_{R\to\infty}[Re^{-ik_0R}\hat{\phi}(X,Y,Z;l)]. \qquad (2.16)$$

Using the formula (2.10), we obtain that

$$K(l) = -\frac{2}{\pi}\hat{D}(\theta_x^{(\text{in})}, \theta_y^{(\text{in})}; l), \tag{2.17}$$

That is, the edge behaviour of the physical problem is represented in terms of the far field of the auxiliary solution.

Next we substitute the relation (2.17) into the embedding formula (2.15), differentiate with respect to z and perform the Fourier transformation in the (x, y)-plane. The result is

$$D(\theta_x, \theta_y; \theta_x^{(\text{in})}, \theta_y^{(\text{in})}) =$$

$$-\frac{\mathrm{i}}{\pi k_0 \left(\cos\theta_x + \cos\theta_x^{(\text{in})}\right)} \int_\Gamma \hat{D}(\theta_x, \theta_y; l)\hat{D}(\theta_x^{(\text{in})}, \theta_y^{(\text{in})}; l) \sin\Theta(l) dl. \tag{2.18}$$

It is interesting to note that another embedding formula emerges by applying the operator H_y and repeating the arguments above:

$$D(\theta_x, \theta_y; \theta_x^{(\text{in})}, \theta_y^{(\text{in})}) =$$

$$\frac{\mathrm{i}}{\pi k_0 \left(\cos\theta_x + \cos\theta_x^{(\text{in})}\right)} \int_\Gamma \hat{D}(\theta_x, \theta_y; l)\hat{D}(\theta_x^{(\text{in})}, \theta_y^{(\text{in})}; l) \cos\Theta(l) dl. \tag{2.19}$$

Note that the arguments remain the same when the boundary conditions on the crack are chosen to be the Dirichlet or impedance ones.

2.4. HIGH FREQUENCY ASYMPTOTICS

One is not limited to dealing with exact or numerical solutions to the eigenstates, it is perfectly viable to adopt an asymptotic approach for, say, high frequencies and utilize this within the embedding framework. So although the embedding formulae themselves are valid for arbitrary ratios of wavelength to the size of the scatterer, it is interesting to construct the short wave/ high frequency approximation and apply it to the embedding formulae. For high frequencies an explicit approximation for the auxiliary function/ eigenstate $\hat{\phi}$ is easy to find and, thus, to write down a complete approximate solution for the diffraction problem. By eigenstate we mean the situation where we take the source to lie precisely at the crack edge.

The embedding formula is in the same form as that arising through the geometric theory of diffraction Keller (1962), Achenbach *et al* (1982); this therefore provides another mathematical route to these asymptotic

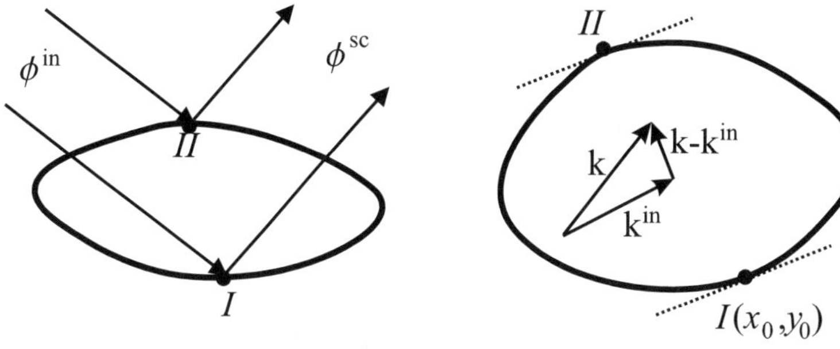

Figure 3. The diffracted rays

solutions, and provides justification for their efficiency and accuracy even at mid to low frequencies when they might be supposed to be poor [Keller, 1962; Achenbach et al., 1982].

The calculation of the function $\hat{D}(\theta_x, \theta_y; l)$ is a very complicated problem. If the wavelength is much smaller than the characteristic size of the scatterer, the crack near the edge is locally a half-plane, and the remote parts of the crack do not play an important role in diffraction. So, it is natural to approximate $\hat{D}$ by the corresponding function for a half-plane crack. Using the exact solution of the half-plane problem with point source incidence,

$$\hat{D}(\theta_x, \theta_y; l) \approx -\sqrt{-\pi \mathrm{i}} \left(\sqrt{k_0^2 - k_\tau^2} - k_\eta \right)^{1/2} e^{\mathrm{i}(k_x x_0 + k_y y_0)} \tag{2.20}$$

where (x_0, y_0) are the coordinates of the point of the edge, k_η and k_τ are the projections of the wavenumber k on the directions normal to Γ and tangential to it, respectively. These values can be calculated using the relations

$$\begin{aligned} k_\eta &= -k_0 \cos\theta_x \sin\Theta + k_0 \cos\theta_y \cos\Theta, \\ k_\tau &= k_0 \cos\theta_x \cos\Theta + k_0 \cos\theta_y \sin\Theta. \end{aligned} \tag{2.21}$$

Substituting the function (2.20) into the embedding formula (2.18) and consider the exponential factor, the integrand oscillates rapidly everywhere except at stationary points of Γ, that is, where the vector dl is orthogonal to the difference of the vectors $\mathbf{k} = (k_0 \cos\theta_x, k_0 \cos\theta_y)$ and $\mathbf{k}^{(\mathrm{in})} = -(k_0 \cos\theta_x^{(\mathrm{in})}, k_0 \cos\theta_y^{(\mathrm{in})})$. These stationary points provide the main terms of the asymptotics of the field.

If we consider the case $\mathbf{k} \neq \mathbf{k}^{(\mathrm{in})}$. There are two stationary points, I and II, at which Γ is orthogonal to $\mathbf{k} - \mathbf{k}^{(\mathrm{in})}$. (there are 2 such points, namely, the points I and II in Fig. 3). At each point we (first, say, for the point I) use the local coordinates η and τ, and calculate the components of the vectors (k_τ, k_η) and $(k_\tau^{(\mathrm{in})}, k_\eta^{(\mathrm{in})})$ using the transformation formulae (2.21). Note that $k_\tau = k_\tau^{(\mathrm{in})}$.

Using the method of stationary phase we obtain

$$D_I \approx D_I^{\mathrm{e}} \times D_I^{\mathrm{a}} \times D_I^{\mathrm{c}}, \tag{2.22}$$

where D_I^{e}, D_I^{a} and D_I^{c} are the exponential, angular and curvature factors, respectively:

$$\begin{aligned}
D_I^{\mathrm{e}} &= \exp\{-\mathrm{i}k_0[x_0(\cos\theta_x + \cos\theta_x^{(\mathrm{in})}) + y_0(\cos\theta_y + \cos\theta_y^{(\mathrm{in})})]\}, \\
D_I^{\mathrm{a}} &= \frac{(\sqrt{k_0^2 - k_\tau^2} + k_\eta^{(\mathrm{in})})^{1/2}(\sqrt{k_0^2 - k_\tau^2} - k_\eta)^{1/2}}{k_\eta - k_\eta^{(\mathrm{in})}}, \\
D_I^{\mathrm{c}} &= \left(\frac{\pi\mathrm{i}}{(k_\eta - k_\eta^{(\mathrm{in})})d\Theta/dl}\right)^{1/2}.
\end{aligned}$$

Analogously, the term corresponding to the stationary point II (and all other stationary points, if there are any others) should be estimated, and the sum over all of them should be taken.

One can see that the expression (2.22) has the structure peculiar to the classical ray asymptotics of the Geometrical Theory of Diffraction [Keller, 1962].

3. Concluding remarks

Overly singular eigenstates/ auxiliary functions are clearly a useful device for extracting directivities using embedding and allows for a physical interpretation in terms of line source incidence. We have demonstrated that embedding is related to high frequency asymptotic techniques. The approach is also useful in combination with Wiener-Hopf techniques and embedding is clearly applicable to elasticity and surface waves, [Craster and Shanin, 2002]. Thus embedding should become a method of choice when solving integral equations in diffraction theory.

Acknowledgements:

The work of RVC was supported by the EPSRC under grant number GR/R32031/01. One of us, AVS, is grateful to Dr. Larissa Fradkin for helpful conversations. The work of AVS was supported by RFBR grants NN 00-15-96530, 01-02-06119 and the program "Universities of Russia".

References

Achenbach, J. D., A. K. Gautesen, and H. McMaken: 1982, *Ray methods for waves in elastic solids.* Pitman.

Biggs, N. R. T. and D. Porter: 2001, 'Wave diffraction through a perforated barrier of non-zero thickness'. *Q. Jl. Mech. appl. Math.* **54**, 523–547.

Biggs, N. R. T. and D. Porter: 2002, 'Wave scattering by a perforated duct'. *Q. Jl. Mech. Appl. Math.* **55**, 249–272.

Biggs, N. R. T., D. Porter, and D. S. G. Stirling: 2000, 'Wave diffraction through a perforated breakwater'. *Q. Jl. Mech. appl. Math.* **53**, 375–391.

Craster, R. V. and A. V. Shanin: 2002, 'Embedding formulae in diffraction theory'. Under consideration.

Junger, M. C. and D. Feit: 1986, *Sound, structures and their interaction.* Acoustical Society of America. second edition.

Keller, J. B.: 1962, 'The geometric theory of diffraction'. *J. Opt. Soc. Amer.* **52**, 116–130.

Martin, P. A. and G. R. Wickham: 1983, 'Diffraction of elastic waves by a penny-shaped crack: analytical and numerical results'. *Proc. R. Soc. Lond. A* **390**, 91–129.

Williams, M. H.: 1980, 'Diffraction by a finite strip'. *Q. Jl. Mech. appl. Math.* **35**, 103–124.

WAVE PROPOGATION AND CRACK DETECTION IN LAYERED STRUCTURES

E. V. GLUSHKOV, N. V. GLUSHKOVA and A. V. EKHLAKOV
Kuban State University
P.O.Box 4102, Krasnodar, 350080, Russia

Abstract. The elastodynamics integral equation method holds an intermediate position between the asymptotic ray approach and direct numerical computation. The paper describes a low-cost implementation of the method in application to the ultrasonic crack detection modeling.

Key words: Embedding, integral equations, acoustics, asymptotics, ray theory

Introduction

The data processing of ultrasonic non-destructive testing is traditionally based on ray methods of general diffraction theory (GDT) [1, 2]. In view of asymptotic nature of the ray approach, it is used in the high-frequency band, when the wavelength of the probing signal is much less than the characteristic dimension of the defect. On the other hand, if the dimensions of the defect are comparable with or less than the wavelength, reliable mathematical models become particularly important, since the reflection in this case gives a very blurred image, which requires special processing to size and shape the defect.

In this case solution of wave problems can be obtained by direct numerical methods like FEM, BEM or Finite Differences, however their application encounters with the well-known difficulties. First of all, these methods cannot be applied to infinite domains (open waveguides). They have to be restricted by introducing artificial absorbing boundary conditions or by using specially constructed infinite elements. Then, the numerical solution does not provide a fine mode structure of the wave field that often is of prime interest for customers. And lastly, these methods are time-consuming, especially if distant points are of interest.

R.V. Goldstein and G.A. Maugin (eds.),
Surface Waves in Anisotropic and Laminated Bodies and Defects Detection, 301–315.

Quite the contrary, semi-analytical wave field representations in terms of boundary integrals [3, 4] are applicable to infinite domains, make it possible to analyze the wave structure (to extract out terms for bulk, surface and channel modes), and generally they yield low-cost codes and efficient analytical asymptotics. These asymptotics are of the same physically descriptive form like the obtained with the ray technique ones, but, in addition, they carry exhaustive information about sources, scatterers and structures.

Thus, the integral approach holds an intermediate position between the ray and direct numerical methods. On the one hand, it provides theoretically exact solution of elastodynamic boundary value problems, on the other, its asymptotics are appropriate for GDT. Therefore, the integral equation methods give the chance to discover and explore fine wave phenomena escaping usually with other approaches. It allowed us, in particular, to investigate the structure of averaged time-harmonic energy flows in layered and stepped waveguides with energy vortices and backward fluxes [5, 6], as well as to clear up their role in resonance extinguishing of surface waves [7, 8].

In this paper we present some results related to sounding of layered structures and to ultrasonic detection of arbitrarily shaped and oriented cracks. Typically, there are two problems of self-dependent interest: 1) calculation of an incident field $\mathbf{u}_0$, excited by a given source, and 2) computation of the scattered field $\mathbf{u}_1$, diffracted by some obstacles (cracks, inclusions, rough surface etc.).

1. The Probe Field

1.1. INTEGRAL REPRESENTATION AND GREEN'S MATRIX

Let us consider an elastic N-layered isotropic half-space occupying volume $-\infty < x, y < \infty$, $-\infty < z \leq 0$ in the Cartesian coordinates $\mathbf{x} = \{x, y, z\}$. Elastic homogeneous layers $S_j : -\infty \leq x, y \leq \infty$, $z_{j+1} \leq z \leq z_j$, $j = 1, 2, ..., N$, $z_1 = 0$, $z_{N+1} = -\infty$, with different material properties are perfectly bonded, i.e. the displacement and stress fields $\mathbf{u} = \{u_x, u_y, u_z\}$ and $\boldsymbol{\tau} = \{\tau_{xz}, \tau_{yz}, \sigma_{zz}\}$ are continuous in the whole volume including the boundary planes $z = z_j$, $j = 2, ..., N$. The half-space may contain an arbitrarily shaped planar crack modeled by an infinitesimally thin material discontinuity along a plane domain Ω with a stress-free inner sides.

The action of a probe (of an ultrasonic transducer) upon this structure is modeled by a given surface load applied to a contact domain D:

$$\boldsymbol{\tau}|_{z=0} = \begin{cases} \mathbf{q}_0(x, y, t), & (x, y) \in D \\ 0, & (x, y) \notin D \end{cases} \tag{1.1}$$

Depending on the type of the transducer, the contact area can be of different form (elliptic, rectangle or even disconnected: $D = \cup_m D_m$ for a system of sources). Function $\mathbf{q}_0$ sets the load distribution in D depending on the source characteristics (longitudinal or transverse, directional, inclined, etc.) [9]. To model a realistic probe, $\mathbf{q}_0$ ought to be chosen in accordance with the law of traction distribution in the interface between the transducer and the tested material when it is caused by an incident plane wave coming from an electrically excited bevelled edge of transducer's piezo-crystal. This law is easily derived if the effect of D finiteness is neglected [9, 10]. Otherwise, $\mathbf{q}_0$ is defined via solution of the Wiener-Hopf type integral equation, to which the contact problem is reduced [4]. Furthermore, we consider $\mathbf{q}_0$ as a known function.

Since any transient pulse $\mathbf{u}(\mathbf{x}, t)$ can be expressed as a linear superposition of the harmonic solutions $\mathbf{u}(\mathbf{x}, \omega)e^{-i\omega t}$:

$$\mathbf{u}(\mathbf{x}, t) = \frac{1}{\pi}\int_0^\infty \mathrm{Re}\left[\mathbf{u}(\mathbf{x}, \omega)e^{-i\omega t}\right]d\omega \tag{1.2}$$

we start from the harmonic steady-state problem with a circle frequency ω, omitting further the harmonic factor $e^{-i\omega t}$.

Due to the linearity of the problem the total field $\mathbf{u}$ in the structure under consideration is combined from a probe field $\mathbf{u}_0$ and a scattered one $\mathbf{u}_1$:

$$\mathbf{u} = \mathbf{u}_0 + \mathbf{u}_1$$

Here $\mathbf{u}_0$ is the solution for the loaded half-space without defect.

Let us $k(\mathbf{x})$ is Green's matrix of the half-space, which columns $\mathbf{k}_j(\mathbf{x})$ are displacement vectors associated with the surface point load $\boldsymbol{\tau}|_{z=0} = \delta(\mathbf{x})\mathbf{e}_j, j = 1, 2, 3$; where $\mathbf{e}_j$ are the unit coordinate vectors for the axes Ox, Oy, Oz respectively. With the matrix k any displacement resulted from a surface load, including the probe field $\mathbf{u}_0$, can be expressed in terms of the convolution integral

$$\mathbf{u}_0(\mathbf{x}) = \iint_D k(\mathbf{x} - \boldsymbol{\xi})\mathbf{q}_0(\boldsymbol{\xi})d\boldsymbol{\xi}, \quad \boldsymbol{\xi} = \{\xi, \eta, 0\} \tag{1.3}$$

The Fourier transform technique allows one to derive Green's matrix in terms of path Fourier integrals:

$$k(\mathbf{x}) = \mathcal{F}^{-1}[K] \equiv \frac{1}{(2\pi)^2}\iint_{\Gamma_1\Gamma_2} K(\alpha_1, \alpha_2, \alpha, z)e^{-i(\alpha_1 x + \alpha_2 y)}d\alpha_1 d\alpha_2 \tag{1.4}$$

where $\alpha = \sqrt{\alpha_1^2 + \alpha_2^2}$, matrix $K = \mathcal{F}[k]$ is the Fourier transform of $k(\mathbf{x})$ over x, y (Fourier symbol); by $\mathcal{F}$ and $\mathcal{F}^{-1}$ we denote direct and inverse transforms. The contours Γ_1 and Γ_2 go in the complex planes α_1, α_2 along the real axes $\operatorname{Im} \alpha_n = 0$, $n = 1, 2$, deviating from them for bypassing real poles and branch points of the matrix K components. The directions of the deviation are governed by the principle of limiting absorption [4].

For an isotropic half-space the symbol K is of the following structure (in the conventional notation introduced by Vorovich and Babeshko [4]):

$$K(\alpha_1, \alpha_2, \alpha, z) = \begin{pmatrix} -i(\alpha_1^2 M + \alpha_2^2 N) & -i\alpha_1\alpha_2(M - N) & -i\alpha_1 P \\ -i\alpha_1\alpha_2(M - N) & -i(\alpha_1^2 N + \alpha_2^2 M) & -i\alpha_2 P \\ \alpha_1 S & \alpha_2 S & R \end{pmatrix} \quad (1.5)$$

Functions M, N, P, R, S depend only on α and z, so that the change of variables

$$\begin{cases} \alpha_1 = \alpha \cos \gamma \\ \alpha_2 = \alpha \sin \gamma \end{cases} \quad \begin{cases} x = r \cos \varphi \\ y = r \sin \varphi \end{cases} \quad \begin{matrix} \alpha = \sqrt{\alpha_1^2 + \alpha_2^2} \\ r = \sqrt{x^2 + y^2} \end{matrix} \quad (1.6)$$

together with the Bessel functions representation

$$2\pi i^n J_n(\alpha r) = \int_0^{2\pi} e^{i\alpha r \cos \gamma - i n \gamma} d\gamma \quad (1.7)$$

brings (1.4) to the one-dimensional path integral form:

$$k(\mathbf{x}) = \frac{1}{2\pi} \int_\Gamma K(i\partial/\partial x, i\partial/\partial y, \alpha, z) J_0(\alpha r) \alpha d\alpha. \quad (1.8)$$

The contour Γ is resulted from Γ_1, Γ_2 in accordance with the change (1.6). It goes in the complex plane α along the real positive axis $\operatorname{Im} \alpha = 0$, $\operatorname{Re} \alpha > 0$ also bypassing real positive poles ζ_m and branch points κ_n of functions M, N, P, R, S.

The multipliers α_1, α_2 are substituted in K by the space derivatives by virtue of the one-to-one correspondence

$$\alpha_1^{p_1} \alpha_2^{p_2} \leftrightarrow \left(i\frac{\partial}{\partial x}\right)^{p_1} \left(i\frac{\partial}{\partial y}\right)^{p_2}, \quad p_1, p_2 = 0, 1, 2$$

Their acting on the Bessel function yields the Bessel functions again [11]:

$$\frac{\partial}{\partial x} J_0(\alpha r) = -\alpha \cos \varphi J_1(\alpha r),$$

$$\frac{\partial^2}{\partial x^2} J_0(\alpha r) = \alpha^2 [(\sin^2 \varphi - \cos^2 \varphi) \frac{J_1(\alpha r)}{\alpha r} + \cos \varphi J_0(\alpha r)], \quad \text{etc.}$$

so that no any derivatives remain in the final integral representation. It should be noted that such representation can also be derived directly from (1.4) without derivatives in (1.8), just substituting $\sin^p \gamma, \cos^p \gamma, p = 1, 2$ into (1.5) accordingly (1.6) in terms of exponents $e^{\pm in\gamma}$, which are accounted then in (1.7).

With arbitrary piecewise continuous dependence of the elastic properties on z matrix K cannot be derived analytically. Moreover, its numerical calculation via certain recurrent matrix algorithms often fails due to exponential behavior of the matrix K components. Therefore, for obtaining K we use a modification of the Thomson-Haskell algorithm, which numerical stability is assured by extracting out such bad terms and factors analytically [12].

For each layer $S_j, j = 1, 2, .., N-1$ and the lower half-space S_N this method yields K in the following form:

$$K = K_j/\Delta, \quad j = 1, 2, ..., N \tag{1.9}$$

$$K_j = \sum_{n=1}^{2} [K_{n,j} e^{\sigma_{n,j}(z-z_j)} + K_{n+2,j} e^{-\sigma_{n,j}(z-z_{j+1})}], \quad j \le N-1$$

$$K_N = \sum_{n=1}^{2} K_{n,N} e^{\sigma_{n,N}(z-z_N)}, \quad j = N$$

Matrices $K_{n,j}(\alpha_1, \alpha_2, \alpha)$ are independent of z and have no poles, while all the poles ζ_m are zeros of the denominator $\Delta(\alpha)$. The branch points can only appear in K due to radicals $\sigma_{n,j} = \sqrt{\alpha^2 - \kappa_{n,j}^2}$, in which $\kappa_{1,j} = \omega/v_{p,j}$ and $\kappa_{2,j} = \omega/v_{s,j}$ are wave numbers of the bulk P and S waves in S_j medium; $v_{p,j}, v_{s,j}$ are their velocities.

The branches of the radicals are fixed in the complex plane α by the cuts $\alpha(t) = \pm\sqrt{\kappa_{n,j}^2 - t^2}, 0 \le t \le \infty$ and conditions $\sigma_{n,j}(\alpha) \to \infty$ as $\alpha^2 \to \infty$, so that no growing exponents occur in any K component as $|\alpha| \to \infty$ or $z \to -\infty$. Simultaneously with the components $K_{n,j}, K_j, \Delta$ and the matrix K itself the code provides explicit values of $\frac{\partial}{\partial z}K$ and $\Delta' = \frac{\partial \Delta}{\partial \alpha}$ entering in the expressions for a traction vector $\boldsymbol{\tau}$ and residuals $\operatorname{res} K|_{\alpha=\zeta_m} = K_j/\Delta'|_{\alpha=\zeta_m}$ used below in the normal mode expansions (1.12) – (1.14).

It might be well to point out that a similar stable algorithm of matrix K calculation works in the case of an anisotropic structure as well. In doing so, we, instead of form (1.5), obtain K with an arbitrary smooth dependence on the polar angle γ, which is expandable, however, in a fast-convergent Fourier series:

$$K = \sum_{n=-\infty}^{\infty} K_n(\alpha, z) e^{in\gamma}$$

Due to the fast convergence, such series may be efficiently approximated by a finite number of terms for $n = 0, \pm1, ..., \pm M$ (in the isotropic case $n = 0, \pm1, \pm2$) so that a finite number of the Bessel functions $J_n(\alpha r)$ enters in the final one-dimensional integral form of k. In other respects the general technique described below is valid for anisotropic materials too.

1.2. FAR-FIELD ASYMPTOTICS

The derived integral representation (1.3), (1.8) is quite applicable for a low-cost direct numerical obtaining of the incident field $\mathbf{u}_0(\mathbf{x})$ in a near-field zone, where a distance from the source $R = |\mathbf{x}|$ is commensurable with a wavelength λ. However, the near-field is of little interest for the crack detection, whereas computing expenses increase dramatically as $R/\lambda >> 1$, up to practical inapplicability at a certain distance. Therefore, the integral representation is used mostly as the starting point for the derivation of far-field asymptotics.

1.2.1. *Normal Modes*

First, consider a contribution of the poles ζ_m obtained using the residual technique. For this purpose contour Γ in (1.8) has to be "unfolded" into the contour σ going along the whole axis $\mathrm{Im}\,\alpha = 0$ from $\mathrm{Re}\,\alpha = -\infty$ to $\mathrm{Re}\,\alpha = \infty$. This is achieved by substituting $J_0(\alpha r) = [H_0^{(1)}(\alpha r) + H_0^{(2)}(\alpha r)]/2$ ($H_0^{(n)}(\alpha r)$ are the Hankel functions) in (1.8) with the change of variable $\alpha = -\alpha$ in the second term:

$$k(\mathbf{x}) = \frac{1}{4\pi} \int_\sigma K(i\partial/\partial x, i\partial/\partial y, \alpha, z) H_0^{(1)}(\alpha r)\alpha d\alpha \tag{1.10}$$

Due to the asymptotic behavior [11]

$$H_0^{(1)}(\alpha r) \sim \sqrt{2/(\pi\alpha r)} \exp{(i\alpha r - i\pi/4)} \text{ as } |\alpha r| \to \infty \tag{1.11}$$

the integrand decreases exponentially as $\mathrm{Im}\,(\alpha r) \to \infty$. This enables to close σ into the upper half-plane $\mathrm{Im}\,\alpha \geq 0$ and to replace integral (1.10) by the sum of residuals plus integrals along the cuts, in line with the Jordan lemma and Cauchy theorem. It brings $\mathbf{u}_0$ to the form

$$\mathbf{u}_0(\mathbf{x}) = \sum_{m=1}^{\infty} \mathbf{u}_m(\mathbf{x}) + \mathbf{u}_{cut}(\mathbf{x}) \tag{1.12}$$

with

$$\mathbf{u}_m(\mathbf{x}) = \iint_D k_m(\mathbf{x} - \boldsymbol{\xi})\mathbf{q}_0(\boldsymbol{\xi})d\boldsymbol{\xi} \tag{1.13}$$

$$k_m(\mathbf{x}) = \frac{i}{2}\text{res}\, K(i\partial/\partial x, i\partial/\partial y, \alpha, z)|_{\alpha=\zeta_m} H_0^{(1)}(\zeta_m r)\zeta_m$$

and $\mathbf{u}_{cut}$ associated with the cut integrals. It should be noted, that only branch points $\kappa_n = \kappa_{n,N}, n = 1,2$ of the underlying half-space S_N contribute in $\mathbf{u}_{cut}$, while all the rest $\kappa_{n,j}, j \leq N-1$ are removable in fact.

By virtue of Hankel's behavior (1.11), the terms $\mathbf{u}_m$ related to the real $\zeta_m, m = 1, 2, ..., M(\omega)$ can be referred to as cylindrical normal modes:

$$\mathbf{u}_m(\mathbf{x}) = \mathbf{b}_m(\varphi, z)e^{i\zeta_m r}/\sqrt{r} + O(r^{-3/2}) \quad \text{as} \quad r \to \infty, \;\; z = \text{const} \tag{1.14}$$

where

$$\mathbf{b}_m = \sqrt{i\zeta_m/(2\pi)}R_m(\varphi, z)\mathbf{Q}_m(\varphi)$$

$$R_m = K_j(-\zeta_m \cos\varphi, -\zeta_m \sin\varphi, \zeta_m, z)/\Delta'(\zeta_m) \text{ for } z \in S_j$$

$$\mathbf{Q}_m = \mathbf{Q}_0(-\zeta_m \cos\varphi, -\zeta_m \sin\varphi)$$

$$\mathbf{Q}_0(\alpha_1, \alpha_2) = \mathcal{F}[\mathbf{q_0}] = \iint_D \mathbf{q}_0(x, y)e^{i(\alpha_1 x + \alpha_2 y)}dxdy.$$

As for cut integrals, it is proved, that

$$\mathbf{u}_{cut} = O(r^{-3/2}) \quad \text{as } r \to \infty, \;\; z = \text{const}$$

hence, they do not contribute into the main part of the far-field asymptotics in a horizontal direction, as well as $\mathbf{u}_m$ related to complex poles $\zeta_m, m \geq M+1$, which exhibit exponential attenuation $O(e^{-\text{Im}\,\zeta_m r})$ as $r \to \infty$.

Contour σ bypasses usually positive real poles ζ_m from below while the negative ones $-\zeta_m$ are rounded from above. This takes place almost for all poles except to that related to irregular parts of the dispersion curves $\zeta_m(\omega)$ featured by the opposite signs of the phase and group velocities of the normal modes $\mathbf{u}_m(\mathbf{x})$: $c_p = \omega/\zeta_m$ and $c_g = d\omega/d\zeta_m$.

Fig. 1 gives an example of the dispersion curves (curves of phase slowness $\zeta_m(\omega)/\omega$) with the backward mode range near the frequency $f = 16$ ($\omega = 2\pi f$) and about such a case at $f = 6$. This is for a two-layered model with a hard elastic half-space covered by a soft layer (the coating of thickness $h_1 = 0.012$ with $v_{p,1} = 0.3$ and $v_{s,1} = 0.16$ and the underlying half-space with $v_{p,2} = 2.5$ and $v_{s,2} = 1.3$; the parameters are in a dimensionless form).

The boundaries of the backward band are determined by the condition $c_g = 0$. Two neighbour real poles merge together at these frequencies resulting in double poles. It leads to infinite values of the residuals (if $\mathbf{q}_0$ does not comply with certain conditions of self-balancing). These frequencies are referred to, therefore, as resonance ones. They are weak (integrable) singularities of the frequency spectrum $\mathbf{u}(\mathbf{x}, \omega)$, determining, thereby, the main frequency components of the time domain solution (1.2).

Fig. 2 gives an example of a signal $\dot{u}_z(t)$ computed in line with eq. (1.2) for a relatively far point $r = 1$ on the surface $z = 0$ of the two-layered structure considered above. It was excited by a short impact point load $\sigma_z = t^2 e^{-bt}\delta(x, y)$ with $b = 100$. The most powerful wave packet arrives to the control point at $t = 6.7$. It is carried by the Rayleigh wave running along the coating with the velocity $v_R = 0.15$. However, due to the underlying hard base with much higher wave velocities, the first arrivals come more than ten times faster.

The spectrum $\mathbf{u}(\mathbf{x}, \omega)$ was computed in several seconds at a PC by numerical path integration for low frequencies ($r\omega < 10$) and by using the mode expansion (1.12) for medium and higher ones.

1.3. BULK WAVES

The real poles ζ_m are always greater than κ_n so that $\sigma_{n,N}(\zeta_m) = \sqrt{\zeta_m^2 - \kappa_n^2} > 0$. In line with the symbol K structure (1.9) the amplitude $\mathbf{b}_m$ in (1.14) diminishes exponentially as z gets deeper in S_N. Therefore, the far-field asymptotics as $z \to -\infty$ is derived from the cut integrals $\mathbf{u}_{cut}$ associated with the continuos spectrum of the problem and the bulk waves. It can also be obtained as a contribution of the stationary points

$$\alpha_{1,n} = -\kappa_n \cos\varphi \sin\psi, \quad \alpha_{2,n} = -\kappa_n \sin\varphi \sin\psi$$

of the oscillating exponential components $\exp(i(\sqrt{\kappa_n^2 - \alpha^2} - \alpha_1 x - \alpha_2 y))$ of the initial double integral (1.4) derived by the steepest descent method [5]:

$$k(\mathbf{x} - \boldsymbol{\xi}) = \sum_{n=1}^{2} k_n(\varphi, \psi) e^{i\kappa_n R}/R + O(R^{-2}) \quad \text{as } R \to \infty, \ z < z_N \tag{1.15}$$

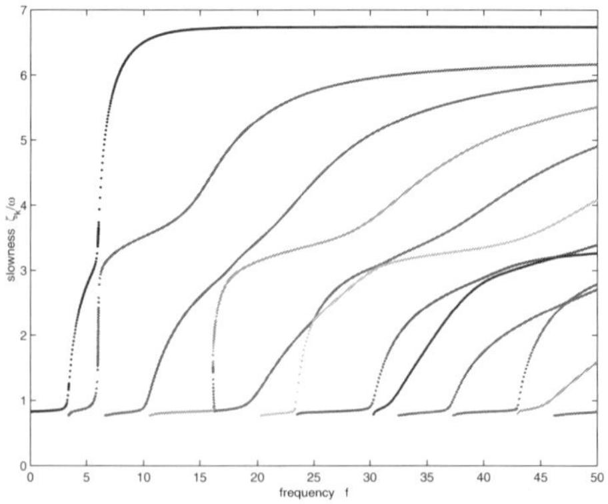

Figure 1. Dispersion curves for a two-layered half-space

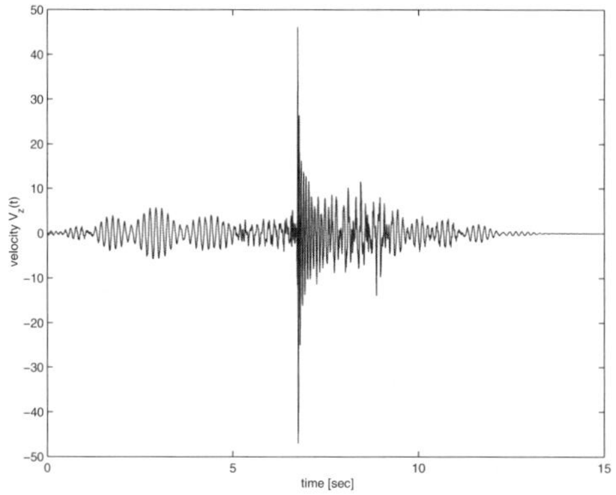

Figure 2. A time-domain signal received at a far distance from the point of impact loading applied to the surface of a coated half-space

$$k_n = i\kappa_n \cos\psi K_{N,n}(\alpha_{1,n}, \alpha_{2,n}, \alpha_n)/(2\pi),$$

$R = \sqrt{(x-\xi)^2 + (y-\eta)^2 + (z-z_N)^2}$, φ, ψ are radius and angles of a spherical coordinate system centered at a current point (ξ, η, z_N)

$$\begin{cases} x - \xi = R\cos\varphi\sin\psi & 0 \le \varphi \le 2\pi \\ y - \eta = R\sin\varphi\sin\psi & \\ z - z_N = R\cos\psi & \pi/2 < \psi \le \pi \end{cases}$$

Two terms of sum (1.15) describe spherical P and S waves for $n = 1$ and $n = 2$ respectively. Replacing convolution integral (1.3) by an approximate cubature formula with nodes $(\xi_k, \eta_k), k = 1, 2, ..., N_{nodes}$, we arrive to the bulk wave asymptotics for $\mathbf{u}_0$ in the lowest half-space:

$$\mathbf{u}_0(\mathbf{x}) \sim \sum_{n=1}^{2} \sum_{k=1}^{N_{nodes}} k_n(\varphi_k, \psi_k)\mathbf{q}_0(\xi_k, \eta_k) s_k e^{i\kappa_n R_k}/R_k \qquad \text{as } R_k \to \infty, \psi > \pi/2 \tag{1.16}$$

in which φ_k, ψ_k, R_k are spherical coordinates of a point $\mathbf{x}$ in the systems centered at points (ξ_k, η_k, z_N); s_k are cubature weight coefficients.

In the ultrasonic crack detection models the normal mode expansion (1.12) is especially convenient for simulating distant inspection of interface and near surface defects, while the bulk waves asymptotics (1.16) are used for deepened flaws [13, 14].

2. The Scattered Field

With the set forth approach the scattered field $\mathbf{u}_1$ is represented in the same explicit integral form like the one known for a buried source in a layered half-space. It is expressed via the unknown crack opening displacement (c.o.d.), which is determined from the integral equations arising when the boundary conditions at the crack sides are satisfied. For deepened cracks, when recurring reflection from far surfaces is neglected, these are Wiener-Hopf equations with a pure difference kernel. For near-surface and surface-breaking cracks the kernel structure is more complicated, however its close analytical representation has been derived for such cases as well.

2.1. HORIZONTAL AND INCLINED CRACKS

Let us give an idea of the integral equation derivation, first, for a horizontal crack situated in a plane $z = -H$. Let, then, $\mathbf{q}_1$ be a vector of traction at this plane associated with the scattered field $\mathbf{u}_1$. This unknown field can be expressed via $\mathbf{q}_1$ in the same manner as $\mathbf{u}_0$ through $\mathbf{q}_0$ using Green's matrices k^+, k^- for the upper and lower media $z > -H$ and $z < -H$. In the Fourier symbols it takes the form

$$\mathbf{U}_1^{\pm}(\alpha_1, \alpha_2, z) = K^{\pm}(\alpha_1, \alpha_2, z)\mathbf{Q}_1(\alpha_1, \alpha_2) \tag{2.1}$$

Let $\mathbf{v} = (\mathbf{u}^+ - \mathbf{u}^-)|_{z=-H}$ be a jump of $\mathbf{u}_1$ at the crack Ω (c.o.d.). Eq. (2.1) allows us to express $\mathbf{Q}_1 = \mathcal{F}[\mathbf{q}_1]$ through $\mathbf{V}(\alpha_1, \alpha_2) = \mathcal{F}[\mathbf{v}]$:

$$\mathbf{Q}_1 = L\,\mathbf{V}, \quad L(\alpha_1, \alpha_2) = (K^+ - K^-)^{-1}|_{z=-H} \tag{2.2}$$

The integral equation with respect to unknown $\mathbf{v}$ follows then from the traction-free boundary condition at crack's sides:

$$(\mathbf{q}_1 + \boldsymbol{\tau}_0)|_{z=-H} = 0, \quad (x, y) \in \Omega \tag{2.3}$$

where $\boldsymbol{\tau}_0 = T_n\mathbf{u}_0$ is a known traction vector at the crack plane related to the incident field $\mathbf{u}_0$. It is obtained simultaneously with $\mathbf{u}_0$ by application of the stress operator $T_n\mathbf{u} \equiv \lambda \mathbf{u}\operatorname{div}\mathbf{u} + 2\mu\partial\mathbf{u}/\partial n + \mu(\mathbf{u} \times \operatorname{curl}\mathbf{u})$ with $\mathbf{n} = \{0, 0, 1\}$ ($\mathbf{n}$ is a unit normal to a stress plane). Eqs. (2.2), (2.3) lead to the Wiener-Hopf integral equation

$$\mathcal{L}\mathbf{v} \equiv \iint\limits_{\Omega} l(\mathbf{x} - \boldsymbol{\xi})\mathbf{v}(\boldsymbol{\xi})d\boldsymbol{\xi} = \mathbf{f}(x, y), \quad (x, y) \in \Omega \tag{2.4}$$

where $l(\mathbf{x}) = \mathcal{F}^{-1}[L]$, $\mathbf{f} = -\boldsymbol{\tau}_0|_{z=-H}$.

To solve eq. (2.4) with a non-classical domain Ω, we use a variational Galerkin scheme with axially symmetric δ-like trial and test functions φ_k set at the nodes (x_k, y_k) of a grid covered Ω with a step h [13, 15]. The axial symmetry allowed us to gain most benefit in reducing numerical costs spent on $\mathbf{v}$ obtaining.

The asymptotical and modal analysis of the $\mathbf{u}_1^{\pm}$ integral representations following from (2.1), gives, as before for $\mathbf{u}_0$, fairly simple formulae for calculating scattered patterns and scattering cross sections (s.c.s.) and for obtaining diffracted signals at the required points where they are recorded. Moreover, the electromechanical reciprocity argument of Auld $\delta\Gamma$, which gives the same information that is measured in practice in pulse-echo scanning [9, 16], can also be easily expressed through $\mathbf{v}$:

$$\delta\Gamma = -\frac{i\omega}{P}\iint_{\Omega} \mathbf{v}\cdot\mathbf{f}\,d\Omega \quad (\mathbf{v}\cdot\mathbf{f} = \sum_{i=1}^{3} v_i f_i)$$

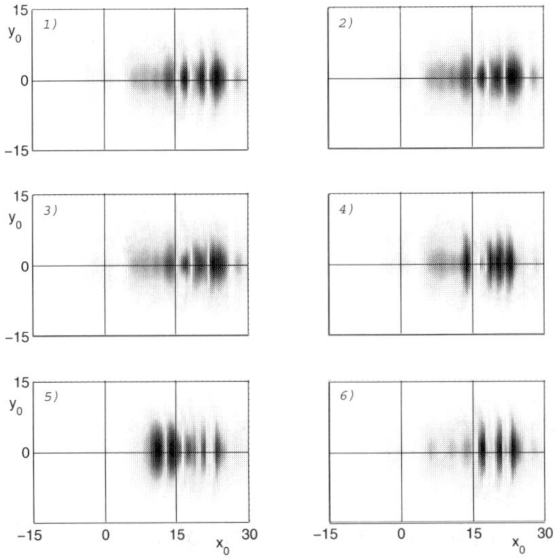

Figure 3. Scan-images of an interface square crack by an inclined 2 MHz 30° P-probe of diameter 10 mm

As an example, Fig. 3 gives scan-images of an interface square crack with the same area like the circle crack of diameter 10 mm considered in Ref. [9]. It is located between a 30 mm coating layer S_1 and a steel half-space S_2 ($v_{p,2}$ = 5.94 and $v_{s,2}$ = 3.23 km/s) with the S- velocities ratio $v_{s,2}/v_{s,1} = 0.5, 0.2, 0.1, 2, 5$, and 10 respectively.

As for inclined cracks, there are two qualitatively different situations. First, the crack may be comparatively far from the surface and boundaries of the layers, so that a contribution of a second-time reflection may be neglected. In this case eq. (2.4) is derived in the local coordinates $\mathbf{x}_1 = \{x_1, y_1, z_1\}$ connected with the crack (O_1z_1 is orthogonal to crack's plane), using Green's matrices of homogeneous half-spaces $z_1 \geq 0$ and $z_1 \leq 0$ as

$K^{\pm}$ in (2.1). Two coordinate systems $\mathbf{x}$ and $\mathbf{x}_1$ are connected by the relation $\mathbf{x}_1 = C(\mathbf{x} - \mathbf{x}_c)$, where C is a transition matrix and $\mathbf{x}_c = \{x_c, y_c, z_c\}$ are coordinates of the center O_1 in the global system $\mathbf{x}$. To calculate $\mathbf{f}$ we use, then, the stress operator T_n with $\mathbf{n} \perp \Omega$ and the matrix C to carry $\boldsymbol{\tau}_0$ from the $\mathbf{x}$- system into crack's one: $\mathbf{f} = C\,T_n \mathbf{u}_0|_{z_1=0}$. In other respects the procedure of scan-image modeling remains the same.

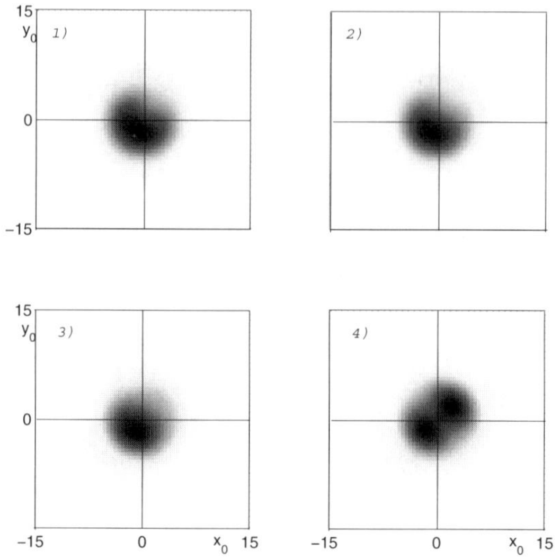

Figure 4. Pulse-echo images of a square crack rotated on $30^o, 45^o, 60^o$ and 90° around the bisecting axis of the x, y quadrant

Fig. 4 gives examples of scan-images of an inclined square crack of the same area like in Fig. 3. It is located with different orientation in a homogeneous steel sample at the depth 30 mm. The plot was obtained with a vertical 2 MHz 0° P-probe. More numerical pulse-echo scan-images, as well as s.c.s. and scattering diagrams, can be found in [13, 14].

In regard to signals recorded at the half-space surface $z = 0$ we have to take into account here, in addition to the scattered field $\mathbf{u}_1$, the field $\mathbf{u}_2$ reflected from this surface: $\mathbf{u} = \mathbf{u}_0 + \mathbf{u}_1 + \mathbf{u}_2$. Since at the far distance from the crack the field $\mathbf{u}_1$ can be described by asymptotics similar to (1.16), it is not difficult to obtain $\mathbf{u}_2$ at the points of recording using the well-known ray formulae for quasi-plane P and S waves reflection from a free surface [10].

Another possibility is to express $\mathbf{u}_2$ through stresses $\mathbf{q}_2 = -T_z \mathbf{u}_1|_{z=0}$, induced at the surface by the coming field $\mathbf{u}_1$:

$$\mathbf{U}_2(\alpha_1, \alpha_2, z) = K(\alpha_1, \alpha_2, z)\mathbf{Q}_2(\alpha_1, \alpha_2) \tag{2.5}$$

and then to derive the asymptotics required from the integrals of the inverse Fourier transform $\mathbf{u}_2(\mathbf{x}) = \mathcal{F}^{-1}[\mathbf{U}_2]$ similar like for $\mathbf{u}_0$, i.e. as a contribution of integrand's poles and stationary points of oscillating exponents. The main problem here is to derive $\mathbf{Q}_2$ explicitly in the global coordinates starting from the expression for $\mathbf{U}_1^{\pm}$ in the local coordinates.

Fourier transformation over z_1 and inversion with respect to the global variable z applied to (2.1) together with the stress operator T_z yield $\mathbf{Q}_2$ in a rather complicate but closed form:

$$\mathbf{Q}_2(\alpha_1, \alpha_2) = \sum_{n=1}^{2} \tilde{M}_n(\alpha_1, \alpha_2)\tilde{\mathbf{V}}(\alpha_1, \alpha_2)\frac{e^{\sigma_n z_c}}{2\sigma_n} e^{i(\alpha_1 x_c + \alpha_2 y_c)} \tag{2.6}$$

where $\tilde{\mathbf{V}} = \mathbf{V}(\gamma_1, \gamma_2), \tilde{M}_n = M_n(\boldsymbol{\gamma})$, $\boldsymbol{\gamma} = C\boldsymbol{\beta}$, $\boldsymbol{\beta} = \{\alpha_1, \alpha_2, -i\sigma_n\}$; $\boldsymbol{\gamma} = \{\gamma_1, \gamma_2, \gamma_3\}$ is a vector of the Fourier transform parameters in the local coordinates $\mathbf{x}_1$, while α_1, α_2 relate to transforms over x, y in the global system $\mathbf{x}$; $\mathbf{V}$ see (2.2), $M_n(\boldsymbol{\gamma}) = T_z(\boldsymbol{\gamma})CK_n^{\pm}(\gamma_1, \gamma_2, \gamma)L(\gamma_1, \gamma_2)$, $T_z(\boldsymbol{\gamma}) = \mathcal{F}[T_z]$. Exponents in (2.6) are used further together with $e^{-i(\alpha_1 x + \alpha_2 y)}$ of the inverse transform to derive $\mathbf{u}_2$ asymptotics as $R = \sqrt{(x - x_c)^2 + (y - y_c)^2 + z_c^2} \to \infty$.

If an inclined crack is in an inner layer $S_j, j \geq 2$, but far enough from its boundaries, the asymptotics of displacements $\mathbf{u}$ at the surface $z = 0$ are derived similarly using an auxiliary vector of tractions $\mathbf{q}_j = -T_z\mathbf{u}_1|_{z=z_j}$ induced by the scattered field at the boundary $z = z_j$. Then, $\mathbf{u}|_{z=0} = \mathcal{F}^{-1}[K(\alpha_1, \alpha_2, 0)\mathbf{Q}_j(\alpha_1, \alpha_2)]$, where K is Green's matrix of the packet of layers overlying S_j.

The second situation, qualitatively different regarding to integral equations, occurs for near-surface cracks, when the reflected field $\mathbf{u}_2$ cannot be neglected. In this case the integral equation has to be derived from the condition

$$(\mathbf{q}_1 + T_n\mathbf{u}_2 + \boldsymbol{\tau}_0)|_{z_1=0} = 0, \quad (x_1, y_1) \in \Omega \tag{2.7}$$

Since in line with (2.5), (2.6) $T_n\mathbf{u}_2$ is also expressed via $\mathbf{v}$, the resulting integral equation becomes of the form

$$(\mathcal{L} + \mathcal{L}_2\mathbf{v}) \equiv \iint_{\Omega} [l(\mathbf{x} - \boldsymbol{\xi}) + l_2(\mathbf{x}, \boldsymbol{\xi})]\mathbf{v}(\boldsymbol{\xi})d\boldsymbol{\xi} = \mathbf{f}(x, y), \quad (x, y) \in \Omega \tag{2.8}$$

This equation takes explicitly into account all recurrent reflections from the crack and boundaries affecting on the crack open displacement $\mathbf{v}$.

If the crack does not touch the surface, the Fourier symbol of l_2 decreases exponentially as $\alpha \to \infty$ due to the $\mathbf{Q}_2$ structure. Therefore, in this case the kernel l_2 is a smooth matrix-function on $\mathbf{x}$ and $\boldsymbol{\xi}$ and the hypersingular kernel $l(\mathbf{x}-\boldsymbol{\xi})$ remains to be the main part of the integral operator, whereas for a surface-breaking crack the addition l_2 becomes also singular. Eq. (2.8), of course, is much more complicate than eq. (2.4), however, the explicit form of the kernels gives the chance to develop and implement low-cost numerical algorithms of its solution as well.

2.2. SCATTERING RESONANCE POLES

Since the inverse problem is ill-posed, the procedures of defect sizing and shaping by processing the scattered signals need certain regularisation to achieve stable numerical convergence. This can be achieved by using additional information about the expected properties of the solution, e.g. about the resonance properties of the object under investigation.

We mean the so-called scattering resonance poles ω_n, $n = 1, 2, ...$ allocated in the lower half-plane of the complex frequency plane ω, which are the poles of a harmonic steady- state scattered field $\mathbf{u}_1(\mathbf{x}, \omega)$, to be considered as an analytical function of the complex variable ω [17]. It is important that they depend only on crack's size and shape and does not depend on its orientation.

To estimate the influence of the crack shape on the poles allocation, the trajectories of several first poles, drifting in the complex ω-plane as crack's shape changed, were traced numerically [18]. Since the poles contribute in the time domain as exponentially decaying harmonic oscillations $\mathbf{u}_n(t) = \mathbf{a}_n e^{-\beta_n t} e^{\pm i\alpha_n t}$, the practical use of the data obtained depends on ability to discern the real and imaginary parts of the poles $\omega_n = \pm\alpha_n - i\beta_n$ from the transient signals recorded.

Conclusion

The integral equation method has proved itself to be a good tool for investigation fine wave phenomena in solids. In combination with asymptotical and numerical calculation of the integrals arisen, it can be used for development fast and efficient computer models for both research and industrial applications not only in NDT, but in designing surface waves devices, coated components, anti-seismic constructions and in many other applications as well.

Acknowledgement. The work was supported by the RFBR and CRDF REC-004 grants.

References

1. Babich, V.M., Buldyrev, V.S., Molotkov, I.A. (1985) *The Space-time Ray Method. Linear and Non-linear Waves*, Leningrad, Leningrad University (in Russian).
2. Chapman, R.K. (1990) A system model for the ultrasonic inspection of smooth planar cracks, *J. Nondestr.Eval.*, **9**, pp. 197–211.
3. Achenbach, J.D. (1973) *Wave Propagation in Elastic Solids*, Elsevier Science B.V., Amsterdam/New York.
4. Vorovich, I.I., Babeshko, V.A. (1974) *Dynamic Mixed Problems of Elasticity for Non-classical Domains*, Nauka, Moscow (in Russian).
5. Babeshko V.A., Glushkov E.V., Glushkova N.V. (1992) Energy vortices and backward fluxes in elastic waveguides, *Wave Motion*, **16**, pp. 183– 192.
6. Glushkov, E.V., Glushkova, N.V., Kirillova, E.V. (1995) Normal mode diffraction in elastic layered waveguides; resonances and energy vortices, *Proceedings of the Third International Conference on Mathematical and Numerical Aspects of Wave Propagation*, Ed. G.Cohen, SIAM, Philadelphia, pp. 604–612.
7. Babeshko, V.A., Glushkov, E.V., Glushkova, N.V. and Kirillova, E.V. (1992) Energy localizaion under high-frequency resonance conditions, *Sov. Phys. Dokl.* **37(8)**, pp. 443–444.
8. Glushkov, E., Glushkova, N. (1997) Blocking property of energy vortices in elastic waveguides, *J. Acoust. Soc. Am.*, **102(3)**, pp. 1356 –1360.
9. Boström, A., Wirdelius, H. (1995) Ultrasonic probe modeling and nondestructive crack detection, *J. Acoust. Soc. Am.*, **(5)**, pp. 2836– 2848.
10. Brekhovskikh, L.M. (1980) *Waves in Layered Media*, 2nd Ed., Academic Press, New York.
11. Abramowitz, M. and Stegun, I.A. (Editors) (1970) *Handbook of Mathematical Functions*, National Bureau of Standards.
12. Babeshko, V. A., Glushkov, E. V. and Glushkova, N. V. (1987) Methods of Green's matrix calculation for a stratified elastic half-space, *Zhurnal Vychislitelnoy Matematiki i Matematicheskoy Fiziki*, **27(1)**, pp. 93–101 (in Russian).
13. Glushkov, Ye.V. and Glushkova, N.V. (1996) Diffraction of elastic waves by three-dimensional cracks of arbitrary shape in a plane, *J. Appl. Maths Mechs*, **60(2)**, pp. 277–283.
14. Glushkov, Ye.V., Glushkova, N.V., Yekhlakov, A.V. (2002) Mathematical model of ultrasonic detection for three-dimensional interface cracks, *J. Appl. Maths Mechs*, **66(1)**, pp. 147–156.
15. Glushkov, E.V., Glushkova, N.V. (2001) On the Efficient Implementation of the Integral Equation Method in Elastodynamics *Journal of Computational Acoustics*, **9(3)**, pp. 889–898.
16. Auld, B.A. (1979) General electromechanical reciprocity relations applied to the calculation of elastic wave scattering coefficients, *Wave Motion*, **1**, pp. 3–10.
17. Alves, C.J.S., Ha Duong, T. (1995) Numerical experiments on the resonance poles associated to acoustic and elastic scattering by a plane crack, *Proceedings of the Third International Conference on Mathematical and Numerical Aspects of Wave Propagation*, Ed. G.Cohen, SIAM, Philadelphia, pp. 544-553.
18. Glushkov, Ye.V. and Glushkova, N.V. (1998) Resonant frequencies of the scattering of elastic waves by three-dimensional cracks, *J. Appl. Maths Mechs*, **62(5)**, pp. 803–806.

Author Index

NATO-ARW held in Moscow, February 7-9, 2002
List of Participants

ALSHITS Vladimir. I.,
Russian Academy of Sciences,
Institute of Crystallography,
LENINSKIY PROSPECT, 59,
11933 MOSCOW, RUSSIA
alshits@ns.crys.ras.ru

AYZENBERG-STEPANENKO Mark,
The Institute for Industrial Mathematics
4 NA-NACHTOM STR.,
84311 BEER-SHEVA, ISRAEL
ayzenbe@math.bgu.ac.il

BABESHKO Vladimir.A.
Kuban State University,
ULITSA STAVROPOLSKAYA, 149,
350040 KRASNODAR, RUSSIA
bva@ksu.kuban.su

BROBERG Bertram,
Department of Mathematical Physics,
University College Dublin
BELFIELD, DUBLIN 4, IRELAND
bertram.broberg@ucd.ie

CHELYUBEEV Dmitry. A.,
Russian Academy of Sciences,
Institute for Problems in Mechanics,
PROSPECT VERNADSKOGO,
DOM 101, KOR. 1,
119526 MOSCOW, RUSSIA
popov@ipmnet.ru

CHEMYSHEV German. N,
Russian Academy of Sciences,
Institute for Problems in Mechanics,
PROSPECT VERNADSKOGO,
DOM 101, KOR. 1,
119526 MOSCOW, RUSSIA
popov@ipmnet.ru

DESCHAMPS Marc,
Laboratoire de Mécanique Physique,
UMR CNRS 5469,
Université de Bordeaux I,
351, COURS DE LA LIBÉRATION
33405 TALENCE, CEDEX, FRANCE
deschamps@lmp.u-bordeaux.fr

ENTOV Vladimir. M,
Russian Academy of Sciences,
Institute for Problems in Mechanics,
PROSPECT VERNADSKOGO,
DOM 101, KOR. 1, 119526 MOSCOW,
RUSSIA.
entov@ipmnet.ru

GLUSHKOV Evgeny V.,
Kuban State University,
ULITSA STAVROPOLSKAYA, 149,
350040 KRASNODAR, RUSSIA
evg@math.kubsu.ru

GLUSHKOVA Natalya V.,
Kuban State University,
ULITSA STAVROPOLSKAYA, 149,
350040 KRASNODAR, RUSSIA
evg@math.kubsu.ru

GOLDSTEIN Robert V.,
Russian Academy of Sciences,
Institute for Problems in Mechanics,
PROSPECT VERNADSKOGO,
DOM 101, KOR. 1,
119526 MOSCOW, RUSSIA,
goldst@ipmnet.ru

KANEL G.I.,
Russian Academy of Sciences,
Institute of Thermal Physics of
Extremal States,
Joint Institute of High Temperatures of RAS,
IJORSKAYA UL. 13/19,
127412 MOSCOW, RUSSIA
kanel@ficp.ac.ru

KAPTSOV Alexander V.,
Russian Academy of Sciences,
Institute for Problems in Mechanics,
PROSPECT VERNADSKOGO,
DOM 101, KOR. 1,
119526 MOSCOW, RUSSIA,
kaptsov@ipmnet.ru

KISELEV Alexey. P.,
St.Petersburg Branch of the
Steklov Mathematical Institute,
Fontanka 27,
191023 St.Petersburg , RUSSIA
kiselev@pdmi.ras.ru

KLINE Ronald,
San-Diego Center For Materials Research,
5500 CAMPANILE DR,
SAN DIEGO, CA,
USA
kline@kahuna.sdsu.edu

KUKUDZHANOV Vladimir. N.,
Russian Academy of Sciences,
Institute for Problems in Mechanics,
PROSPECT VERNADSKOGO, DOM 101, KOR. 1, 119526 MOSCOW, RUSSIA,
kukudjanov@ipmnet.ru

KUZNETSOV Sergey. V.,
Russian Academy of Sciences,
Institute for Problems in Mechanics,
PROSPECT VERNADSKOGO,
DOM 101, KOR. 1,
119526 MOSCOW, RUSSIA
svkuznec@ipmnet.ru

LOWE Michael,
Department of Mechanical Engineering,
Imperial College,
EXHIBITION ROAD,
LONDON, SW7 2BX, UK,
m.lowe@ic.ac.uk

MARCHENKO Alexey. V.,
Russian Academy of Sciences,
GENERAL PHYSICS INSTITUTE OF THE RAS,
38, VAVILOV STREET,
MOSCOW, 117942, RUSSIA,
amarch@orc.ru

MAUGIN Gérard,
Laboratoire de Modélisation en Mécanique,
University Pierre et Marie Curie, Case 162
4 PLACE JUSSIEU
75252 PARIS CEDEX 05,
France
gam@ccr.jussieu.fr

MOROSOV Nikita. F.,
St.Petersburg University,
Faculty of Mathematics and Mechanics,
Department of Theory of Elasticity,
198904, ST.PETERSBURG,
PETRODVOREC, BIBLIOTECHNAYA PL. 2,
RUSSIA
morozov@gamma.math.spbu.ru

MOZHAEV Vladimir.G.,
Moscow State Lomonosov University,
FACULTY OF PHYSICS,
MOSCOW STATE UNIVERSITY,
117234 MOSCOW, RUSSIA,
V.Mozhaev@ifw-drezden.de

PARKER David. F.,
Department of Mathematics and Statistics,
University of Edinburgh,
THE KING'S BUILDINGS,
EDINBURGH, EH9 3JZ, U.K,
D.F.Parker@ed.ac.uk

PETROV Yury. V.,
St.Petersburg University,
Faculty of Mathematics and Mechanics,
Department of Theory of Elasticity,
198904, ST.PETERSBURG,
PETRODVOREC,
BIBLIOTECHNAYA PL. 2, RUSSIA
yp@yp1004.spb.edu

POPOV Alexander. L.,
Russian Academy of Sciences,
General Physics Institute,
PROSPECT VERNADSKOGO,
DOM 101, KOR. 1,
119526 MOSCOW, RUSSIA,
popov@ipmnet.ru

SALGANIC Rafael. L.,
Russian Academy of Sciences,
Institute for Problems in Mechanics,
PROSPECT VERNADSKOGO,
DOM 101, KOR. 1,
119526 MOSCOW, RUSSIA,
salganic@ipmnet.ru

SHANIN Andrew. V.,
Moscow State University,
Faculty of Physics, Physical Department,
MOSCOW STATE UNIVERSITY,
117234 MOSCOW, RUSSIA,
shanin@ok.ru

SHEMYAKIN Evgeny. I.,
Department of wave dynamics
Moscow State Lomonosov University,
FACULTY OF PHYSICS,
MOSCOW STATE UNIVERSITY,
117234 MOSCOW, RUSSIA,
schemyakine@mail.ru

SHITIKOVA Marina. V.,
Voronezh State University
of Architecture and Civil Engineering,
UL. KIROVA 3-75,
314018 VORONEZH RUSSIA,
shitikova@vmail.ru

SIMONOV Igor V.,
Russian Academy of Sciences,
General Physics Institute,
PROSPECT VERNADSKOGO,
DOM 101, KOR. 1,
119526 MOSCOW, RUSSIA,
simonov@ipmnet.ru

TING Thomas. C.T.,
Department of Mechanical Engineering
Stanford University,
DURAND BLDG 262
STANFORD, CA 94305-4040
USA,
tting@uic.edu

TOKMAKOVA Svetlana.P.,
N.N.Andreev Acoustics Institute,
SHVERNIKA, 4,
117036 MOSCOW, RUSSIA,
sveta@akin.ru

USTINOV Konstantin.B. E.,
Russian Academy of Sciences,
Institute for Problems in Mechanics,
PROSPECT VERNADSKOGO,
DOM 101, KOR. 1,
1199526 MOSCOW, RUSSIA,
ustinov@ipmnet.ru

WILDE Maria,
Saratov State University,
Mathematics and Mechanical Faculty,
Department of Theory of Elasticity
and Biomechanics,
ASTRAHANSKAYA, 83, SARATOV,
410026 RUSSIA,
Kossovichlu@info.sgu.ru

ZAKHAROV Dmitrij D.,
Russian Academy of Sciences,
Institute for Problems in Mechanics,
PROSPECT VERNADSKOGO,
DOM 101, KOR. 1,
119526 MOSCOW, RUSSIA
zakhard@sbu.ac.uk